Bacterial Endotoxin

verlag
chemie

Bacterial Endotoxin

Chemical, Biological and Clinical Aspects

Edited by
J. Y. Homma, S. Kanegasaki, O. Lüderitz,
T. Shiba and O. Westphal

verlag chemie
Weinheim
Deerfield Beach, Florida
Basel

Production Manager: Dipl.-Wirt.-Ing. (FH) Myriam Nothacker

This book contains 157 figures and 110 tables.

Deutsche Bibliothek Cataloguing-in-Publication Data

Bacterial endotoxin: chem., biolog. and clin. aspects /
ed. by J.Y. Homma ... – Weinheim; Deerfield Beach, Florida;
Basel: Verlag Chemie, 1984.
   ISBN 3-527-26164-8 (Weinheim, Basel)
   ISBN 0-89573-228-9 (Deefield Beach, Florida)
NE: Homma, J.Y. [Hrsg.]

© Verlag Chemie GmbH, D-6940 Weinheim, 1984
All rights reserved (including those of translation into other languages). No part of this book may be reproduced in any form – by photoprint, microfilm, or any other means – nor transmitted or translated into a machine language without written permission from the publishers.
Registered names, trademarks, etc. used in this book, even when not specifically marked as such, are not to be considered unprotected by law.
Printing: betz-druck gmbh, D-6100 Darmstadt
Bookbinding: Josef Spinner, D-7583 Ottersweier
Printed in the Federal Republic of Germany

Preface

When the term endotoxin was first introduced by Richard PFEIFFER early this century, definition was based simply on the toxicity of this principle which, in contrast to exotoxins, was found to be a characteristic component of, and firmly bound to, the cells of Gram-negative bacteria. Only much later was the toxin isolated from bacterial cultures as a high-molecular "endotoxic complex". It was then found that endotoxin does not only exert toxic effects in animals and man, but also causes other biological manifestations at non-toxic dose levels, such as fever, changes of the white blood cell count, activation of hormonal and enzymatic systems or stimulation of immune responses. It was observed that bacterial endotoxin, in sensitive animals and man, acts as a potent biological signal and induces a series of "endotoxic reactions". In clinical medicine, purified preparations were found to induce non-specific changes of the body's defense mechanisms against infections, certain types of cancer, lethal irradiation and other noxious irritations. Taking all these findings together, it is no wonder that endotoxin has attracted many investigators each of whom focussed on his aspect of the "endotoxin problem".

Although endotoxin has been a matter of world-wide interest for about half a century, many important questions still have not been answered. Moreover, research work is still growing, new research groups enter the field, and it is increasingly difficult for many endotoxicologists to keep informed about progress.

This book describes the state of the art in those fields of endotoxin research which are of interest from chemical, analytical, biological and biomedical points of view. Progress has been made and is being made in particular by Japanese research groups, most of which work in international cooperation. It is for this reason that Japanese contributions to this book are prominent.

The most sensitive test for endotoxin presently is the Limulus test. It makes use of the gelation of the hemolymph of this animal after incubation with tiny amounts of endotoxin. In the sea around Japan Limulus species abound and provide material for many investigations, including clinical tests, for example in studies of acute endotoxemia. Japanese researchers have elucidated the mechanism of this test. For the quantitation of bacterial endotoxin activity a standard preparation is needed. This problem is dealt with by some of the papers in this book.

After it had been established that "lipid A" is the biologically active center for endotoxicity in the endotoxic complex, analytical chemists, mainly in Germany and Japan, have elucidated the complex and unique structure of lipid A so that organic chemists in Japan could start to work on the partial and total synthesis. The substances obtained were also biologically analyzed as part of an attempt to evaluate the relationship between substructures and biological activities.

With regard to clinical investigations, interest has recently focussed (again) on the mechanisms of action of endotoxin on tumors. The search for endotoxin preparations eliciting tumor hemorrhague, as well as work – especially in the Memorial Sloan Kettering Cancer Center, New York, – on endogenous mediators induced by endotoxin, are delt with in this book. Also the question of any direct action of endotoxin on tumors is raised. Clinical problems of "endotoxemia", an old and ever new task, are discussed by several research groups. Moreover, biological effects in animals and their significance for the understanding of clinical effects of endotoxin are reported.

The authors of this book have discussed endotoxin problems at several occasions. It was eventually proposed to summarize their thoughts and the results in the form of this progress report.

O. WESTPHAL, J.Y. HOMMA

Contents

Preface V

Endotoxin: General Introduction 1
Otto Westphal

Lipid A: Structure and Synthetic Approach

Bacterial Lipopolysaccharides and Their Lipid A Component 11
Ernst Th. Rietschel, Hubert Mayer, Horst-Werner Wollenweber, Ulrich Zähringer, Otto Lüderitz, Otto Westphal and Helmut Brade

Characterization and Biological Activities of the Lipid A Precursors from an E. coli Mutant Deficient in Phosphatidylglycerol 23
Masahiro Nishijima, Fumio Amano, Akemi Minoura, Yuzuru Akamatsu and Christian R.H. Raetz

How Can Chemical Synthesis Approach the Entity of Lipid A? 29
Tetsuo Shiba, Shoichi Kusumoto, Masaru Inage, Haruyuki Chaki, Masahiro Imoto, Tetsuo Shimamoto

Synthetic Studies on Lipid A and Related Compounds 39
Makoto Kiso, Akira Hasegawa

Synthesis of Diglycosyl Phosphate via a Phosphite Intermediate: Partial Structure of Lipid A 51
Tomoya Ogawa and Akinori Seta

Endotoxic Activities of Lipid A and Synthetic Analogues

Biological Activities of Synthetic Lipid A Analogues 61
Motohiro Matsuura, Yasuhiko Kojima, J. Yuzuru Homma, Yoshio Kumazawa, Yoshiyuki Kubota, Tetsuo Shiba and Shoichi Kusumoto

Biological Activities of Synthetic Lipid A Analogues 73
Ken-ichi Tanamoto, Gery McKenzie, Chris Galanos, Otto Lüderitz, Otto Westphal, Ulrich Zähringer, Ulrich Schade, Ernst Th. Rietschel, Shoichi Kusumoto and Tetsuo Shiba

Shwartzman Activities of Synthetic Lipid A Analogues and Their Effects on Hepatic Enzyme Activities 81
Nobuhiko Kasai, Kiyoshi Egawa, Junichi Mashimo, Tetsuo Shiba and Shoichi Kusumoto

Bone Marrow Reactions Produced with Synthetic Lipid A Analogues 89
Masao Yoshida, Michimasa Hirata, Nobuko Tsunoda, Katsuya Inada, Tetsuo Shiba and Shoichi Kusumoto

Biological Activities of Synthetic Lipid A Analogues in Artificial Membrane Vesicles 101
Shiro Kanegasaki, Tatsuji Yasuda, Toru Tsumita, Takushi Tadakuma, J. Yuzuru Homma, Shoichi Kusumoto and Tetsuo Shiba

Contents

Comparative Studies on the Immunobiological Activities of Synthetic Lipid A Analogues and Lipophilic Muramyl Peptides 111
Shozo Kotani, Haruhiko Takada, Masachika Tsujimoto, Tomohiko Ogawa, Yoshihide Mori, Tetsuo Shiba, Shoichi Kusumoto, Masaru Inage and Nobuhiko Kasai

Further Chemical and Biological Studies on Some Bacterial Lipopolysaccharides

Comparison of the Biological Activities of Lipopolysaccharides Complexed with Outer Membrane Protein 149
Shiro Kanegasaki, Tatsuji Yasuda, Sonoko Kobayashi, Ken-ichi Tanamoto, Kazushige Hirasaw, Yasuhiko Kojima, Ernst Th. Rietschel, HiaMI Yamada and Shoji Mizushima

Relationship between Biological Activities and Chemical Structure of Pseudomonas aeruginosa Endotoxin 159
J. Yuzuru Homma and Ken-ichi Tanamoto

Biological Activities of Klebsiella O3 Lipopolysaccharide 173
Izumi Nakashima, Michio Ohta, Nobuo Kido, Yasuaki Fujii, Takashi Yokochi, Fumihiko Nagase, Takaaki Hasegawa, Masashi Mori, Ken-ichi Isobe, Kenji Mizoguchi, Mitsuru Saito, Nobuo Kato and Nobuhiko Kasii

Lipopolysaccharides of the Family Vibrionaceae 187
Kazuhito Hisatsune, Seiichi Kondo, Takehiro Iguchi, Fumihiro Yamamoto, Makoto Inaguma, Shigeru Kokubo and Shigeri Arai

Action on Endotoxin of Tumors and Related Aspects

Antitumor Effects of Endotoxin: Possible Mechanisms of Action 205
Herbert F. Oettgen, Lloyd J. Old, Michael K. Hoffmann, Malcolm A.S. Moore

Induction and Properties of the Tumor Necrosis Factor 223
Akihiro Yamamoto, Barbara Williamson, Elizabeth C. Richard, Nancy Fiore and Lloyd J. Old

Effects of Endotoxin Administration on Tumors and Host: An Experimental Observation on Tumor-Bearing Rabbits 235
Naoto Aoki, Wataru Mori

Tumor Regression Induced by Endotoxin Combined with Trehalose Dimycolate 251
Kazue Fukushi, Hiroshi Asano and Jin-ichi Sasaki

Hemorrhagic Tumor Necrosis Induced by Endotoxin 261
Nikolaus Freudenberg, Kensuke Joh, Chris Galanos, Marina A. Freudenberg and Otto Westphal

Endotoxin Target Cells and Endogenous Mediators: The Role of Macrophages and Arachidonic Acid Metabolites 269
Ulrich Schade, Thomas Lüderitz and Ernst Th. Rietschel

Recovery of Immune Response in Immunodeficient Mice after Administration of Lipopolysaccharide: LPS-Induced Non-Specific Resistance in Beige Mice with Chèdiak-Higashi Syndrome 281
Masayasu Nakano, Kazuyasu Onozuka, Tatsuo Saito-Taki and Nagahiro Minato

Distribution, Degradation and Elimination of Intravenously Applied Lipopolysaccharide in Rats 295
Marina A. Freudenberg, Bernhard Kleine, Nikolaus Freudenberg and Chjris Galanos

Endotoxemia: Clinical Aspects

Clinical Aspects of Endotoxemia 307
Masaru Ishiyama, Chiyuku Watanabe and Shoetsu Tamakuma

Factors and Mechanisms of Clinical Endotoxemia 315
Masahiko Onda, Kenji Adachi and Akiro Shirota

Clinical Aspects of Endotoxemia in Liver Diseaes 327
Kyuichi Tanikawa and Masataka Iwasaki

Complement Levels in Endotoxemia 339
Tetsuro Takaoka, Akishige Nakamura, Nagao Shinagawa and Jiro Yura

Endotoxemia and Blood Coagulation: Procoagulant Activity of Mouse Bone Marrow Cells 351
Michimasa Hirata, Masao Yoshida, Nobuko Tsunoda and Katsuya Inada

Limulus Test, Endotoxin Standard Preparation

The Limulus Coagulation System Sensitive to Bacterial Endotoxins 365
Sadaaki Iwanaga, Takashi Morita, Toshiyuki Miyata, Takanori Nakamura, Masuyo Hiranaga and Sadami Ohtsubo

Basic and Applied Studies of the Limulus Test 383
Makoto Niwa, Masaru Umeda and Kunihiro Ohashi

A Candidate for Japanese Reference Lipopolysaccharide in Control of Limulus Amoebcyte Lysate Testing 395
Kiyoto Akama, Kazuo Kuratsuka, Reiko Homma, Seizaburo Kanoh, Makoto Niwa, Sadaaki Iwanaga and Chizuko Nakahara

Isolation and Purification of a Standardized Lipopolysaccharide from Salmonella abortus equi 409
Chris Galanos, O. Lüderitz and O. Westphal

Endotoxin: General Introduction

Otto Westphal

Max Planck-Institut für Immunbiologie, 7800 Freiburg i.Br. FRG

Endotoxin is a high molecular complex firmly bound to the cell envelope (or cell wall) of Gram-negative bacteria. It was Richard PFEIFFER, a pupil of Robert KOCH, who in 1904 after studies on Vibrio cholerae introduced the term endotoxin to distinguish this heat-stabile toxic principle from exotoxin as the heat-labile toxic material which is being released from live bacteria in growing cultures or during acute infection. Endotoxin, in contrast, is only released when the bacteria undergo disintegration.

In the following decades real interest in the nature of endotoxin developed gradually. Many biological and clinical effects of live or killed Gram-negative bacteria were discovered and could be attributed to one and the same principle: endotoxin. With the growing pharmaceutical industry unpredictable and highly undesired biologically active contaminations in ampoules, ready for injection or infusion, created increasing problems - such as the well known pyrogenic effects. It was Dr. Florence SEIBERT in the United States in the twenties who pioneered by introducing a sensitive test for pyrogenicity in rabbits (1).

In a big search for the chemical nature of such pyrogenic contaminants, Florence SEIBERT ended up with such minute amounts of undefined residual matter that she gave up, speaking finally only of her "little blue devil", whatever it might be ... However, in the following two decades - starting from quite a different angle - it became clear that SEIBERT's pyrogenic contaminations were of bacterial origin and, in fact, traces of endotoxin from small numbers of autolyzed organisms ever present in water or air.

It was Florence SEIBERT who organized what we may call the first Endotoxin Conference. It took place in New York under the patronage of the New York Academy of Sciences. Dr. William WINDLE, a neurophysiologist, acted as chairman and as the most important contributor, because he had just discovered the activation of nerve-regeneration by bacterial endotoxin in higher animals (2). The great excitement then created, however, shifted to great disappointment: WINDLE himself and others were unable to confirm these findings. Today we know that this excellent scientist lost his image not because he was wrong, but because the industrial preparation of endotoxin he was using changed from batch to batch in an unpredictable way! We ourselves, a few years later - together with the physiologists Professor von MURALT and Dr. BAMMER in Bern/Switzerland - repeated such investigations by measuring the recurrence of the corneal reflex in the rabbit's eye in which the light-sensitive corneal nerves had been cut surgically (3). After the intravenous injection of defined doses of purified endotoxin there was a dramatic restauration of the flashlight-induced corneal reflex compared to untreated controls!

Thus, already in these early days the question of reproducible and standardized endotoxin preparations was seriously raised; but it took another quarter of a century until the problem could be clearly formulated and practically solved. C. Galanos et al. will deal with this important matter.

Many endotoxin conferences followed after 1946, in the average more than one per year, so that people spoke of an "endotoxin explosion". Conferences were held in many parts of the United States as well as in Western and Eastern Europe. If one follows the items of these meetings, it seems fair to say that there were always exciting news to be presented - but there also remained certain crucial questions unanswered and, thus, stimulating future research. We can make the statement that endotoxin in its many basic scientific and applied biomedical aspects has attracted investigators for almost half a century. There is little doubt that a future generation of endotoxicologists will not suffer unemployment; in the contrary, the field is still continuously growing.

About twenty years ago, at the occasion of an endotoxin conference in New Brunswick, a pioneer in endotoxin research, Ivan BENNETT, said (4):
"Endotoxins possess an intrinsic fascination that is nothing less than fabulous. They seem to have been endowed by Nature with virtues and vices in the exact and glamorous proportions needed to render them irresistible to any investigator who comes to know them. - They intrigue the chemist. The molecular basis for their biological action seems always on the verge of discovery but somehow just eludes detection".

Twenty years later, today, we are only a little further in this respect. BENNETT then elaborated that endotoxin equally intrigues biologists and clinicians. It is indeed true that
"endotoxins possess a range of biological activity that seems designed to cut across all of the categories into which research and researchers are presently classified".

In experimental animals and in man endotoxin can affect the structure and function of numerous organs and cells, activate or inhibit many enzyme systems, hormonal systems and mediators. These will reflect on changes of the white blood count in tissues and blood, changes of body temperature, changes of resistance to bacterial and viral infections and other irritations as well as changes in the direct or indirect destructive capacity of cells of the immune system towards cancer. This makes endotoxin research so attractive. This book in several sections, will deal with many of these aspects and phenomena. Approaches in many different fields are needed. <u>We still absolutely need cooperation</u> within the "endotoxin family". We gladly notice the growing contributions of the japanese wing of that family to exiting and important new chapters in the "story of endotoxin".

In the thirties André BOIVIN in Paris and Walter MORGAN in London were the first to introduce more generally applicable extraction procedures for - what they called - the <u>O-antigenic complex</u> of Enterobacteriaceae, such as from Salmonella and Shigella strains. It was found that their extracts exerted not only strong <u>anti-bacterial immunogenicity</u> (due to their O-antigen), but also <u>endotoxicity</u> - which, of course complicated the use of such materials as vaccines. The complexes were mainly composed of <u>polysaccharide</u>, <u>protein</u>, and more or less undefined <u>lipoidic material</u>. We ourselves, somewhat later, developed the phenol/water method for the extraction of protein-free lipopolysaccharides (LPS) from Gram-negative bacteria. Together with Dr. Anne Marie STAUB at the Pasteur Institute in Paris it could then clearly be established that the polysaccharide component contained the serologically active determinants in the form of defined oligosaccharides in the repeating units of the species-specific bacterial O-antigen.

There were several endotoxin conferences, in 1965 and 1967, organized by the later Nobel laureate Dr. LURIA, under the headline of <u>"Microdermatology"</u>. This would mean aspects of the fine structure of the bacterial "skin", especially the many variations in sugar composition of the polysaccharide component and the hundreds or thousands of combinations of serological determinants along the O-specific chain, as expressed (for example) in the numbers of the Kauffmann-White scheme for Salmonella species (5). Much insight into the question about the size of oligosaccharide determinants and the reflexion of slight defined structural changes in serological specificity brought about by <u>lysogenic phage conversions</u> and the underlying biochemical events in Salmonella O-antigens were vigorously discussed during these conferences (6).

On the other hand, the lipid component - termed <u>lipid A</u> - was claimed to contain the very structures responsible for endotoxicity (7). This claim was originally based on the fact that lipid A appeared to be ubiquitous for many, otherwise different lipopolysaccharides extracted from quite a variety of bacterial species. But it took some time until final proof could be achieved.

Of great help in this direction was the study of non-pathogenic <u>rough mutants</u> in comparison with their pathogenic parent smooth strains (8). In a short time many R-mutants were selected, extracted applying a specially developed technique, the PCP-method (9), and their LPS' analyzed. The various R-mutants were thus shown to be deficient in the biosynthesis of their complete (smooth) O-antigenic polysaccharide. The most deficient strains - so-called <u>Re-mutants</u> - produced only a trisaccharide composed of an unusual C_8 sugar acid, keto-deoxy-octonic acid (KDO), bound to lipid A. Such low-molecular glycolipids exerted full endotoxicity, thus proving that the long-chain polysaccharide component of LPS is not essential for endotoxicity. Finally, free lipid A could be produced from Re-glycolipids by mild acid treatment. After dispersion in water lipid A was shown to exert practically all known endotoxic activities (10).

Extended structural analyses of lipid A preparations, especially in joint German-Japanese efforts, and ever since continued, not only allowed evolutionary speculations about endotoxin, but also brought an insight into the general unique chemical make up of the endotoxic principle and furnished ideas about possible minimal structural requirements for certain endotoxic reactions. At this stage the time appeared to be ripe for synthetic approaches. Several chapters of this book will inform about the present state of investigations on the fine structure of lipid A and relations to biological activity. In the same context the work of other active japanese colleagues on the synthesis of lipid A, lipid A-like structures and substructures and first results about endotoxic activities will be described.

The synthetic approach, as always in the analysis of natural products, will lead to a final understanding of structure and biological activity. Of course, it should provide ample proof of the anticipated structure, specific details of which can easily be checked by testing for any of the pronounced activities of endotoxin, such as pyrogenicity (in the order of ngs/kg in sensitive animals or man). Synthesis will further allow the production of lipid A analogues or partial structures which Nature cannot provide. We are looking forward to raising again the old question whether or not some of the many typical endotoxic activities can be chemically differentiated in an approach to more selective primary endotoxic target reactions.

Referring to William WINDLE's work on endotoxin-mediated nerve regeneration, one may be reminded of another scientific hero who, around the same time, was much engaged with the biochemical analysis of the answer of higher animals to irritation. Animal and man react in a characteristic manner, and irritating agents, such as adequate doses of endotoxin, are percieved by the body as biological signals. I am speaking of the russian-born pathologist Valy MENKIN who was working in the United States, mainly in Philadelphia, on mechanisms of local inflammation (11). He arrived at the conclusion that exogenous irritants, such as toxins, would react with highly affine receptors on target cells, the latter ones in turn answering by the activation of processes - he thought mainly of proteolysis - that lead to the production of more specifically acting mediators.

A pioneer in endotoxin research, Lewis THOMAS in New York, once stressed the concept that it is not endotoxin which is so highly active or toxic per se; it is the higher animal with its affine cells and their receptors and the following sequences of endogenous reactions via mediators that make a species more or less sensitive to endotoxin. May I give you one example: some primates, like the baboon, are about 10.000 to 100.000 fold less sensitive to endotoxin compared to the human (12)! So, lipopolysaccharide or lipid A are only "endotoxically" active in animals where they find the respective environment to percieve their signals.

The mediator theory of Valy MENKIN at his time was not accepted (or otherwise ignored) by his american colleagues. They argued: "You have too much imagination and fantasy" .... But what would Science be without these attributes of human thinking? Today, 25 years after MENKIN's too early death, one of the most important fields of investigations is the <u>search for mediators</u> elicited by endotoxic signals and for the type of cell producing such highly active secondary products, with the hope to finally purify and identify and even synthesize these biologically most interesting agents. - Soon we hope to understand the chemical nature of the signals (lipid A structures) and that of the receptors as well as the resulting mediators.

We should, however, be aware of the fact that highly purified endotoxin preparations and, of course, synthetic products - biologically speaking - are <u>artifacts</u>. They would never occur in such degree of chemical purity during bacterial infection or by release from the flora of the gut. Endotoxin as it is released from the bacterial cell is in a dynamic state of continuous degradation. All these products produce, so to say, <u>"biological noise"</u>, a mixture of signals. In the puristic attitudes of modern Science we are trying to select <u>"pure music"</u> from the mass of noise, and this is what we hope to be able by testing highly purified and structurally well defined materials.

This brings us to <u>clinical aspects</u> of endotoxin. When I was in Japan in 1981 I was impressed by the mass of clinical investigations about <u>endotoxemia</u> as expressed by the <u>Limulus test</u>. Clinicians reported about continuous estimations of endotoxin levels in patients under various clinical conditions, based on quantitative figures of the Limulus test. It seems to me that this is a very important approach which, however, raises a series of questions: does a positive test indicate the presence of only endotoxin? Would a negative test really indicate the absence of endotoxin, at least below a very tiny threshold dose? Are we certain about the specificity of the Limulus reaction? What clinical implications would a positive Limulus test have in patients without typical symptoms? Could such positive tests have diagnostic implications? In the New Brunswick Conference of 1962, Ivan BENNETT pointed to similar problems (4):

"We have to determine the actual significance of endotoxins in human health and disease, not in the controlled environment of the research laboratory but in daily life, in the field, in nature as we have modified it. Endotoxin can cause fever, but how many human fevers are endotoxic? Endotoxin can cause shock, but how often is shock in man endotoxic? Endotoxin can modify resistence to infection, but how often does endotoxin influence the susceptibility of man?" ....

It is known that it is experimentally impossible to establish completely "<u>endotoxin-free life</u>". The question has been raised as to whether the ever-present endotoxin, to which higher animals have adapted during millions of years of evolution, may exert the function of an "<u>exo-hormone</u>" - in the sense that a continuous stimulation of important biological processes by very small but definite amounts of endotoxin is essential for the higher animals life? - We need a highly sensitive and specific test to handle such exciting problem and to answer these questions.

Speaking of clinical aspects we are, of course, faced with questions of relative <u>doses</u> of endotoxin. We are faced with untoward effects of high endotoxin doses, for example in endotoxemia, but also with many beneficial - that means therapeutic - effects of low doses of the same principle. Many of these approaches remained only phenomenological for a long time, because the available <u>techniques</u> were not sufficiently sensitive, for example to follow the extremely low quantities of purified LPS which exert already some typical treshold activities. Our own standard-LPS (13) (Novopyrexal) produces a definite granulocytosis of about 200-300% within 4 hours without any other reactions after the intravenous injection of 0.03 µg <u>totally</u> into healthy adult volunteers. Doses above 0.05 - 0.06 µg would just be slightly pyrogenic. It appears to be highly desirable to follow such tiny quantities of material where even radio-labeling would not be sufficient for that purpose. Whether a sensitive enough radio-immuno-assay, using monospecific (i.e. monoclonal) anti-lipid A antibody, can be established is at present under investigation.

From the clinical point of view endotoxin can be looked at in a threefold manner: as a <u>diagnostic agent</u> to test the acute state of certain defense mechanisms - as a <u>therapeutic agent</u> in the treatment of many, predominantly chronic diseases (including cancer) - and as a <u>toxic agent</u> if higher doses are occurring in the circulation due to release of endotoxin from the gut or during enterobacterial infections.

I like to add one word on bacterial products, such as the <u>muramyl peptides and related synthetic compounds</u> - a speciality of very active groups in France and Japan. Some of these products exert biological activities comparable to endotoxic reactions. It seems that the combination of the two bacterial principles will bring new clinical (and especially therapeutic) outlooks. The subject is now about 90 years old, since Dr. William B. COLEY introduced the so-called Coley-Vaccine for the treatment of certain cancers, especially sarcoma (14). This vaccine was a mixture of a Gram-positive (Streptococcus) and a Gram-negative strain (B. prodigiosus or Serratia marcescens). Dr. OETTGEN will refer on that matter.

May I finish with a short story around the second endotoxin conference which was organized in August 1958 in Freiburg, the home town of many German endotoxicologists. We had invited some 40 to 50 world widely known scientists, all experts in endotoxin research. Hotel accomodation and conference room etc. were fixed long ahead in the most fashionable hotel near the center of the town. Two weeks before our conference should start, King Ibn Saud of Saudi Arabia with a great staff arrived in Freiburg to let the general health of many of his sons check in the Medical Faculty. A famous member of the faculty had once saved the most beloved daughter of Ibn Saud. Since that time Freiburg was a kind of Medical Mecca to the Saudi Arabians. The whole group occupying our hotel was scheduled for their flight home a few days before the start of our conference. One day before that time Ibn Saud decided to stay longer. It was high summer season, all other hotels were completely occupied - and nobody dared to prompt the Saudis to go out in time.

Within twenty four hours we were forced to arrange accomodation for all our participants in many far-away Black Forest hotels. By courtesy of a photoshop we were invited to use his presentation room with good facilities in the cellar for our sessions. Our guests dispersed over the Black Forest by taxis and were brought to the sessions again by taxis.

After the first shock, as one can imagine, our guests began to enjoy spending beautiful nights and mornings in the mountains. The whole affair finally created great amusement, and our sessions were full of good spirit. One evening session was held in a hotel on top of the second highest mountain of the Black Forest. When we just started the subject (it was "endotoxin and the properdin system") an enormous thunderstorm with heavy lightnings blew out all electricity, so that we continued the session with candle light and without projection of slides. Every participant still today remembers the great spirit of this early conference with all its improvisations at the last moment.

The long story of endotoxin, thus, has always been full of scientific and social excitement. The world-wide family of endotoxin researchers has grown, its ever increasing members will grant further collegial cooperation.

References

(1) F.B. Seibert, Am. J. Physiol. 67 (1923) 90-104; 71 (1925) 621-652.

(2) W.F. Windle, Trans. N. Y. Acad. Sci. 14 (1952) 159.

(3) H. Bammer, V. Martini, Pflügers Arch. 257 (1953) 308.

(4) J. Bennett, Introduction. Bacterial Endotoxin, M. Landy, W. Braun edit., Proc. Symp. Rutgers Univ., Inst. of Microbiol., Rutgers Univ. Press, New Brunswick (1963).

(5) F. Kauffmann, Serological diagnosis of Salmonella species, Munksgaard, Copenhagen (1971).

(6) Review see O. Lüderitz, A.M. Staub, O. Westphal, Bact. Rev. 30 (1966) 193-255.

(7) O. Westphal, O. Lüderitz, Angew. Chemie 66 (1954) 407-417.

(8) O. Lüderitz, O. Westphal, Angew. Chemie 78 (1966) 172-185.

(9) C. Galanos, O. Lüderitz, O. Westphal, Eur. J. Biochem. 9 (1969) 245-249.

(10) C. Galanos, O. Lüderitz, E.T. Rietschel, O. Westphal, Int. Rev. Biochem., Biochem. of Lipids II, 14 (1977) 239-335; O. Westphal, K. Jann, K. Himmelspach, Progr. Allergy 33 (1983) 9-39.

(11) V. Menkin, Modern Views of Inflammation, Int. Arch. Allergy appl. Immunol. 4 (1953) 131-168; V. Menkin, Biochemical Mechanisms of Inflammation, 2nd edit., C.C. Thomas, Springfield, Ill. (1956).

(12) O. Westphal, Int. Arch. Allergy appl. Immunol. 40 (1975) 1-43; see also B.M. Sulzer, Infect. Immunol. 5 (1972) 107-113.

(13) C. Galanos, O. Lüderitz, O. Westphal, Zbl. Bakt. I. Orig. A 243 (1979) 226-238; see also C. Galanos et al., Chapter in this book.

(14) W.B. Coley, Am. J. Med. Sci. 105 (1893) 487-511; J. Am. Med. Assoc. 31 (1898) 389-395.

Lipid A:
Structure and Synthetic Approach

Bacterial Lipopolysaccharides and Their Lipid A Component

Ernst Th. Rietschel[1], Hubert Mayer[2], Horst-Werner Wollenweber[2], Ulrich Zähringer[1], Otto Lüderitz[2], Otto Westphal[2], and Helmut Brade[1]

[1] Forschungsinstitut Borstel, D-2061 Borstel, FRG
[2] Max Planck-Institut für Immunbiologie, D-7800 Freiburg i.Br., FRG

1 INTRODUCTION

In the mediation of pathophysiological changes, such as fever and shock, which occur in the course of severe gramnegative infection, the involvement of a bacterial toxin has been postulated. Since this toxin was found to be a constituent of the bacterial cell wall and not to be released into the surrounding environment it was termed endotoxin (1). The endotoxin could be enriched by extraction of bacterial cells with phenol-water, a procedure which led to preparations which were shown to consist of a heteropolysaccharide (O-specific chain and core) and a covalently bound lipid component (lipid A). Chemically, therefore, endotoxins represent lipopolysaccharides (2) (Fig. 1).

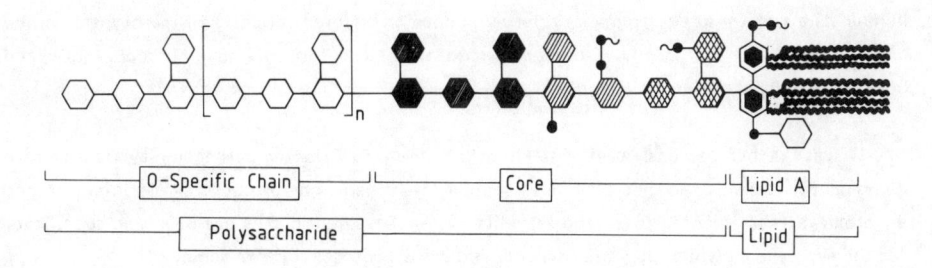

Fig. 1: Schematic Representation of the Structure of a Salmonella Lipopolysaccharide (3)
The structure of the lipid A proximal core region (KDO-trisaccharide) is presently being reinvestigated (10). The number of fatty acid residues (∿∿∿) is arbitrary.
∿ , Phosphorylethanolamine; ● , phosphate; ○ , sugar residue.

Because of their biological significance bacterial lipopolysaccharides (LPS) were studied both from a chemical and biological point of view by a great number of investigators. Some findings and developments in this field were i) the discovery of genetically defective mutants which synthesize LPS with an incomplete polysaccharide portion; ii) the finding that the polysaccharide portion of LPS consists of a relatively invariable core oligosaccharide and a highly variable polysaccharide moiety which is responsible for the O-antigenicity; iii) the discovery that the endotoxic properties of LPS are anchored in the lipid A portion; iv) the development of methods to obtain purified preparations being essentially free of other cell wall components and possessing defined physicochemical properties; v) the elucidation of the biosynthesis of LPS; vi) the analysis and finally vii) the chemical synthesis of lipid A structures.

The main interest of our laboratories presently concerns the establishment of relationships between chemical structure and biological activity of LPS (3). These studies have focussed on both the polysaccharide and the lipid A portion of the endotoxin molecule. In the following the chemical architecture of these LPS components, with emphasis on lipid A, is described briefly and some biological aspects are discussed.

2 POLYSACCHARIDE COMPONENT (O-SPECIFIC CHAIN AND CORE OLIGOSACCHARIDE)

LPS are amphipathic molecules, the hydrophilic portion being represented by the O-specific chain and the core oligosaccharide. The O-specific chain is built up by repeating oligosaccharide units (Fig. 1). Many types of sugars have been found as constituents of repeating units including neutral sugars with 5-7 carbon atoms, deoxy and amino sugars, uronic and aminuronic acids, O-methyl, O-acetyl, and phosphate-substituted sugars (2, 4). This variability of constituents and the different types of linkages cause an immense diversity of the O-specific chains.

The O-polysaccharide is responsible for the O-antigenicity of LPS, which is determined by different O-factors, each representing a distinct immunodeterminant group. Thus, by different combinations of these determinants a large number of different O-antigens (serotypes) can be constructed. Many of these have been shown to occur in nature and have been characterized both chemically and immunologically (2, 4).

It was further demonstrated that O-antigens may function as pathogenicity factors of a given bacterial genus and that antibodies against O-antigens may protect against infection and endotoxin effects such as fever and lethality (5, 6). These antibodies are of a high specificity for the serotype by which they are induced, and cross-protection is not achieved.

Oligosaccharide determinants in the O-chain may function as receptors for bacteriophages which are able to degrade the polysaccharide by means of specific phage-born enzymes. These phages and the isolated enzymes have become powerful tools in investigations of LPS structure (7).

The discovery of rough (or R)-mutants opened a new field for chemical, biological and genetic investigations on LPS. LPS of R-form bacteria are devoid of the O-polysaccharide and some lack parts of the core oligosaccharide (core-defective mutants). According to present knowledge, the lipid A portion of LPS derived from different R-mutants does not differ from that of the parent strain. The mutants most defective in polysaccharide biosynthesis, still multiplying under laboratory conditions are the Re-mutants. This type of mutant has been isolated from different enterobacterial genera (Salmonella, Shigella, Escherichia coli, Proteus, Yersinia) and it produces a LPS which is made up by lipid A and 3-deoxy-D-mannooctulosonic acid (KDO) as the sole sugar constituent of the core (2).

The core consits of a branched heterooligosaccharide. In Enterobacteriaceae at least six individual core types have been described which show only minor differences in their composition. Thus, the core region of LPS expresses much less diversity than the O-polysaccharide (2, 3). In general, two core regions are recognized: the inner core which is made up by oligosaccharides of KDO and L-glycero-D-mannoheptose and the outer core being composed of D-glucose, D-galactose and, often, N-acetyl-D-glucosamine. The inner core is variably substituted by phosphate, phosphorylethanolamine, and in some core types (R2, K-12) by neutral sugars (Fig. 1). Most of the linkages between the sugar residues, in different core types, have been determined (for summary see refs. 2-4). However, the linkages between the KDO residues, their anomeric configuration and even the number of KDO units present, are not yet completely known. Recently, in our laboratories a new antigenic specificity was described which is present in all examined LPS except in those of the Re-type (8, 9). This determinant resides in the inner core region and is different from the so far known core specificities and the lipid A specificity. In the course of studies concerning the chemical structure of this antigenic determinant a KDO disaccharide was isolated from LPS of different Salmonella minnesota rough-mutants without cleaving the polysaccharide-lipid A linkage (10). This finding is not compatible with the previously proposed structure of the inner core region (a branched KDO trisaccharide, compare Fig. 1). The chemical structure of this region, therefore, is presently being reinvestigated in our laboratories.

With the structural classification of different core types, the biological relevance of these structures has been elucidated. Like the O-specific chain, core structures represent receptors for certain phages, and one way to classify R-mutants is by determination of the phage pattern (11). This is of great importance, as the highly lipophilic R-form bacteria cannot be typed by agglutination with specific antisera (they agglutinate spontaneously in physiological saline).

Moreover, the core structures represent antigens and induce core-specific antibodies which can be detected by the passive hemagglutination and the passive hemolysis test. These antibodies have attracted the interest of many scientists working on the biology of LPS and gramnegative infection. As mentioned above, there exists only a limited diversity of core structures. This opens the possibility of preparing vaccines with a broad cross-reactivity. Notably, antisera raised against Re- and Rc-mutants of Salmonella and E. coli species have been discussed as potential candidates for the induction of cross-immunity against LPS and bacteria of different genera. So far, however, the reported results are conflicting (for review of this topic compare ref. 6).

One explanation for this discrepancy may be the presence of the aforementioned new antigenic specificity. A prominent biological feature of this antigen is that it is exposed on the bacterial cell surface and that all mammalian species tested (including man) possess, in high titers, a natural complement-dependent hemolytic activity directed against the determinant (8, 12). The biological relevance of these natural antibodies, however, as well as the potential use of the new antigen as a vaccine has yet to be elucidated. Thus, in biological analyses of endotoxins one should be aware that in all in vivo and many in vitro test systems this antigen and the corresponding antibody are present and may influence the parameters measured.

3 LIPID A COMPONENT

The polysaccharide component is covalently bound to lipid A, which represents the hydrophobic portion of LPS. Lipid A is responsible for the beneficial or harmful biological properties of LPS and it participates in a number of functions exerted by LPS in bacterial cells (2-4). Further, lipid A represents an antigen which, upon immunization, leads to the production of lipid A-specific antibodies (13).

In order to elucidate the biochemical basis of its endotoxic, antigenic and other biologically significant properties the structure of lipid A has been studied intensively in various laboratories. These investigations show that lipid A's of endotoxins obtained from different bacterial sources are structurally closely related. As a representative of this group the structure of Salmonella lipid A will be discussed in some detail below. Essentially distinct lipid A structures have been detected in LPS which do not induce typical endotoxin effects. In the present paper also these lipid A structures will be briefly discussed.

4 SALMONELLA LIPID A

According to present knowledge Salmonella lipid A consists of two ß(1'-6) interlinked 2-amino-2-deoxy-D-glucopyranosyl residues. This glucosamine disaccharide carries two phosphoryl groups, one being bound to the hydroxyl group in position 4' of the non-reducing glucosaminyl residue (GlcN II), the other being α-linked to the glycosidic hydroxyl group of the reducing glucosaminyl residue (GlcN I). This structure which is termed the lipid A backbone carries substituents at the phosphoryl, amino and hydroxyl groups. The phosphate group at position 4' is substituted by a ß-linked 4-amino-4-deoxy-L-arabinopyranosyl residue while the phosphoryl group at position 1 (GlcN I) is substituted by phosphorylethanolamine. Both substituents are present in non-molar amounts (for literature compare refs. 14, 15).

The amino groups of the backbone carry (R)-3-hydroxytetradecanoic acid residues which (in S. minnesota) are, at their 3-hydroxyl group, acylated by dodecanoic and hexadecanoic acid. Recent studies show that the amino group of GlcN II is substituted by (R)-3-dodecanoyloxytetradecanoic and the amino group of GlcN I by (R)-3-hexadecanoyloxytetradecanoic acid (16, 17).

Attached to hydroxyl groups of the backbone are the polysaccharide component (in intact lipopolysaccharide) and two mol equivalents of acyl groups. According to recent investigations it is the primary hydroxyl group in position 6' (and not that in position 3') of GlcN II which represents the attachment site of the polysaccharide portion (14, 15, 18 and F. Unger, personal communication). Hence in free lipid A (devoid of polysaccharide) this hydroxyl group is not substituted. This assumption is supported by the demonstration that in free Salmonella lipid A the hydroxyl group in position 6' can be methylated (39) by diazomethan in the presence of SiO_2 (28). In this experiment a second glucosamine derivative was obtained which carried a methoxy group in position 4. This finding indicates that in free lipid A of Salmonella the hydroxyl group in position 4 (GlcN I) is unsubstituted (for direct proof of this assumption in E. coli free lipid A see below and ref. 19). These findings suggest that the two moles of ester-bound acyl groups are linked to the hydroxyl groups in positions 3 and 3'. In Salmonella the O-acyl groups comprise 1 mol of each (R)-3-hydroxytetradecanoic and (R)-3-tetradecanoyloxytetradecanoic acid. In this latter structure approximately 10% of tetradecanoic acid is present in the α-oxidized form (i.e. (S)-2-hydroxytetradecanoic acid). However, the sum of (R)-3-tetradecanoyloxy- and (R)-3-(S)-2-hydroxytetradecanoyloxytetradecanoic acid equals 1 mole equivalent. Very recently, it was demonstrated that (R)-3-tetradecanoyloxytetradecanoic acid is bound to GlcN II and (R)-3-hydroxytetradecanoic acid to GlcN I. This was shown by subjecting free Salmonella lipid A fractions to positive fast atom bombardment mass spectrometry (20) and by analysing dephosphorylated free lipid A by laser desorption mass spectrometry (17). Collectively these data show that (R)-3-tetradecanoyloxytetradecanoic acid is linked to GlcN II (presumably to the hydroxyl group at position 3') and that (R)-3-hydroxytetradecanoic acid is bound to GlcN I (position 3).

The above data are summarized in Fig. 2 which represents a proposal for the chemical structure of Salmonella minnesota lipid A. The calculated molecular weight of the acylated glucosamine disaccharide lacking 4-amino-4-deoxy-L-arabinose, phosphorylethanolamine and phosphate would have a theoretical value of 1876. When hydrogen fluoride treated (i.e. dephosphorylated) S. minnesota free lipid A was analysed by laser desorption mass spectrometry, quasi molecular ions $(M+K)^+$ were seen at m/z = 1915 $\pm$ 2 and 1931 $\pm$ 2 (17). Correction of these values for K^+ (= 39) yields 1876 and 1892, respectively. The former figure correlates with the calculated molecular weight of the acylated Salmonella lipid A backbone shown in Figure 2. The latter value most likely corresponds to a molecular species which contains 2-hydroxytetradecanoic acid instead of tetradecanoic acid.

It is emphasized that the structural proposal made in Fig. 2 only partially reflects the genuine lipid A structure. This is because some of the lipid A components are not present in

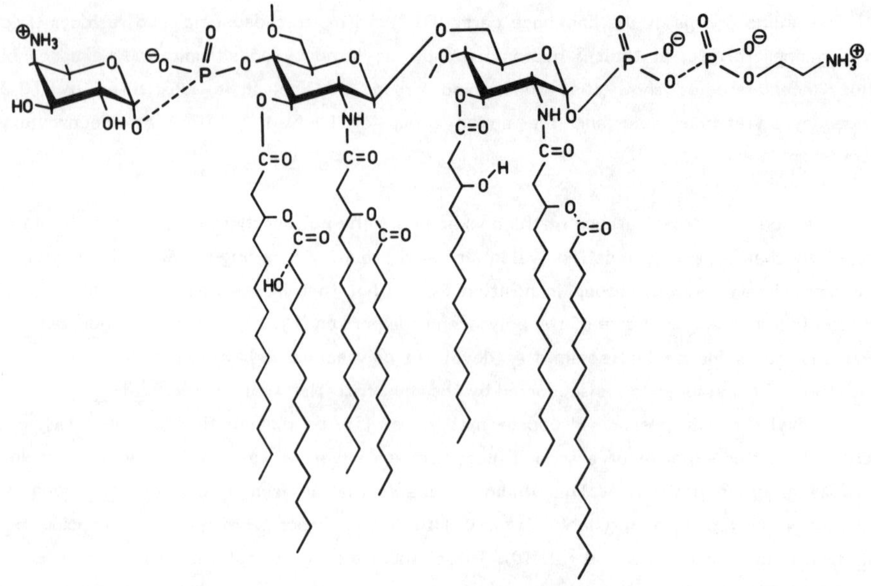

Lipid A: Structure and Synthetic Approach

Fig. 2: Proposed Chemical Structure of the Lipid A Component of Salmonella minnesota Lipopolysaccharide (18)

Dotted bonds indicate nonmolar substitution. The polysaccharide component is linked to the primary hydroxyl group of the nonreducing glucosaminyl residue. In free lipid A the hydroxyl group in position 4 of the reducing glucosaminyl residue is probably not substituted.

molar amounts relative to the lipid A backbone. Thus, approximately only 50% of the phosphoryl groups are substituted by 4-amino-4-deoxy-L-arabinose and phosphorylethanolamine (21). Further, ester and amide-bound 3-hydroxytetradecanoyl residues may only be partially 3-O-acylated (22). In addition, the hydroxyl groups in position 3 and 3' of the backbone may not be quantitatively substituted by 3-hydroxytetradecanoyl or 3-tetradecanoyloxytetradecanoic acid residues (23), and in the latter structure part of the tetradecanoic acid may be present in the α-oxidized form. Therefore, in a lipid A preparation several molecules are present which possess an identical backbone which differ, however, in the degree of substitution of this backbone. It is probable that this structural diversity forms the basis for the well known intrinsic heterogeneity of lipid A.

5 SALMONELLA-RELATED LIPID A TYPES

The β(1'-6)-linked D-glucosamine disaccharide identified in Salmonella also forms the backbone of a large variety of lipid A's from other bacterial sources. It was identified in other Enterobacteriaceae, in Pseudomonadaceae, Vibrionaceae and a number of other bacterial groups (compare refs. 3, 14, 18). In E. coli and Proteus mirabilis the substitution pattern of the backbone has been studied more intensively and these strains, therefore, will be discussed in more detail below.

In P. mirabilis, the ester-linked phosphate group in position 4' (GlcN II) is quantitatively substituted by 4-amino-4-deoxy-L-arabinopyranose, in E. coli it is free (24, 25). On the other hand, the glycosidic phosphoryl group carries, in E. coli, a phosphoryl group while in P. mirabilis it appears to be non-substituted. By nuclear magnetic resonance (NMR, 26, 27) and chemical analysis (24) it was demonstrated that in both strains the saccharide portion in LPS is linked to the primary hydroxyl group at position 6' of the lipid A backbone. In both E. coli and P. mirabilis lipid A, 2 moles of (R)-3-acyloxytetradecanoic and 2 moles of (R)-3-hydroxytetradecanoic acid are present. Of each of these groups 1 mol is involved in amide linkage, the 3-acyloxytetradecanoyl residue being linked to GlcN II and 3-hydroxytetradecanoic acid to GlcN I. Also, of each group 1 mol is present in ester-linkage, again the 3-acyloxytetradecanoyl residues being bound to GlcN II and the non 3-O-acylated fatty acid to GlcN I (16, 17). In a major E. coli lipid A fraction the linkage sites of the two ester-bound acyl groups were, by means of two dimensional (2D)-NMR analysis, unequivocally identified as positions 3 and 3' of the lipid A backbone (19). In this study it was also demonstrated that in this free lipid A fraction the hydroxyl groups in positions 4 and 6' are free.

These data and the results of chemical investigations on endotoxic lipid A's of other gramnegative genera suggest that these lipid A's are structurally closely related but not identical. Thus, all biologically active free lipid A's contain an 1,4'-diphosphorylated ß(1'-6)-linked D-glucosamine disaccharide with free hydroxyl groups in position 4 and 6'. To this backbone, in general, 4 moles equivalents of (R)-3-hydroxy fatty acids are bound, two of which occupy amino functions and two of which are linked to backbone hydroxyl groups (positions 3 and 3'). Both amide-bound and one of the ester-linked (R)-3-hydroxy fatty acids may be acylated at their 3-hydroxyl group.

This structure has not been identified so far in other natural compounds and, hence, it is unique and characteristic for lipid A. Because of its exclusive occurrence in lipid A it is tempting to assume that this structural element may form or contain the structure determining the endotoxic and antigenic properties of lipid A.

Comparison of structures of lipid A derived from different bacterial origin and obtained under different conditions of growth shows that they may differ from each other by (i) the presence and nature of polar head groups substituting the backbone phosphate residues (e.g. 4-amino-4-deoxy-L-arabinose, D-glucosamine, phosphate, phosphorylethanolamine); (ii) the presence of other backbone substituents such as D-glucosamine in Rhodospirillum tenue (position 4); (iii) the facultative 3-O-acylation of ester- and amide-linked 3-hydroxy fatty acids by other fatty acids; (iv) the nature (normal, branched, hydroxylated) and chain lengths of fatty acids acylating 3-hydroxy fatty acids, and (v) the chain lengths and number of 3-hydroxy fatty acid substituting the backbone.

Variation in these parameters creates a large number of related but not identical chemical structures. The term lipid A, therefore, does not describe one defined molecule but it rather designates a family of lipids which are characteristic components of bacterial endotoxin.

6 UNUSUAL LIPID A TYPES

The notion that the lipid A portion of LPS is a rather conservative structure applies to the majority of lipid A's studied so far. However, some species have been recognized in which the lipid A component is structurally different from that of the Salmonella type lipid A. These species are still few in number and they may be characterized as containing unusual lipid A types. Such unusual lipid A types have been identified in phototrophic bacteria but they obviously also occur in Brucella, Bacteroides and other species. Interestingly, the LPS of these bacteria is, in general, of low endotoxicity.

Systematic chemotaxonomic studies carried out on various species of the photosynthetic family Rhodospirillaceae revealed that the chemical composition of lipid A may show important and characteristic differences (29). Different strains of one species usually share, however, the same characteristic constituents, such as D-glucosamine, amide-linked fatty acids and phosphate. The so far investigated species may be arranged in four lipid A groups: group 1 with Rhodopseudomonas gelatinosa, Rhodospirillum tenue and Rhodocyclus purpureus, group 2 with Rhodopseudomonas capsulata and Rhodopseudomonas sphaeroides, group 3 with Rhodomicrobium vannielii, and group 4 with Rhodopseudomonas viridis, Rhodopseudomonas palustris, and Rhodopseudomonas sulfoviridis.

Lipid A's of groups 1 and 2 contain the Salmonella type backbone structure, i.e. the 1,4'-diphosphorylated β(1'-6)-D-glucosamine disaccharide. This backbone is substituted in species of group 1 in a characteristic way by additional sugar units (30). Lipid A's of group 2 strains contain in amide linkage 3-oxotetradecanoic acid which is replacing 3-hydroxytetradecanoic acid either partly (R. sphaeroides) or totally (some strains of R. capsulata) (31).

Lipid A of R. vannielii group 3 also contains the β(1'-6)-linked glucosamine disaccharide which lacks, however, phosphate (32). This disaccharide is substituted to about 30% by non-acylated D-mannopyranosyl-units which are linked to the hydroxyl group at C-4' (GlcN II), i.e. to the position usually occupied by the ester-linked phosphate group (18).

Group 4 lipid A lacks glucosamine. The backbone is formed by a 2,3-diamino-2,3-dideoxy-D-glucopyranosyl residue, a rare sugar not found elsewhere in nature. It may carry one (R. viridis) or two amide-linked fatty acids (33).

The free lipid A's of the species investigated (R. gelatinosa, R. sphaeroides and R. vannielii) carry free hydroxyl groups in position 6' and 4 as was demonstrated by silica gel-catalyzed methylation. It, therefore, could be speculated that KDO is, as was shown in enteric lipid A's (18, 19, 24, 26, 27) linked to the hydroxyl group in position 6' of GlcN II. It might also be assumed that in lipid A's of group 4, KDO is linked to the primary hydroxyl group of the diaminoglucose.

Only the LPS and free lipid A of members of group 1 exhibit endotoxic activity, those of groups 2 and 4 are non-toxic and non-pyrogenic (34). Low toxicity is found with R. vannielii (group 3) (32). Lipid A's of groups 1 and 2 are serologically cross-reacting with each other and with Salmonella free lipid A. Members of group 4 cross-react with each other but not with free Salmonella lipid A or with members of other lipid A groups (29).

The grouping of strains according to their lipid A composition is in good correlation with data on the phylogenetical relationship of purple bacteria, as recently demonstrated by comparison of the 16S-RNA sequence of members of Rhodospirillaceae and Chromatiaceae (35). These studies show that a rather close phylogenetic relationship exists between i) R. sphaeroides and R. capsulata, ii) R. viridis, R. vannielii and R. palustris and iii) R. gelatinosa, R. tenue and R. purpureus (35). With the exception of R. vannielii, which forms a separate lipid A group, these data on genealogical clusters fit well with the chemical lipid A classification.

7 CONCLUDING REMARKS

The chemical analysis of structural parts of the LPS molecule, O-polysaccharide, core and lipid A were, in part, performed with the aim to elucidate the determinants responsible for the biological activities of these portions. In many cases these structures could be identified. Ideally such structural proposals derived from analytical work should be verified by chemical synthesis. Indeed many examples exist where such groups were analytically characterized and chemically synthesized, and where both the bacterial and synthetic structure exhibited comparable biological activity. Thus, O-specific oligosaccharides have been analysed and synthesized and the artificial structure proved to be serologically as active as the natural antigen. Such synthetic products may be of importance for serological investigations and for the preparation of artificial vaccines (for literature compare ref. 14).

Approaches to the chemical synthesis of core part structures have also been initiated (36). Such structures are of great interest because of their potential to serve as immunogens leading to cross protection against various pathogenic gramnegative bacteria.

More recently, the chemical synthesis of lipid A has been approached (37, and other contributions of this volume) and some of the part structures and analogues available are being presently analysed for biological activities i.e. endotoxicity and serological specificity (14, 38, and other contributions of this volume). These investigations should prove or disprove the validity of the proposals made with regard to the lipid A structure. They should further help to decide whether lipid A activities are due to one or several determinants and to define those minimal structural requirements which are important for endotoxic activities, serological specificity, and physiological functions of lipid A.

8 REFERENCES

(1) O. Westphal, U. Westphal, Th. Sommer, In: Microbiology-1977 (D. Schlessinger, ed.) Am. Soc. Microbiol., Washington DC (1977) 221-238.

(2) O. Lüderitz, M.A. Freudenberg, C. Galanos, V. Lehmann, E.Th. Rietschel, D.W. Shaw, In: Current Topics in Membranes (C.S. Razin, S. Rottem, eds.), Microbiological membrane lipids, Academic Press, New York, Vol. 17 (1982) 79-151.

(3) E.Th. Rietschel, C. Galanos, O. Lüderitz, O. Westphal, In: Immunopharmacology and the Regulation of Leukocyte Function. (D. Webb, ed.), Marcel Dekker Inc., New York (1982) 183-229.

(4) C. Galanos, O. Lüderitz, E.Th. Rietschel, O. Westphal, In: International Review of Biochemistry, Biochemistry of Lipids II. (T.W. Goodwin, ed.), University Press, Baltimore, Vol. 14 (1977) 239-335.

(5) C.S. Young, In: Principles and Practics of Infectious Diseases (G.L. Mandell, R.G.Douglas, Jr., J.E. Bennet, eds.), John Wiley, New York, Vol. 1 (1979) 571-608.

(6) C.A. Johnston, S. Greismann, In: Pathophysiology of endotoxin (L. Hinshaw, ed.) Elsevier /North Holland, Biomedical Press (1983) in press.

(7) A.A. Lindberg, In: Surface Carbohydrates of the procaryotic Cell (J.W. Sutherland, ed.) Academic Press, New York (1977) 289-306.

(8) H. Brade, C. Galanos, J. Med. Microbiol. 16 (1983) 203-210.

(9) H. Brade, C. Galanos, Infect. Immun. 42 (1983) 250-256.

(10) H. Brade, C. Galanos, O. Lüderitz, Eur. J. Biochem. 131 (1983) 201-203.

(11) A.A. Lindberg, Ann. Rev. Microbiol. 27 (1973) 205-241.

(12) H. Brade and L. Brade, unpublished results.

(13) C. Galanos, O. Lüderitz, O. Westphal, Eur. J. Biochem. 24 (1971) 116-122.

(14) E. Th. Rietschel, U. Zähringer, H.-W. Wollenweber, K. Tanamoto, C. Galanos, O. Lüderitz, S. Kusumoto, T. Shiba, In: Handbook of Natural Toxins, Vol. II Bacterial Toxins (A. Tu, W.H. Habig, M.C. Hardegree, eds.), Marcel Dekker Inc., New York (1983) in press.

(15) E. Th. Rietschel, H.-W. Wollenweber, Z. Sidorczyk, U. Zähringer, O. Lüderitz, In: Bacterial Lipopolysaccharides: structure, synthesis, biological activities (L. Anderson and F. Unger, eds.), Am. Chem. Soc., Washington DC 231, (1983) 195-218.

(16) H.-W. Wollenweber, K. Broady, O. Lüderitz, E. Th. Rietschel, Eur. J. Biochem. 124 (1982) 191-198.

(17) B. Lindner, H.-W. Wollenweber, U. Zähringer, E. Th. Rietschel, U. Seydel, Biomed. Mass Spect., in preparation.

(18) E. Th. Rietschel, H.-W. Wollenweber, R. Russa, H. Brade, U. Zähringer, Rev. Infect. Dis. (1984) in press.

(19) M. Imoto, S. Kusumoto, T. Shiba, H. Naoki, T. Iwashita, E. Th. Rietschel, H.-W. Wollenweber, C. Galanos, O. Lüderitz, Tetrahed. Lett. 24, (1983) 4017-4020.

(20) N. Qureshi, K. Takayama, D. Heller, C. Fenselau, Rev. Infect. Dis. (1984) in press.

(21) P.F. Mühlradt, V. Wray, V.A. Lehmann, Eur. J. Biochem. 81 (1977) 193-203.

(22) H.-W. Wollenweber, S. Schlecht, O. Lüderitz, E. Th. Rietschel, Eur. J. Biochem. 130 (1983) 167-171.

(23) N. Qureshi, K. Takayama, E. Ribi, J. Biol. Chem. 257 (1982) 11808-11815.

(24) Z. Sidorczyk, U. Zähringer, E. Th. Rietschel, Eur. J. Biochem., submitted.

(25) M.R. Rosner, H.G. Khorana, A.C. Satterthwait, J. Biol. Chem. 254 (1979) 5918-5925.

(26) S.M. Strain, S.W. Fesik, J.M. Armitage, J. Biol. Chem. 258 (1983) 2906-2910.

(27) S.M. Strain, S.W. Fesik, J.M. Armitage, Rev. Infect. Dis. (1984) in press (abstract).

(28) K. Ohno, H. Nishiyama, H. Nagase, Tetrahed. Lett. 45 (1979) 4405-4406.

(29) J. Weckesser, G. Drews, H. Mayer, Ann. Rev. Microbiol. 33 (1978) 215-239.

(30) R.N. Tharanathan, J. Weckesser, H. Mayer, Eur. J. Biochem. 84 (1978) 385-394.

(31) P.V. Salimath, J. Weckesser, W. Strittmatter, H. Mayer, Eur. J. Biochem. 136, (1983) 195-200..

(32) O. Holst, D. Borowiak, J. Weckesser, H. Mayer, in preparation.

(33) N.M. Ahamed, H. Mayer, H. Biebl, J. Weckesser, FEMS Microbiol. Lett. 14 (1982) 27-30.

(34) W. Strittmatter, J. Weckesser, P.V. Salimath, C. Galanos, J. Bacteriol. (1983) in press.

(35) J. Gibson, E. Stackebrandt, L.B. Zablen, R. Gupta, C.R. Woese, Microbiol. 3 (1979) 59-66.

(36) F.M. Unger, Adv. Carbohydr. Chem. Biochem. 38 (1981) 323-382.

(37) S. Kusumoto, M. Inage, H. Chaki, M. Imoto, T. Shimamoto, T. Shiba, In: Bacterial Lipopolysaccharides: structure, synthesis and biological activities (L. Anderson and F. Unger, eds.), Am. Chem. Soc., Washington DC (1983) in press.

(38) O. Lüderitz, K. Tanamoto, C. Galanos, O. Westphal, U. Zähringer, E.Th. Rietschel, In: Bacterial Lipopolysaccharides: structure, synthesis and biological activities (L. Anderson and F. Unger, eds.), Am. Chem. Soc., Washington DC (1983) in press.

(39) H.-W. Wollenweber, R. Russa, O. Lüderitz, E.Th. Rietschel, unpublished results.

ACKNOWLEDGEMENTS

We thank H. Moll for mass spectrometric analyses and I. Keller for expert technical assistance. We also thank M. Lohs and B. Köhler for illustrations and photographs, and M. Buth and G. Dettki for typing this manuscript.

Characterization and Biological Activities of the Lipid A Precursors from an E. coli Mutant Deficient in Phosphatidylglycerol

Masahiro Nishijima, Fumio Amano, Akemi Minoura, Yuzuru Akamatsu
Department of Chemistry, National Institute of Health, Kamiosaki, Shinagawa-ku, Tokyo 141, Japan

Christian R. H. Raetz
Department of Biochemistry, University of Wisconsin-Madison, Madison, Wisconsin 53706, U.S.A.

Abstract

A temperature-sensitive mutant of E. coli, deficient in phosphatidylglycerol at 42°C, was found to accumulate two glycolipids, designated X and Y. These substances were purified by DEAE-cellulose column and preparative thin layer chromatographies. Compositional analyses demonstrate that X and Y contain 1 mol of phosphate / mol of glucosamine, suggesting that they are derivatives of lipid A, probably precursors of lipid A. In addition, X contains approximately one amide-linked and one ester-linked 3-hydroxymyristic acid moieties per mol of glucosamine. Y is similar but is further esterified with palmitate. Neither X nor Y possesses detectable 3-deoxy-D-mannooctulosonic acid, and both are deficient in esterified laurate and myristate when compared with mature lipopolysaccharide. As for the biological activities of these compounds, we tested their abilities for the macrophage activation. Both X and Y were shown to activate macrophages, judging from their morphological changes and also their killing activity for tumor cells.

1 Introduction

   Lipopolysasccharide is an important constituent of the outer surface membrane of gram-negative bacteria and elicits an extraordinary variety of distinct biological effects such as pyrogenecity, adjuvanticity, macrophage activation, B-lymphocyte mitogenecity, and tumor regression (1). A hydrophobic part of the lipopoysaccharide molecule called lipid A has been shown to be responsible for these broad biological activities (1).
Recently we have isolated a temperature-sensitive E. coli mutant in which the combined level of phosphatidylglycerol and cardiolipin could be dramatically reduced at the nonpermissive temperature (2,3). In characterizing the lipid composition of the mutant, we discovered that novel lipopolysaccharide precursors accumulate, particularly under nonpermissive conditions, when phosphatidylglycerol levels are low (2,3). We now report the purification and initial biochemical characterization of these glycolipids, which are designated as X and Y (4). The structural studies suggest that X and Y are closely related metabolites, most likely precursors of lipid A. We also report the preliminary results for the biological activities of these substances.

2 Accumulation of two novel glycolipids in the mutant (strain 11-2) deficient in phosphatidylglycerol

   Fig. 1 shows two-dimensional thin layer chromatograms of the chloroform extracts from cultures of the parent and the mutant (strain 11-2). These cells were labeled with ^{32}P-phosphate at the nonpermissive temperature. The major phospholipids in wild-type cells were phosphatidylethanolamine, phosphatidylglycerol and cardiolipin. However, in the cells of this mutant, the contents of phosphatidylglycerol and cardiolipin were so much reduced. In addition to this striking decrease in the phosphatidylglycerol and cardiolipin contents, two compounds designated X and Y appeared in substantial quantities, particularly if the phospholipids were extracted under acidic conditions.
   These unexpected compounds were efficiently labeled with ^{32}P-phosphate and ^{14}C-N-acetyl-glucosamine, but they were not labeled with ^{3}H-glycerol. These results suggested that both X and Y were not phospholipids, but might be intermediates in the generation of lipopolysaccharide.

Lipid A Precursors

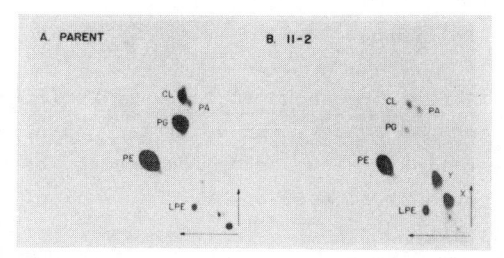

Fig. 1. Two-dimensional thin layer chromatograms of the chloroform extract from mutant (strain 11-2) and parent.

3 Extraction and purification of X and Y

In order to know more detailed characters of these compounds, we purified these substances. First of all, mutant cells were grown at 30°C; when the optical density arrived at about 0.2, the culture was shifted to 42°C, and the cells were harvested after 3 hours at 42°C. Next, X and Y, along with all major phospholipids, were extracted into chloroform by the method of Bligh and Dyer under acidic conditions. When the extraction was conducted at pH 7.4, Y was also in chloroform phase, but X was almost completely in methanol-water phase.

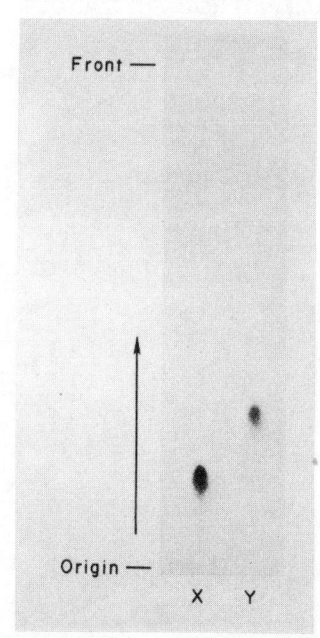

After extraction, X and Y were further purified by DEAE-cellulose column and preparative thin layer chromatograms. Purified X and Y migrated as well defined spots upon analytical Silica Gel 60 thin layer chromatographies, when visualized with either by autoradiography (Fig. 2) or by phosphate staining with the Dittmer-Lester reagent. On DEAE-cellulose thin layer plates, X and Y also migrated as single spots, although Y was somewhat broader than X, suggesting further microheterogeneity of Y.

Fig. 2. Autoradiogram of ^{32}P-labeled X and Y purified from strain 11.2.

4 Compositional anlyses of X and Y

Some chemical characters of X and Y are shown in Table 1. Purified X and Y both contain glucosamine and phosphate in an equimolar ratio, but neither has measurable KDO. Ester-linked fatty acids, determined by a spectrophotometric method, were present at 0.99 and 1.79 mol/mol of glucosamine in X and Y, respectively.

Table 1. Composition of X and Y.

Compound	X	Y
	molar equivalents (a)	
Glucosamine	1.00	1.00
Phosphate	0.99	1.06
KDO	0.02	0.01
Fatty acid ester	0.99	1.79

(a) Analytical values are given in molar ratios to glucosamine.

Table 2. Fatty acid composition of X and Y.

Component and treatment	X 0.1 N NaOH	X 4 N KOH	Y 0.1 N NaOH	Y 4 N KOH
	molar equivalents			
Glucosamine	1.00	1.00	1.00	1.00
Laurate	_(a)	-	-	-
Myristate	Trace	0.03	0.07	0.12
Palmitate	0.07	0.12	0.65	0.73
3-Hydroxymyristate	0.92	1.92	0.89	1.63
Total fatty acid	0.99	2.17	1.61	2.48

(a); not detected.

The fatty acid composition was determined by gas chromatography. Upon mild alkaline hydrolysis (0.1 NaOH, Table2), 0.92 mol of 3-hydroxymyristate and a small amount of palmitate were released from X per mol of glucosamine, while Y released 0.65 mol of palmitate in addition to 0.89 mol of 3- hydroxymyristate and a small amount of myristate per mol of glucosamine. On the other hand, when treated with strong alkali (4 N KOH), both X and Y liberated, respectively, 1.00 mol and 0.74 mol of additional 3-hydroxymyristate, presumably reflecting the presence of amide linkages. No lauric acid was present in either compound. The total amounts of fatty acid esters (mild alkali-labile) in X and Y determined by gas chromatography were in good agreement with the spectrophotometric values shown in Table 1.

Based on these and other analytical studies, initially we speculated that the backbone structure of X was β-1,6-glucosamine disaccharide. However, Takayama et al. have recently demonstrated that the backbone structure of X was not β-1,6-glucosamine disaccharide but glucosamine-1-phosphate (5). Fig 3 shows the proposed structure of X and a possible structure of Y.

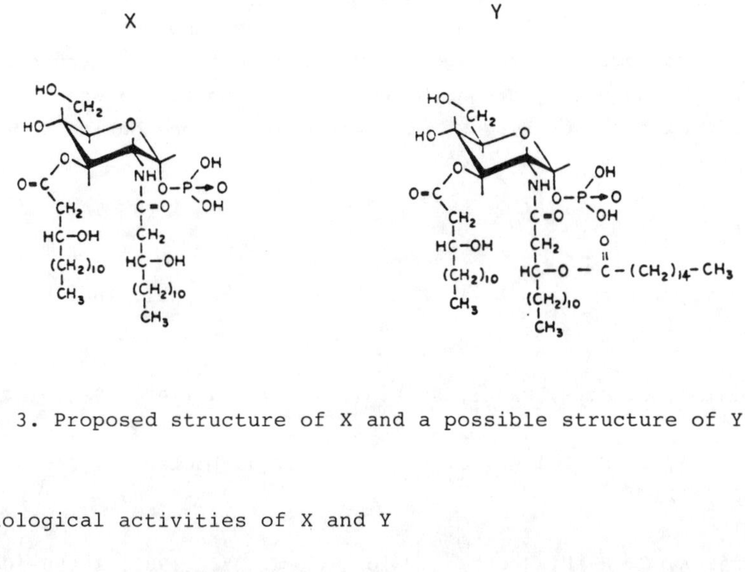

Fig. 3. Proposed structure of X and a possible structure of Y.

5 Biological activities of X and Y

As for the biological activities of X and Y, we tested their abilities for the macrophage activation. For that purpose we examined whether X and Y were able to induce morphological changes of macrophages, using a macrophage-like cell line called J774.1. The cells spread as shown in Fig. 4, when treated with lipopolysaccharide. The microscopic observation revealed that both X and Y were able to induce the morphological changes of the macrophage-like cells.

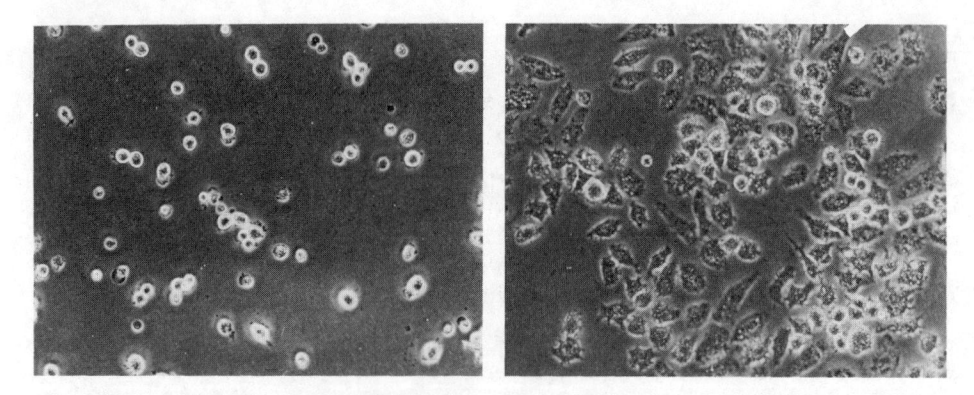

Fig. 4. Morphological change of J774.1 cells by lipopolysaccharide.
Left, - LPS; right, + LPS.

In addition to the induction of the morphological changes of the macrophages, X and Y were also shown to induce the tumor cell killing activity in macrophages; macrophages from ICR mice that had previously received i.p. injection of peptone became fully tumoricidal when incubated with either X or Y, just like with lipopolysaccharide and lipid A (manuscript in preparation).

6 References

(1) C. Galanos, O. Lüderitz, E. T. Rietschel, O. Westphal, Int. Rev. Biochem. 14 (1977) 239-335.

(2) M. Nishijima, C. R. H. Raetz, J. Biol. Chem. 254 (1979) 7837-7844.

(3) M. Nishijima, C. E. Bulawa, C. R. H. Raetz, J. Bacteriol. 145 (1981) 113-121.

(4) M. Nishijima, C. R. H. Raetz, J. Biol. Chem. 256 (1981) 10690-10696.

(5) K. Takayama, N. Qureshi, P. Mascagni, M. A. Nashed, L. Anderson, C. R. H. Raetz, J. Biol. Chem. in press.

How Can Chemical Synthesis Approach the Entity of Lipid A?

Tetsuo Shiba[*], Shoichi Kusumoto, Masaru Inage, Haruyuki Chaki, Masahiro Imoto, Tetsuo Shimamoto

Department of Chemistry, Faculty of Science, Osaka University, Toyonaka, Osaka 560, Japan

Abstract

In order to elucidate a relationship between chemical structure and multiple biological activities of lipid A, thirteen acylated derivatives of β-1,6-D-glucosamine disaccharide with or without phosphate groups based on the proposed structure for the natural lipid A have been prepared. In this study, new methods for the asymmetric synthesis of (R)-3-hydroxytetradecanoic acid as well as the phosphorylation by use of butyllithium were exploited.

From biological activities of these synthetic compounds so far tested in many collaborated research groups, the noteworthy conclusions were drawn that the presence of acyloxyacyl type fatty acid attached to the amino groups and also l-phosphate group seem to be particularly important for exhibition of pyrogenicity, lethal toxicity, Shwartzman reaction, and immunoadjuvant activity.

1 Introduction

Since the chemical structure for lipid A of Salmonella type was proposed by Westphal, Lüderitz and their colleagues as shown in (1) (1,2), many attentions have been paid to a real structural entity responsible for biological activities of endotoxin. However, extreme difficulties in purification of the natural lipid A due to the intrinsic heterogeneity have much disturbed for many investigations to confirm the correct structure of lipid A. Although the structural skeleton of β-1,6-D-glucosamine disaccharide was reconfirmed by Khorana et al. in their study on lipid A of E. coli type (3,4), exact positions and definite structures of fatty acids in lipid A are still not yet fixed.

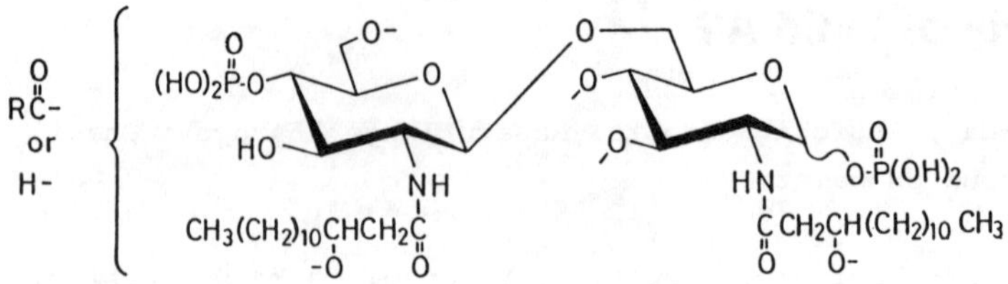

1

In this situation, a progressive approach from the side of organic synthesis was expected to give us very important and valuable informations to solve the structure problem and to establish a structure-activity relationship in lipid A chemistry.

2 Syntheses of dephosphoacyldisaccharides

A fundamental skeleton of β-1,6-glucosamine disaccharide corresponding to the proposed structure of lipid A was constructed by coupling of two moieties (2)(3) of N-acetyl-D-glucosamine according to Königs-Knorr method. In order to carry out the selective acylation through versatile method which will give any acylated derivatives requested, we investigated a series of the reaction processes leading to the dephosphoacyldisaccharide, i.e., 6-O-(2-deoxy-2-tetradecanoylamino-6-O-tetradecanoyl-β-D-glucopyranosyl)-2-deoxy-2-tetradecanoylamino-3,4-di-O-tetradecanoyl-D-glucopyranose (4)(5) as shown below.

In this case, only tetradecanoic acid was used as an acyl function. However, by application of the above synthetic procedure, other acyldisaccharides (5) which possess different acyl functions at hydroxyl and amino groups were also prepared (6)(7). (R)-3-Hydroxytetradecanoic acid as a major fatty acid component in natural lipid A was synthesized by use of novel asymmetric hydrogenation of methyl 3-oxotetradecanoate over Raney nickel modified with (R,R)-tartaric acid (8).

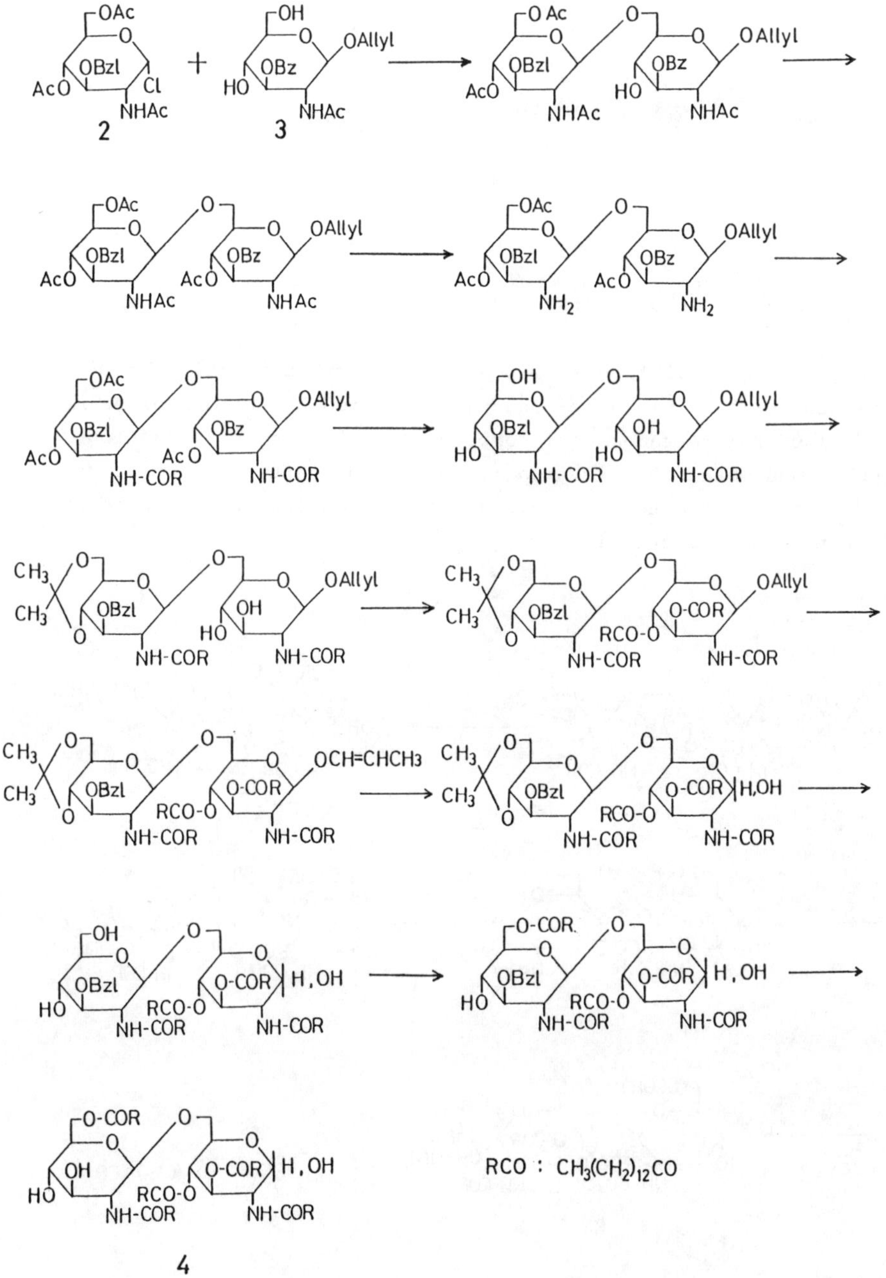

Synthetic Approach to Lipid A

RCO : $CH_3(CH_2)_{12}CO$

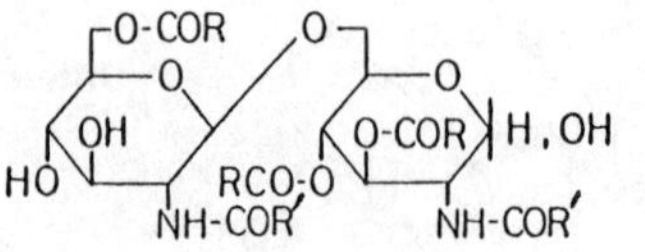

RCO : $CH_3(CH_2)_{12}CO$

R'CO : $CH_3(CH_2)_{10}CH(OH)CH_2CO$

5

3 Syntheses of phosphoacyldisaccharides

For the phosphorylation at 1 position on acyl derivative of D-glucosaminyl-β(1,6)-D-glucosamine was performed by the oxazoline method in the case of N-tetradecanoyl compound (6)(9). In this reaction, α-phosphate was obtained as a predominant product after prolonged reaction time due to the thermodynamical conversion from β-phosphate which was first formed kinetically (9).

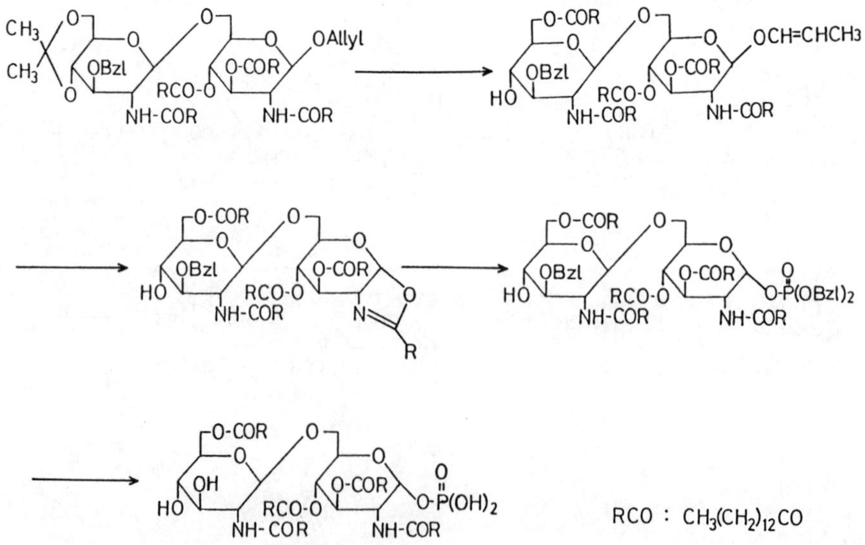

RCO : $CH_3(CH_2)_{12}CO$

6

For preparation of 4'-phosphate (7) of the disaccharide, phenyl-phosphate–dicyclohexylcarbodiimide method afforded us the best result so far tried (9). Combination of both phosphorylation techniques afforded 1-α,4'-diphosphate (8) as shown below (9).

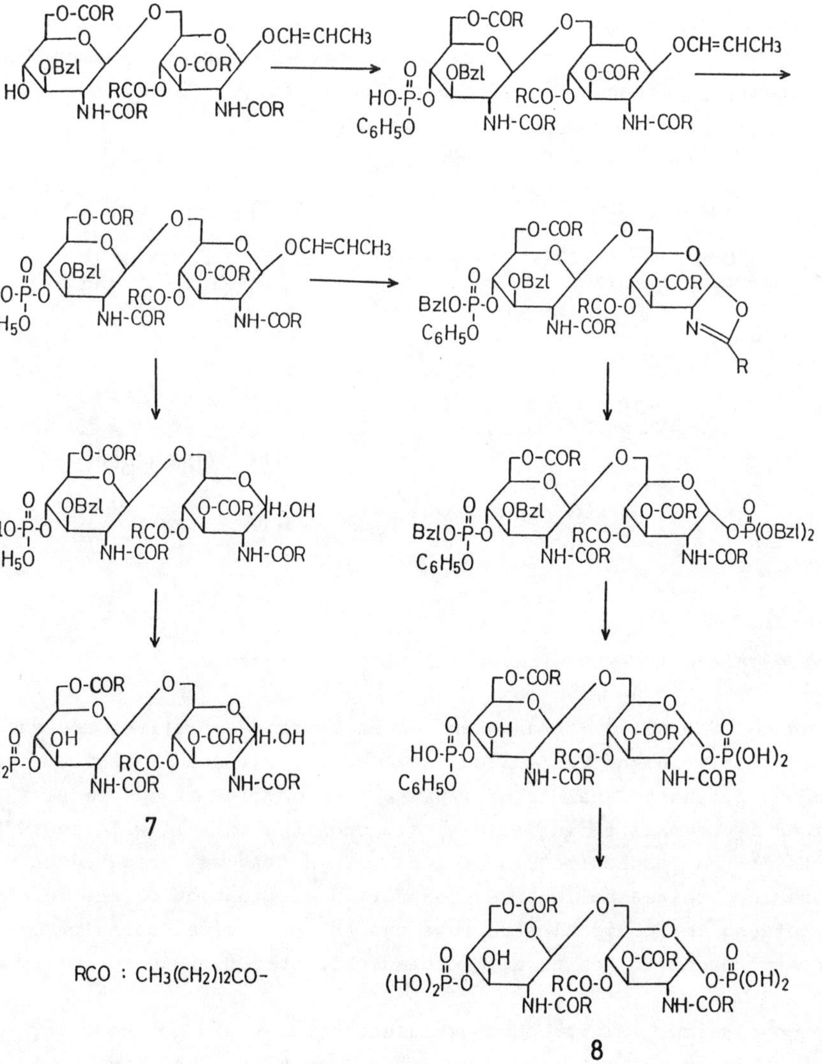

RCO : CH₃(CH₂)₁₂CO-

The oxazoline method was found to be not applicable to an N-3-hydroxyacyl derivative as in the natural compound since β-elimination occurred at the 3-hydroxyl group in the fatty acid part to give an α,β-unsaturated acyl derivative. To overcome this defect, we exploited a new phosphorylation method by use of dibenzyl phosphorochloridate and butyllithium (10). The reaction proceeded smoothly at -70°C, and the desired 1-phosphate was obtained after hydrogenation. This method was not only satisfactorily effective to the preparation of 1-phosphate of N-3-hydroxyacyl disaccharide (9) but also readily applicable to the phosphorylation at other positions on sugar moieties (7).

$RCO : CH_3(CH_2)_{12}CO$

$R'CO : CH_3(CH_2)_{10}\overset{OBzl}{C}HCH_2CO$

$R''CO : CH_3(CH_2)_{10}\overset{OH}{C}HCH_2CO$

4 Biological activities of synthetic lipid A analogs

Many collaborative research groups whose representatives are shown in parenthesis in the Table 1 carried out the various biological tests using our synthetic analogs of lipid A. An outline of the results obtained is summarized in Table 1, although the details will be left to reports of the respective researchers. Such list was prepared only for purpose to obtain a collective prospect in elucidation of the relationship between the chemical structure and the biological activity of lipid A in order to search a more plausible structure as an entity of lipid A.

In conclusion, all dephosphoacyldisaccharides did not show any specific biological activities of lipid A such as pyrogenicity, lethal toxicity and Shwartzman reaction while they manifested some activities related to the immunity more or less. Particularly, the lipid A

Table 1. Biological activities of synthetic lipid A analogs

C_{14} : tetradecanoyl,
C_{14}-OH : (R)-3-hydroxytetradecanoyl,
C_{14}-O-(C_{14}) : (R)-3-tetradecanoyloxytetradecanoyl,
P : $P(OH)_2=O$

Sample number	R^1	R^2	R^3	R^4	Pyrogenicity (Lüderitz)(Homma)	Toxicity (Lüderitz)	Limulus test (Iwanaga)	Shwartzman reaction (Kasai)	Modification of hepatic enzyme act. (Kasai)	Antigenicity (Lüderitz)
301	C_{14}	C_{14}	H	H	− −	−	−	−	−	−
302	C_{14}	C_{14}-OH	H	H	− −	−	−	−	−	−
307	H	C_{14}-OH	H	H	− −	−	−	−	−	−
311	C_{14}-OH	C_{14}-OH	H	H	− −	−	−	−	−	−
314	C_{14}-OH	C_{14}-O-(C_{14})	H	H	− −	−	−	−	±	+
304	C_{14}	C_{14}	P	H	− ±	− −	−	−	−	−
321	C_{14}	C_{14}-OH	P	H	± −	− −	−	−	± +	−
312	C_{14}-OH	C_{14}-OH	P	H	− −	−	−	−	−	−
315	C_{14}-OH	C_{14}-O-(C_{14})	P	H	− −	−	−	−	−	−
303	C_{14}	C_{14}	H	P	− ±	− +	−	−	− ‡	−
316	C_{14}	C_{14}-OH	H	P	± −	+	−	+	− +	+
305	C_{14}	C_{14}	P	P	− −	−	−	−	−	−
317	C_{14}	C_{14}-OH	P	P	− +	+	−	±	− +	+

Lipid A: Structure and Synthetic Approach

Table 1. Biological activities of synthetic lipid A analogs (continued)

C_{14} : tetradecanoyl,
C_{14}-OH : (R)-3-hydroxytetradecanoyl,
C_{14}-O-(C_{14}) : (R)-3-tetradecanoyloxytetradecanoyl,
$P : P(OH)_2 \!=\! O$

Sample number	R^1	R^2	R^3	R^4	Polyclonal B-cell activation (Homma)	B-cell mitogenicity (Kotani)	Monocyte migration enhancement (Kotani)	Adjuvant activity Liposome antiTNP PFC (Kanegasaki)	Adjuvant activity Ab (Kotani)	Adjuvant activity W/O DTH (Kotani)	IF induction (Homma)	Tumor suppression (Dai-ichi)
301	C_{14}	C_{14}	H	H	−	−	−	−	−	−	−	±
302	C_{14}	C_{14}-OH	H	H	−	+	−	−	±	+	−	−
307	H	C_{14}-OH	H	H	−	−	++	−	−	−	−	
311	C_{14}-OH	C_{14}-OH	H	H	−	−	+	−	+	+	−	−
314	C_{14}-OH	C_{14}-O-(C_{14})	H	H	−	+	++	±	++	++	−	
304	C_{14}	C_{14}	P	H	−	−	−	−	−	−	−	+
321	C_{14}	C_{14}-OH	P	H	+	±	+	−	−	+	−	
312	C_{14}-OH	C_{14}-OH	P	H		−	++	+±	++	+	++	
315	C_{14}-OH	C_{14}-O-(C_{14})	P	H					++	++	+	
303	C_{14}	C_{14}	H	P	+	−	+	+±	−	−	+	++
316	C_{14}	C_{14}-OH	H	P	+	+	+	+±	±±	±	+	
305	C_{14}	C_{14}	P	P	±	−	++	+±	++	+	−	+
317	C_{14}	C_{14}-OH	P	P	±	±	++	+±	++	++	−	+±

activities were significantly recognized in the synthetic compounds which have (R)-3-hydroxytetradecanoyl groups at amino functions on both glucosamine moieties and phosphate groups either at 1 (compound 316) or 1,4' (compound 317) positions. Furthermore, these two compounds exhibited the remarkable activities of immunoadjuvancy as well as antitumor effect. Therefore, our synthetic approach seems to be very closed to a final goal to reach a real structure of natural lipid A. However, much lower activities in the synthetic compounds compared with those in the natural substance still remain as an inevitable question to be solved (11).

On the other hand, it should be pointed out that, for example, the compound 312 indicated remarkable activities in the interferon-induction and antitumor effect whereas it did not show undesirable toxicity or pyrogenicity. This fact may amply suggest us that the multiple biological activities in lipid A could be separated depending on each fragment structure. Therefore one can expect a future possibility for creative production of a new compound which has selective pharmacological activity without any hazardous effects.

5 References

(1) O. Westphal, O. Lüderitz, Angew. Chem. 66 (1954) 407-417.

(2) C. Galanos, O. Lüderitz, E. T. Rietschel, O. Westphal, in T. W. Goodman (Ed.), International Review of Biochemistry, Biochemistry of lipids. II. Vol. 14, University Park Press, Baltimore 1977, pp. 239-

(3) M. R. Rosner, J. -Y. Tang, I. Barzilag, H. G. Khorana, J. Biol. Chem. 254 (1979) 5906-5917.

(4) M. R. Rosner, H. G. Khorana, A. C. Statterthwait, J. Biol. Chem. 254 (1979) 5918-5925.

(5) M. Inage, H. Chaki, S. Kusumoto, T. Shiba, Tetrahedron Lett. 21 (1980) 3889-3892.

(6) M. Inage, H. Chaki, S. Kusumoto, T. Shiba, Chemistry Lett., 1980, 1373-1376.

(7) M. Inage, H. Chaki, M. Imoto, T. Shimamoto, S. Kusumoto, T. Shiba, Tetrahedron Lett. 24 (1983) 2011-2014.

(8) A. Tai, M. Nakahata, T. Harada, Y. Izumi, S. Kusumoto, M. Inage, T. Shiba, Chemistry Lett., 1980, 1125-1126.

(9) M. Inage, H. Chaki, S. Kusumoto, T. Shiba, Tetrahedron Lett. 22 (1981) 2281-2284.

(10) M. Inage, H. Chaki, S. Kusumoto, T. Shiba, Chemistry Lett. 1982, 1281-1284.

(11) According to our recent structural study with 2D-NMR method on the chemically pure sample of E. coli Re lipid A, it was strongly suggested that 1,4'-bisphospho-β(1,6)-glucosamine disaccharide in lipid A was acylated at 3 and 3' hydroxyl groups as well as two amino groups, leaving free at 6'-hydroxyl position. We are now engaged in the synthetic study of this newly proposed structure of lipid A. M. Imoto, S. Kusumoto, T. Shiba, H. Naoki, M. Iwashita, E. Th. Rietschel, N. -W. Wollenweber, C. Galanos, O. Lüderitz, Tetrahedron Lett., in submission.

Synthetic Studies on Lipid A and Related Compounds

Makoto Kiso, Akira Hasegawa

Department of Agricultural Chemistry, Gifu University, Gifu 501-11
Japan

Abstract

2-Acylamino-2-deoxy-α-D-glucopyranosyl phosphates and related compounds were synthesized by treatment of the 1-halides with the new phosphorylating agent, $Bu_3SnOP(O)(OR^1)(OR^2)$. Quantitative 4-O-phosphorylation was performed by using diphenyl phosphorochloridate in the presence of 4-dimethylaminopyridine. By applying these methods, a variety of monosaccharide and disaccharide analogs of lipid A were synthesized.

1 Introduction

In the course of studies (1-3) on cell-surface complex carbohydrates having a variety of interesting immunomodulating activities, we have developed some new synthetic routes to the lipid A and related compounds (4). Especially, the successful applications of the ferric chloride-catalyzed glycosylation (5) as well as a new phosphorylation method (6) developed by us, stimurated our recent approach to the active center of endotoxin.

2 Monosaccharide analogs of lipid A and related compounds

The outlines for the preparation of monosaccharide analogs of lipid A are shown in Fig. 2. The ferric chloride-catalyzed glycosylation was conducted by treating compound <u>1</u> (R^2=NHAcyl or NPhth; 1 mol equiv.)

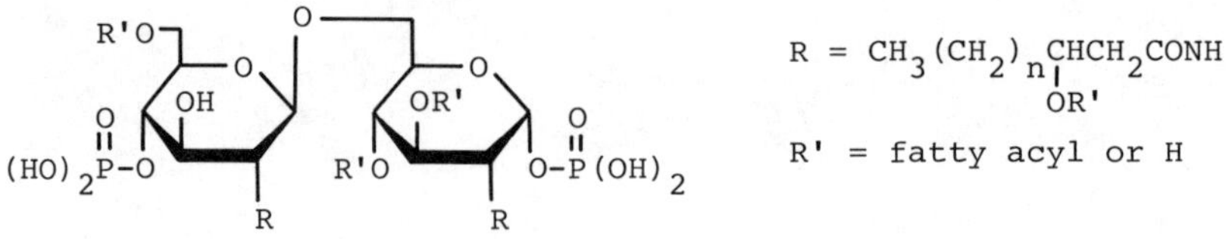

Fig. 1. Proposed structure of lipid A.

$R = CH_3(CH_2)_n\underset{OR'}{CHCH_2CONH}$

R' = fatty acyl or H

with benzyl or allyl alcohol (2-5 mol equiv.) in the presence of anhydrous ferric chloride (1.5 mol equiv.) in dichloromethane (5, 7-8) to give the corresponding β-glycosides 2 (R^2=NHAcyl or NPhth) in 75-90% yield. The N-benzyloxycarbonyl derivatives of 2 (R^2=NHZ) were prepared by adding a catalytic amount of silver perchlorate to a mixture of 5 (X=Cl; 1 mol equiv.), allyl or benzyl alcohol (3 mol equiv.), silver carbonate (1.3 mol equiv.) and anhydrous calcium sulfate in dichloromethane. The β-glycosides 2 (R^2=NHAcyl) were first converted into 2-deoxy-2-(R- and S-3-hydroxytetradecanoylamino)-D-glucose (4) by the selective protections of the primary and secondary hydroxyl groups (9).

The conversion of the 1-β-acetates 1 (R^2=NHAcyl) into the corresponding oxazolines 14 was accomplished by the method of Matta and Bahl (10). The resulting oxazolines were crystallized from ether-petroleum ether (11), and their phosphorylation was carried out by treating the oxazolines with dibenzyl phosphate by the procedure reported by Khorlin et al. (12) to give the desired dibenzyl phosphates 12 (R=Bn) (11). However, the oxazolines having $CH_3(CH_2)_{10}CH(OR)CH_2$ (R=H, Ac, levulinoyl) as R' group did not yield phosphate derivatives by Khorlin's method. The common, major product was 2-(E-1-tridecenyl)-(3,4,6-tri-O-acetyl-1,2-dideoxy-α-D-glucopyrano)-[2,1-d]-2-oxazoline (15).

Since the synthesis of the 1-phosphates of the 3-hydroxytetradecanoyl derivatives by the oxazoline method was unsuccessful as just described, we examined a new procedure using 1-halides as the starting materials. 3,4,6-Tri-O-acetyl-2-benzyloxycarbonylamino-2-deoxy-α-D-glucopyranosyl chloride (or bromide) 5 (1 mol equiv.) was treated with dibenzyl tributylstannyl phosphate (1.1-1.2 mol equiv.) in dichloromethane containing tetraethylammonium chloride (0.5-2.0 mol equiv.) for 3-8 hr at 40-45°C to yield two major glycosyl phosphates, 7(α), needles, m.p. 101-103°C, $[\alpha]_D$ +75.1° (chloroform), and 7(β) which decomposed during the chromatography. Tetraethylammonium halides are known to catalyze the anomerization of α- and β-glycosyl halide (13). As shown in Fig. 3., in fact, the isolated yield of 7(α) as well as the total yield increased with an increasing ratio of the catalyst (Et_4NCl) to the glycosyl chloride 5.

However, since the selective, hydrogenolytic removal of the benzyl-

Synthetic Studies on Lipid A

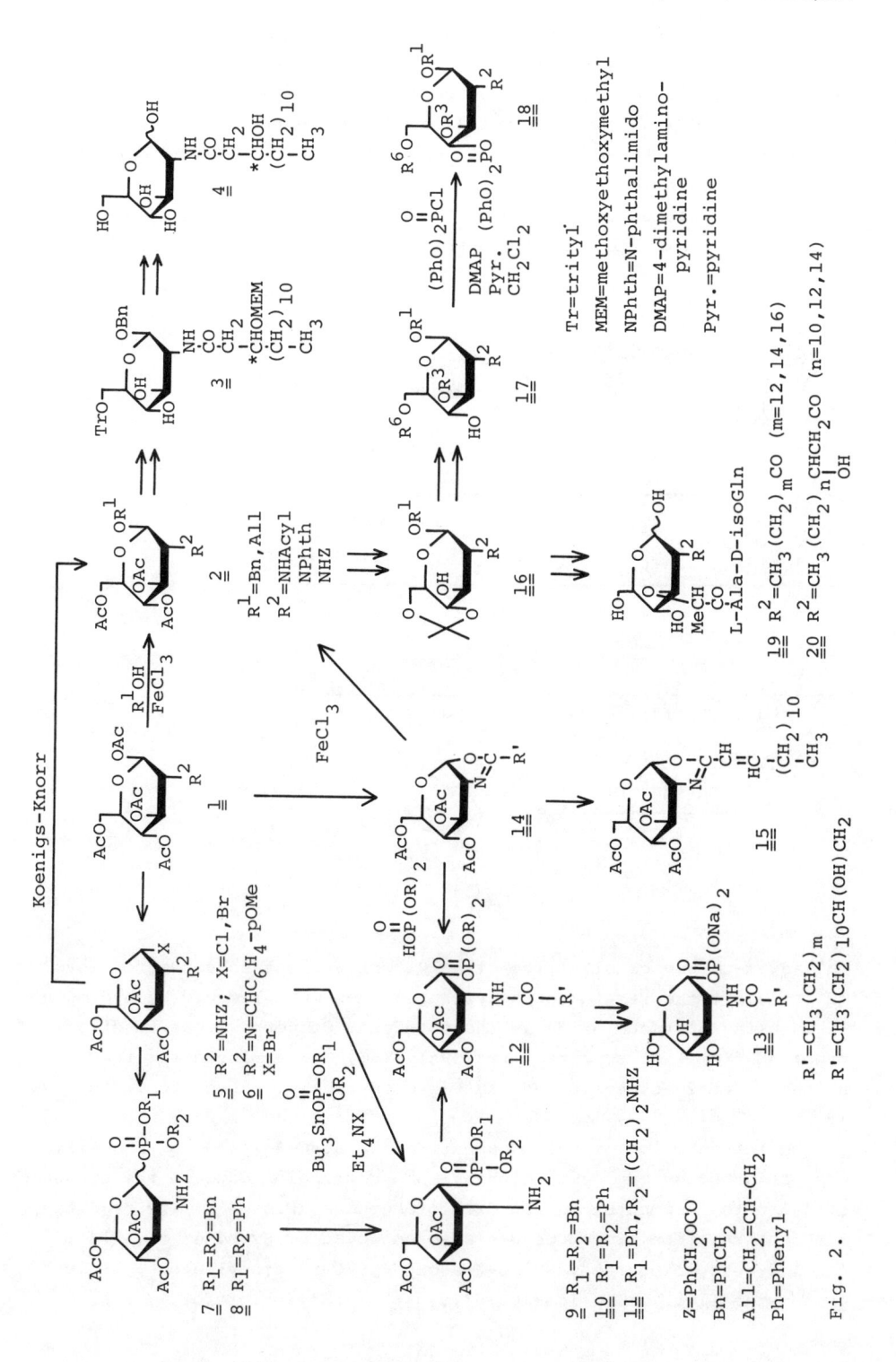

Fig. 2.

41

oxycarbonyl(Z) group of 7(α) was unsuccessful, the new phosphorylation method was further examined for the preparation of diphenyl phosphoryl derivative 8(α). The reaction was conducted at 50°C by adding 2 mol equiv. of Et$_4$NCl as the catalyst (Fig. 3). Although the reactivity of diphenyl tributylstannyl phosphate was slightly lower than that of the dibenzyl derivative, the desired 8(α), amorph., [α]$_D$ +54.3° (chloroform), was obtained in 65% yield. The benzyloxycarbonyl(Z) group on nitrogen atom was cleaved by hydrogenolysis in the presence of 10%

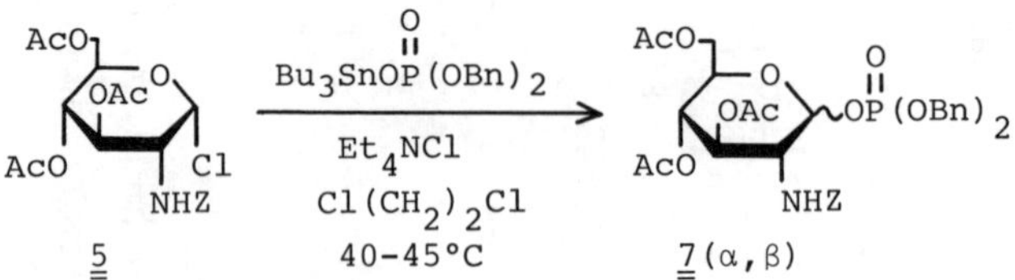

Et$_4$NCl Chloride	7(α) (%)[a]	7(β) (%)[b]	Total (%)
0.5	40	30	70
1.0	60	15	75
2.0	64	15	79

[a]Isolated. [b]T.l.c.

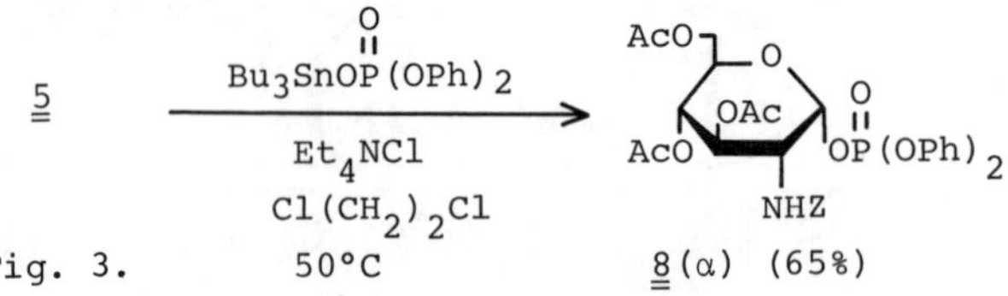

Fig. 3.

palladium-carbon catalyst, to give 10, which was then converted into 12. Finally, the remaining phenyl(Ph) groups were removed by hydrogenolysis using platinum oxide as the catalyst. Compound 9 was readily prepared from 6 as reported previously (6), and employed for the synthesis of 2-acylamino-2-deoxy-D-glucopyranosyl phosphate derivatives 13.

The introduction of the phosphate group at C-4 (16 → 18; Fig. 2), was performed by employing diphenyl phosphorochloridate in the presence of 4-dimethylaminopyridine in dichloromethane (Fig. 4). Very recently, the same reaction procedure was also employed by Szabó et al. (14). Allyl 2-benzyloxycarbonylamino-2-deoxy-4,6-O-isopropylidene-β-D-glucopyranoside (16a), m.p. 165-165.5°, [α]$_D$ -40.3° (chloroform), was

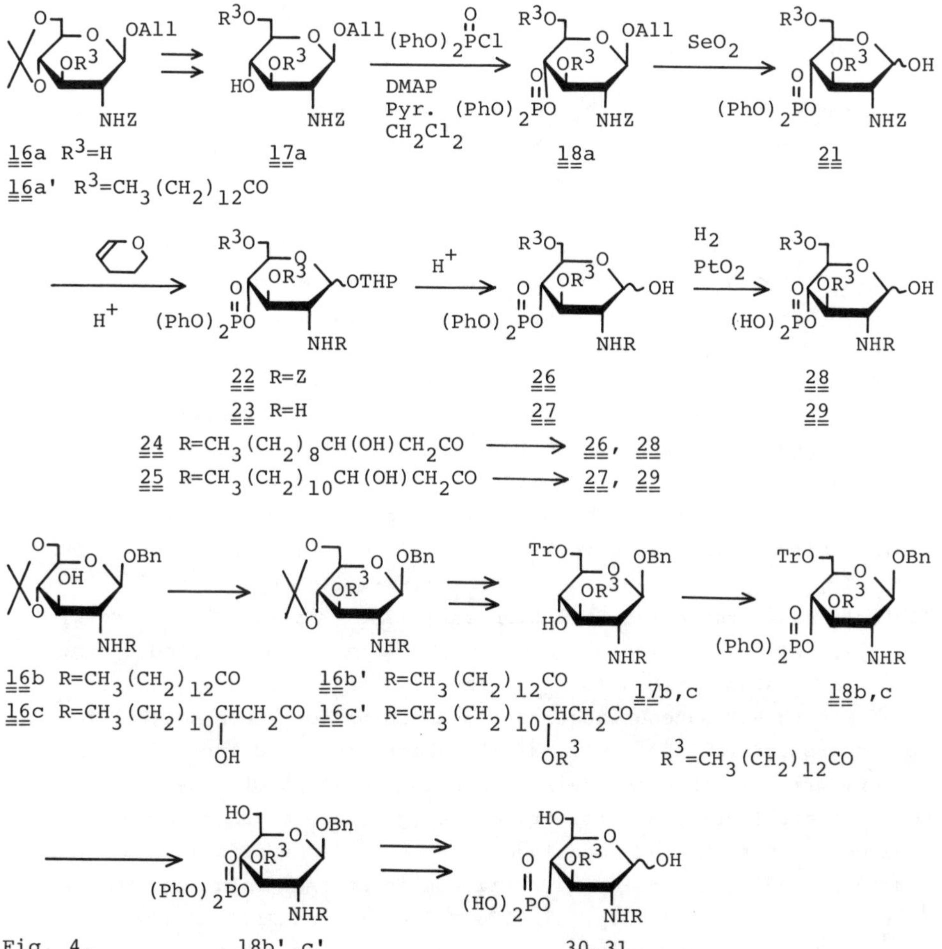

Fig. 4.

esterified with tetradecanoyl chloride in pyridine to give 16a', $[\alpha]_D$ -19.3° (chloroform), and mild hydrolysis of the isopropylidene group, followed by selective tetradecanoylation, afforded 17a, m.p. 91.5-92°C, $[\alpha]_D$ -15° (chloroform). The phosphorylation of 4-OH was conducted by treating 17a (1 mol equiv.) overnight with diphenyl phosphorochloridate (3 mol equiv.) in dichloromethane containing 4-dimethylaminopyridine (2 mol equiv.) and a small amount of dry pyridine at room temperature, to yield 18a, amorph., $[\alpha]_D$ ~0° (chloroform), in almost quantitative yield. Next, compound 18a (1 mol equiv.) was heated at 85° with selenium dioxide (15) (1.1 mol equiv.) and acetic acid (1.5 mol equiv.) in dry 1,4-dioxane (3 mL/100 mg of 18a) for 12 hr to afford 21, $[\alpha]_D$ +28.4° (chloroform), in 70-80% isolated yield. 1-O-Pyranylation and hydrogenolytic removal of benzyloxycarbonyl(Z) group gave 23, syrup,

$[\alpha]_D$ +54.9° (chloroform), which was then converted into 24 or 25 by treating with 3-hydroxydodecanoic acid or 3-hydroxytetradecanoic acid in the presence of dicyclohexylcarbodiimide (DCC) and N-hydroxysuccinimide (HOSu) in dry 1,4-dioxane. Mild hydrolysis of THP group gave 26, amorph., $[\alpha]_D$ +36° (chloroform), and 27, amorph., $[\alpha]_D$ +39° (chloroform). Finally, the phenyl(Ph) groups were hydrogenolyzed in the presence of platinum oxide in ethanol solution to yield 28, amorph., $[\alpha]_D$ +35° (chloroform), and 29, amorph., $[\alpha]_D$ +36° (chloroform).

For the synthesis of 6-OH free, monosaccharide analogs 30 and 31, we employed the corresponding benzyl 2-(acylamino)-2-deoxy-4,6-O-isopropylidene-β-D-glucopyranoside as the starting material: 16b, $[\alpha]_D$ -67° (chloroform); and 16c, $[\alpha]_D$ -44° (chloroform), which were prepared stepwise by O-deacetylation of the benzyl 3,4,6-tri-O-acetyl-2-(acylamino)-2-deoxy-β-D-glucopyranoside 2 (R^2=NHAcyl) (5,16), and subsequent 4,6-O-isopropylidenation. Esterification of 16b,c with tetradecanoyl chloride in 2:1 dichloromethane-pyridine containing a trace of 4-dimethylaminopyridine gave 16b',m.p. 112°C, $[\alpha]_D$ -42.7° (chloroform), and 16c', syrup, $[\alpha]_D$ -30° (chloroform). O-Deisopropylidenation of 16b',c' and subsequent tritylation gave 17b, $[\alpha]_D$ -31.2° (chloroform), and 17c, $[\alpha]_D$ -22.5° (chloroform), which were phosphorylated by the procedure just described for 17a. The resulting 18b and 18c were treated with 80% aqueous acetic acid in chloroform-methanol solution at 50°C to afford 18b', $[\alpha]_D$ -37.1° (chloroform) and 18c', $[\alpha]_D$ -28° (chloroform), which might be suitable as acceptors of 3-deoxy-D-mannooctulosonate (KDO). The protecting, benzyl and phenyl groups were cleaved stepwise by hydrogenolyses in the presence of palladium-carbon and then of platinum oxide as catalysts in ethanol solutions to yield 30 [R=R^3=$CH_3(CH_2)_{12}CO$], m.p. 171-173°C (dec.), and 31 [R=$CH_3(CH_2)_{10}CH$-(OR^3)CH_2CO, R^3=$CH_3(CH_2)_{12}CO$], m.p. 159-162°C (dec.), after lyophilization from 1,4-dioxane (Fig. 4).

Some N-acylmuramyl-L-alanyl-D-isoglutamine derivatives, such as 19 and 20 (Fig. 2), were also prepared (16) and their biological activities examined (17).

3 Disaccharide analogs of lipid A and related compounds

The glycosylation procedure employing oxazoline derivatives 14 was first adopted for the synthesis of disaccharide analogs of lipid A as outlined in Fig. 5 (18). The oxazoline method was also employed for the synthesis of 38 (Fig. 5) (19). Although the oxazoline method is a general procedure for the synthesis of β-glycosides, the acylamino groups at C-2 of the starting materials are unchangeable to others.

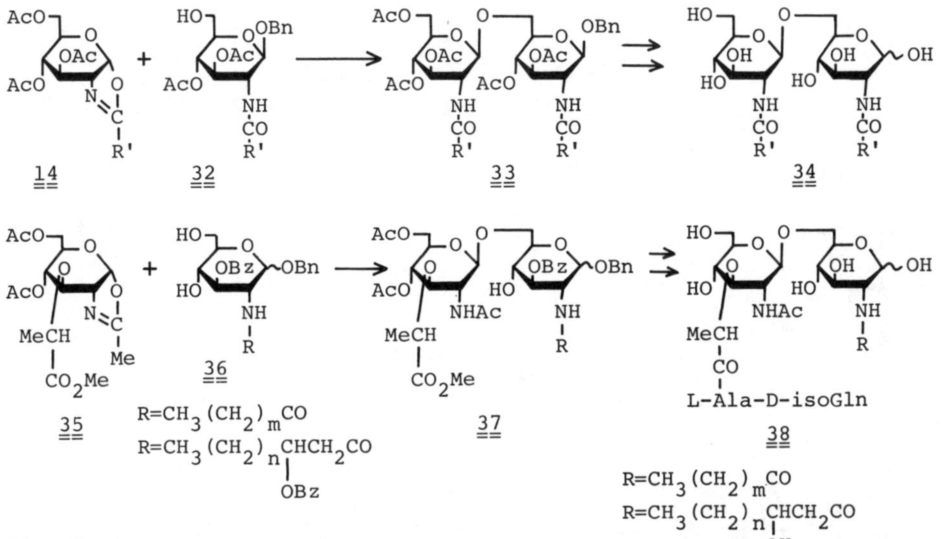

Fig. 5.

We next examined a more flexible method by which a variety of fatty acyl residues can be introduced after formation of disaccharides (Fig. 7). The glycosyl halides, 5(X=Cl) and 4,6-di-O-acetyl-3-O-benzyl-2-deoxy-2-phthalimido-β-D-glucopyranosyl chloride (44), were employed as the glycosyl donors, whereas allyl 3-O-benzoyl-2-deoxy-2-phthal-imido-β-D-glucopyranoside (45) and benzyl 3-O-benzoyl-2-benzyloxy-carbonylamino-2-deoxy-β-D-glucopyranoside (47) as the glycosyl acceptors.

As shown in Fig. 6, benzylation of allyl 2-deoxy-4,6-O-isopropyl-idene-2-phthalimido-β-D-glucopyranoside (16; R^1=All, R^2=NPhth), m.p. 109°C, $[α]_D$ -29.6° (chloroform), to give 39, m.p. 104.5°C, $[α]_D$ +55° (chloroform), and O-deisopropylidenation, followed by acetylation, afforded 41, m.p. 94°C, $[α]_D$ +56.8° (chloroform). The allyl group was then cleaved by selenium oxide to give 42, m.p. 151-151.5°C, $[α]_D$ +83.5° (chloroform). Acetylation of 42 and subsequent chlorination with anhydrous hydrochloric acid in dichloromethane gave 44, $[α]_D$ +80.2° (chloroform). Benzoylation of 16(R^1=All, R^2=NPhth) or 16(R^1=Bn, R^2=NHZ), m.p. 139.5-140°C, $[α]_D$ -61.3° (chloroform), to give 40, m.p. 141.5°C, $[α]_D$ +79.1° (chloroform) or 46, m.p. 157.5-158°C, $[α]_D$ -4.5° (chloroform), respectively, and mild hydrolysis of the isopropylidene group gave the desired glycosyl acceptors, 45, $[α]_D$ +101.2° (chloroform) and 47, m.p. 131°C, $[α]_D$ +28.8° (chloroform).

The coupling of 44 (2 mol equiv.) to 45 (1 mol equiv.) was achieved (Fig. 7) by using silver carbonate and silver perchlorate as the

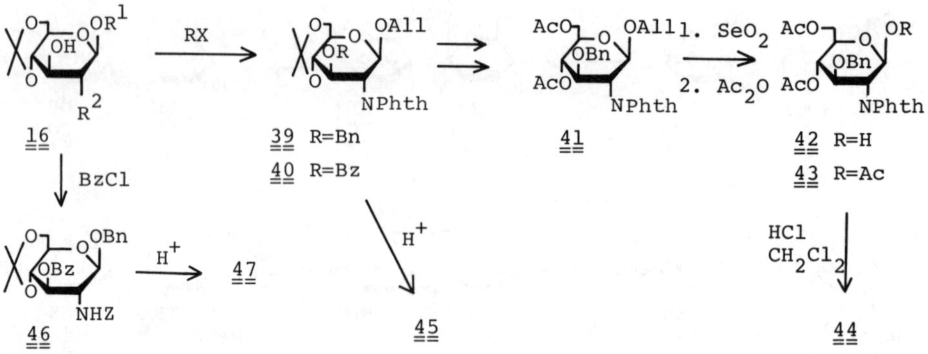

Lipid A: Structure and Synthetic Approach

Fig. 6.

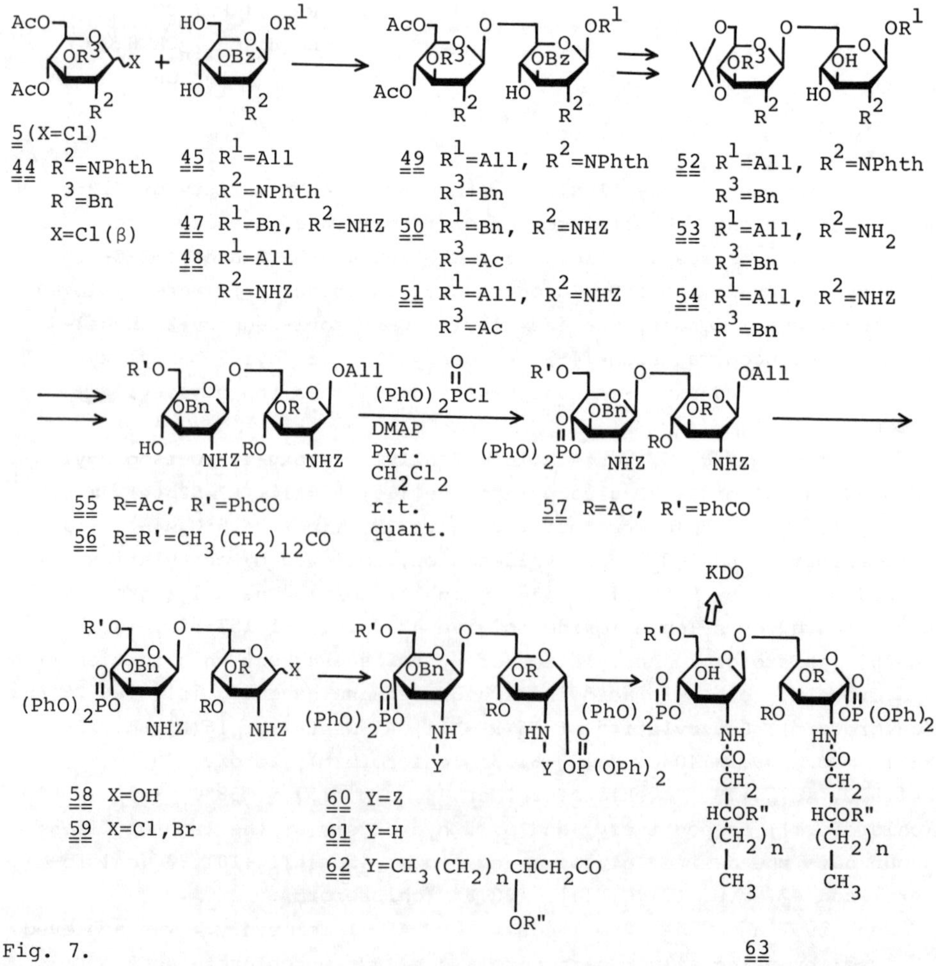

Fig. 7.

catalysts in dichloroethane to give 49, $[\alpha]_D$ +70° (chloroform), in 75%
yield (this reaction was complete within 2 hr). The use of CF_3SO_3Ag-
tetramethylurea as the catalysts (20) reduced the isolated yield of 49.
O-Deacetylation of 49 and 4',6'-O-isopropylidenation with 2,2-dimeth-
oxypropane in 1,4-dioxane afforded 52, $[\alpha]_D$ +4.1° (chloroform), in 83%
yield. The cleavage of phthalimido groups was carried out by treating
52 (1 mol equiv.) with hydrazine hydrate (8 mol equiv.) in ethanol at
the reflux temperature, to give 53, which was promptly benzyloxy-
carbonylated to 54, $[\alpha]_D$ -23.7° (chloroform) [lit. -24.3° (21)]. The
glycosylation procedure just described was also employed for the
coupling of 5(X=Cl) with 47 to give 50, m.p. 203°C, $[\alpha]_D$ +0.1° (chloroform) in high yield. As already described, the phosphorylation of 4-OH
(see Fig. 4) was accomplished by using diphenyl phosphorochloridate in
dichloromethane containing 4-dimethylaminopyridine and a small amount
of dry pyridine. When the same procedure was employed for the conversion of 55, $[\alpha]_D$ -8.2° (chloroform), into 57, $[\alpha]_D$ +22° (chloroform),
the yield was quantitative and the single product 57 was readily
isolated just by conventional extractive processing.

In our strategy for the total synthesis of lipid A (56 → 63; Fig. 7),
the selective protection of the amino, hydroxy, and phosphate functions
by benzyloxycarbonyl (Z), benzyl (Bn), and phenyl (Ph) groups, which
can be selectively cleaved by hydrogenolysis using palladium-carbon or
platinum oxide as the catalyst, is a most important feature. Very
recently, however, a slight modification of the proposed structure of
lipid A (Fig. 1) has been suggested (21), especially for the combining
site of KDO or O-acyl group in the sugar skeleton. Therefore, an
alternative route, employing 50 or the corresponding allyl glycoside
derivative 51 as the disaccharide intermediate, might be useful for the
synthesis of the "true lipid A" and its analogs.

4 Acknowledgements

We thank Mr. H. Nishiguchi, Mr. K. Nishihori. Mr. S. Murase, Mr. H.
Ishida, and Mr. T. Kito for thier efforts which had led to the results
described in this article.

5 References

(1) A. Hasegawa, M. Kiso, I. Azuma, Y. Yamamura in Y. Yamamura et al.
(Ed.), Immunomodulation by microbial products and related
synthetic compounds, Excerpta Medica, Amsterdam 1982, p. 155.

(2) M. Kiso, A. Hasegawa, H. Okumura, I. Azuma in Y. Yamamura et al. (Ed.), Immunomodulation by microbial products and related synthetic compounds, Excerpta Medica, Amsterdam 1982, p. 281.

(3) H. Okumura, I. Saiki, K. Kamisango, Y. Tanio, I. Azuma, M. Kiso, A. Hasegawa in Y. Yamamura et al. (Ed.), Immunomodulation by microbial products and related synthetic compounds, Excerpta Medica, Amsterdam 1982, p. 163.

(4) M. Kiso, A. Hasegawa, ACS symposium: Carbohydrates in bacterial lipopolysaccharides--Synthesis and biomedical significance, Kansas City, 1982 (ACS SYMPOSIUM SERIES, in press).

(5) M. Kiso, H. Nishiguchi, A. Hasegawa, Carbohydr. Res. 81 (1980) c13-c15.

(6) M. Kiso, K. Nishihori, A. Hasegawa, H. Okumura, I. Azuma, Carbohydr. Res. 95 (1981) c5-c8.

(7) M. Kiso, L. Anderson, Carbohydr. Res. 72 (1979) c12-c14.

(8) M. Kiso, L. Anderson, Carbohydr. Res. 72 (1979) c15-c17.

(9) M. Kiso, H. Nishiguchi, S. Murase, A. Hasegawa, Carbohydr. Res. 88 (1981) c5-c9.

(10) K. L. Matta, O. P. Bahl, Carbohydr. Res. 21 (1972) 460-464.

(11) M. Kiso, K. Nishihori, A. Hasegawa, Agric. Biol. Chem. 45 (1981) 545-548.

(12) A. Ya. Khorlin, S. E. Zurabyan, T. S. Antonenko, Tetrahedron Lett. (1970) 4803-4804.

(13) R. U. Lemieux, K. B. Hendriks, R. Y. Stick, K. James, J. Am. Chem. Soc. 97 (1975) 4056-4062.

(14) P. Szabó, S. R. Sarfati, C. Diolez, L. Szabó, Carbohydr. Res. 111 (1983) c9-c12.

(15) K. Kariyone, H. Yazawa, Tetrahedron Lett. (1970) 2885-2888.

(16) M. Kiso, Y. Goh, E. Tanahashi, A. Hasegawa, H. Okumura, I. Azuma,

Carbohydr. Res. 90 (1981) c8-c11.

(17) H. Okumura, K. Kamisango, I. Saiki, Y. Tanio, I. Azuma, M. Kiso, A. Hasegawa, Y. Yamamura, Agric. Biol. Chem. 46 (1982) 507-514.

(18) M. Kiso, H. Nishiguchi, K. Nishihori, A. Hasegawa, I. Miura, Carbohydr. Res. 88 (1981) c10-c13.

(19) A. Hasegawa, M. Ozaki, Y. Goh, M. Kiso, I. Azuma, Carbohydr. Res. 100 (1982) 235-245.

(20) S. Hanessian, J. Banoub, Carbohydr. Res. 53 (1977) c13-c16.

(21) M. Inage, H. Chaki, S. Kusumoto, T. Shiba, Chem. Lett. (1980) 1373-1376.

(22) E. T. Rietschel, H.-W. Wollenweber, U. Zähringer, O. Lüderitz, ACS symposium: Carbohydrates in bacterial lipopolysaccharides-- Synthesis and biomedical significance, Kansas City, 1982 (ACS SYMPOSIUM SERIES, in press).

Synthesis of Diglycosyl Phosphate via a Phosphite Intermediate: Partial Structure of Lipid A

Tomoya Ogawa and Akinori Seta

The Institute of Physical and Chemical Research,
Wako-shi, Saitama, 351, Japan

Abstract

An approach to the synthesis of diglycosyl phosphates via the intermediacy of diglycosyl phosphites is developed and the application of this approach to the synthesis of a part structure of the lipid A of Salmonella Lipopolysaccharide is described.

1 Introduction

Phosphates linked to anomeric carbon atoms have been reported to be present at cell surface of such microbes as Staphylococcus lactis, Saccharomyces cerevisiae, Micrococcus lysodeikticus, and Staphylococcus pneumoniae, and to be a part structure of glycan chains of glycoproteins. Lipid A moiety of LPS also contains glycosyl phosphate units.

In 1976, Letsinger and Lunsford[1] reported a new synthetic approach toward oligonucleotide via phosphite triester intermediate. As part of our project on the synthetic studies on oligosaccharides, we have reported[2] an approach to the synthesis of glycosyl phosphates through the intermediacy of glycosyl phosphite. From synthetic view points, this approach is expected to be efficient due to the high reactivity of phosphorochloridite as a phosphorylating reagent. We describe first an example of this approach to the synthesis of glycosyl phosphate and then an application to the synthesis of a part structure of Lipid A.

Lipid A: Structure and Synthetic Approach

Scheme 1 shows a strategy for the synthesis of glycosyl phosphate via phosphite intermediate. Route (a) indicate a deprotection of phosphite triester and subsequent oxidation to phosphodiester. Route (b) shows an oxidation of phosphite to phosphate triester and subsequent deprotection. After several experiments, we chose to follow route (b).

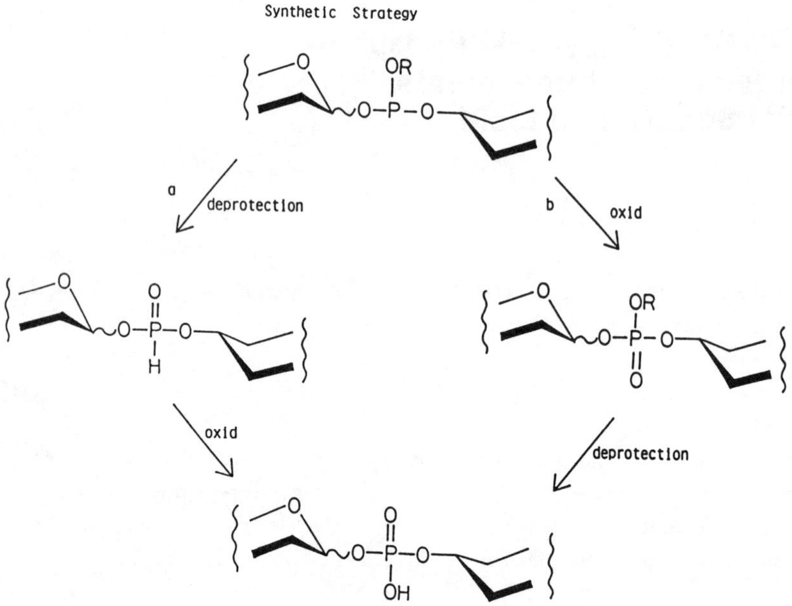

Scheme 1

An example of this approach was illustrated in scheme 2. 2,3,4,6-Tetra-O-benzyl-α-D-mannopyranose 1 was treated with trichloroethyl-phosphorodichloridite at -78° to give monochloridite intermediate 2 which, without isolation, was treated with methyl 2,3,4-tri-O-benzyl-α-D-mannopyranoside 3 to afford an 81% yield of phosphite triester 4. Air oxidation of 4 in the presence of AIBN in benzene gave phosphate triester 5 in 75% yield. Trichloroethyl group was removed by treating 5 with Zn-Cu in DMF-acetylacetone-Et$_3$N to give 6 in 72% yield. Further deprotection of benzyl ethers under Birch reduction and purification afforded target structure 7 in 36% yield.

Having established above mentioned phosphite approach, now we turned to its application to the synthesis of a part structure of Lipid A. Scheme 3 denote the proposed structure of Lipid A of Salmonella LPS by Rietschel et al in 1982[3]. It is to be noted that this structure 8

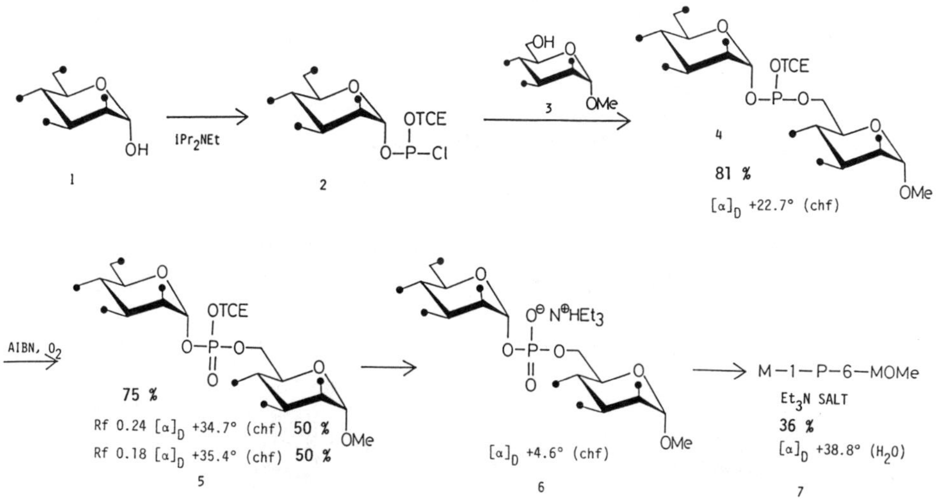

Synthesis of Diglycosyl Phosphate

Scheme 2

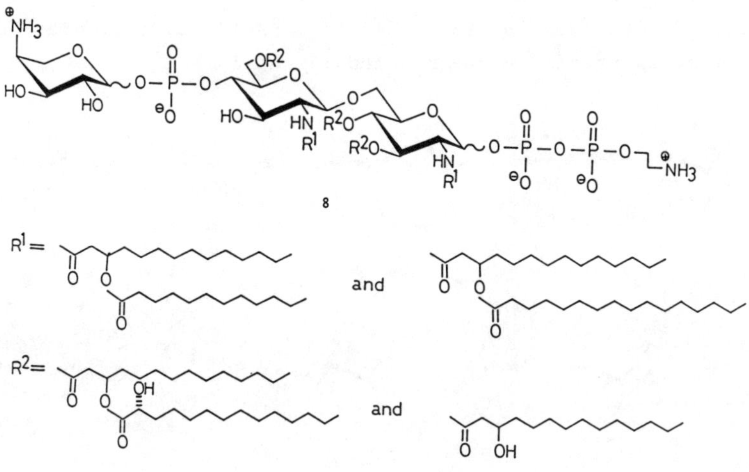

Scheme 3

contains a phosphodiester linkage as well as a pyrophosphate. In order to see the applicability of our approach to the synthesis of these complex structures, we design a simpler part structure 9 of 8. Retrosynthetic analysis shows two monosaccharide synthons 10 and 11 which are required to reconstruct the target structure 9.

Lipid A: Structure and Synthetic Approach

Target and Synthetic Plan

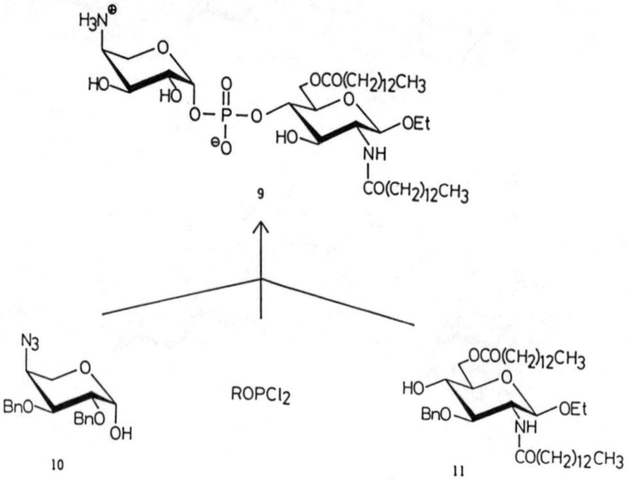

Scheme 4

Synthesis of the monosaccharide synthon 10.

A regioselective activation through trialkylstannylation[4] of methyl β-D-xylopyranoside 12 and subsequent tritylation[5] gave a 70% yield of monotrityl derivatives 15, 16 and 17 in a ratio 25:25:1, via the intermediacy of partially stannylated 13 and 14.

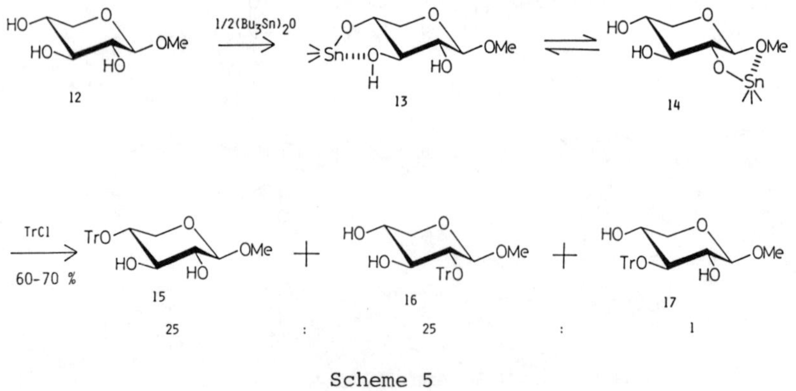

Scheme 5

Benzylation of 15, detritylation of 18, mesylation of 19, and S_N2 displacement of mesyl group of 20 afforded 21 in good yield as described in Scheme 6.

Acetolysis of 21, however, did not give 22 even though many experimental conditions were examined. Therefore, an alternative route to 10 was studied. Mesylate 20 was subjected to acetolysis to give a diastereomeric mixture of acyclic diacetates 23. Upon treatment with NaN$_3$ in DMF 23 was converted into 24 without any neighbouring participation from C-5 acetate group. Zemplen deacetylation of 24 afforded the desired synthon 10.

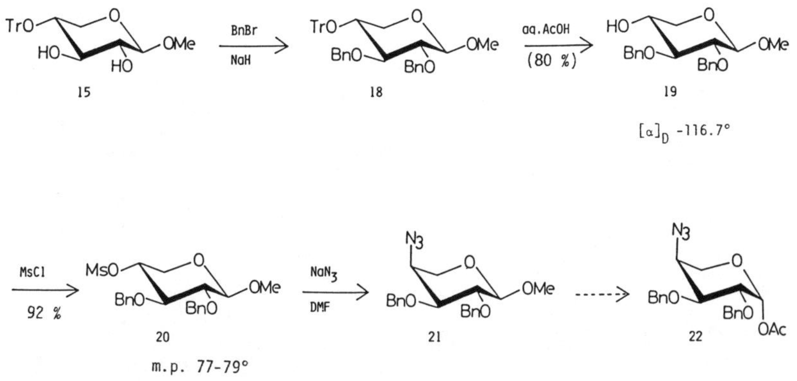

Scheme 6

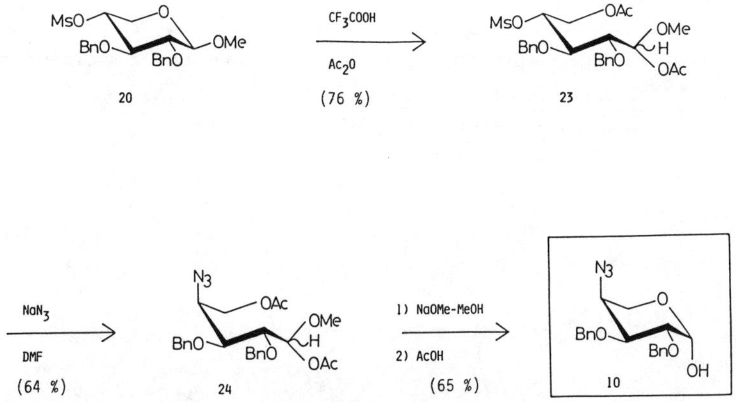

Scheme 7

Synthesis of Monosaccharide Synthon 11

Ethyl 3-O-benzyl-4,6-O-benzylidene-2-deoxy-2-phthalimido-β-D-glucopyranoside 26, readily prepared from 25 was treated with hydrazine hydrate to give amine 27. Condensation of 27 with tetradecanoic acid in the presence of DCC gave 28, which was subsequently solvolysed in aq.AcOH to give a 75% yield of diol 29. Selective acylation of 29 with tetradecanoyl chloride in pyridine afforded a desired synthon 11 as crystals in 78% yield.

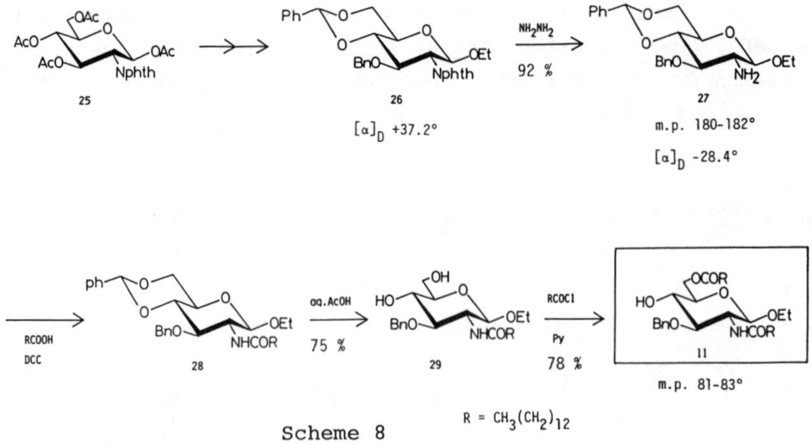

Lipid A: Structure and Synthetic Approach

Scheme 8

R = CH$_3$(CH$_2$)$_{12}$

Synthesis of Phosphodiester linkages

First, a model reaction of 11 using cyclohexanol was studied as shown in Scheme 9. Treatment of 11 with trichloroethylphosphorodichloridite and subsequent reaction of 30 with cyclohexanol, followed by Silica gel chromatography afforded phosphite triester 32 in about 66% yield. Extractive work up of this reaction mixture gave very low yield of the desired 32, and tlc examination showed the formation of more polar products such as 34. Oxidation of 32 was effected in the presence of AIBN to give phosphate triester 33 in 67% yield.

Model Reaction

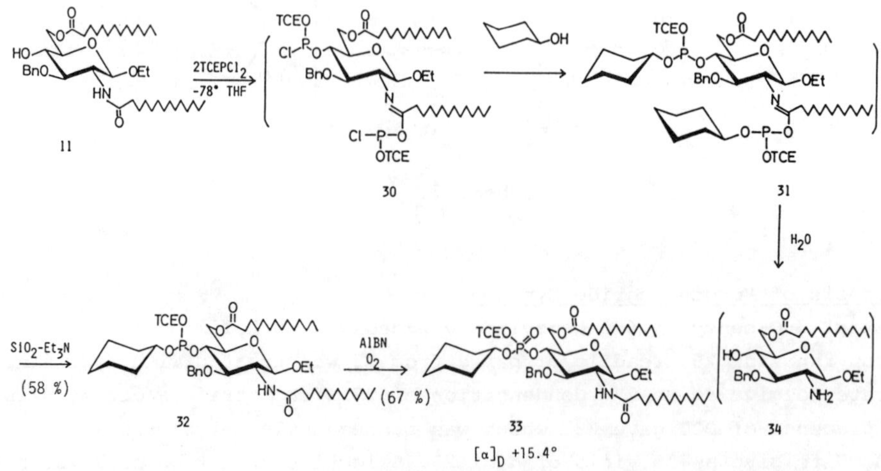

Scheme 9

As the model sequence was well established, the synthesis of the target structure 9 was now studied. The same sequence of the reactions with 11 using 10 in place of cyclohexanol afforded a 69% yield of the desired phosphite triester 36 which was further oxidized into 37 in 60% yield. The structure of 37 was supported by IR date (2100 cm^{-1} for N_3, 1720 cm^{-1} for RCO_2, 1640 cm^{-1} for RCONH), and by ^1H NMR data which showed the presence of benzyl group and tetradecanoyl groups. Further transformation of 37 into 9 in now under investigation.

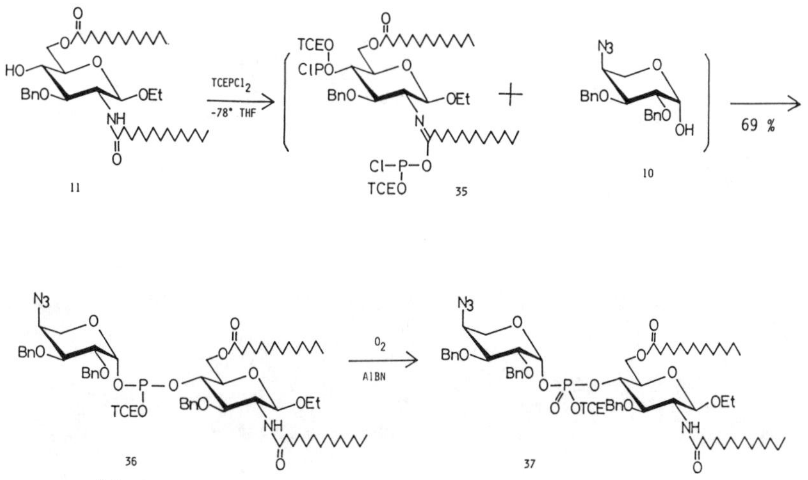

$[\alpha]_D$ +15.4°

IR 2100, 1730, 1650 cm^{-1}

Scheme 10

3 References

(1) R. R. Letsinger and W. B. Lunsford, J. Am. Chem. Soc., 98 (1976) 3655-3661.

(2) T. Ogawa and A. Seta, Carbohydr. Res., 110 (1982) C1-C4.

(3) H-W. Wollenweber, K. W. Broady, O. Luderitz, and E. T. Rietschel, Eur. J. Biochem., 124 (1982) 191-198.

(4) T. Ogawa and M. Matsui, Carbohydr. Res., 51 (1976) C13-C18; Tetrahedron 37 (1981) 2363-2369.

(5) T. Ogawa, T. Nukada and M. Matsui, Carbohydr. Res., 101 (1982) 263-270.

Endotoxic Activities of Lipid A and Synthetic Analogues

Biological Activities of Synthetic Lipid A Analogues

Motohiro Matsuura, Yasuhiko Kojima, J. Yuzuru Homma

The Kitasato Institute, Shirokane, Minato-ku, Tokyo 108, Japan

Yoshio Kumazawa, Yoshiyuki Kubota*

School of Pharmaceutical Sciences and *Research and Development Center of Hygienic Science, Kitasato University, Shirokane, Minato-ku, Tokyo 108, Japan

Tetsuo Shiba, Shoichi Kusumoto

Department of Chemistry, Faculty of Science, Osaka University, Toyonaka, Osaka 560, Japan

Abstract

Synthetic lipid A analogues solubilized with triethylamine and complexed with bovine serum albumin were assayed for biological activities such as mitogenic, polyclonal B cell activation (PBA), interferon-inducing, proclotting enzyme of horseshoe crab activation and macrophage activation activities. Analogues which exhibited mitogenic, PBA and interferon-inducing activities simultaneously were C-1 or C-4' monophosphorylated analogues. Ester- and amide-bound fatty acid substituents of the C-1 analogues were tetradecanoyl (C_{14}) or (R)-3-hydroxytetradecanoyl (C_{14}-OH) groups and those of the C-4' analogues were C_{14}-OH or (R)-3-tetradecanoyloxytetradecanoyl (C_{14}-O-(C_{14})) groups. Only mitogenic activity was shown by a nonphosphorylated analogue with C_{14}-OH and C_{14}-O-(C_{14}) groups as ester- and amide-bound substituents and by C-4' mono- and C-1,4' di-phosphorylated analogues with C_{14} and C_{14}-OH as ester- and amide-bound substituents. None of the other activities were detectable in all the analogues tested.

It is suggested from the results that phosphate residues play an
important role for manifesting some of the biological activities of
lipid A and that ester- and amide-bound substituents also participate
in manifestation of the activities.

1 Introduction

Lipid A moiety of lipopolysaccharide has been shown to be essential
for manifesting various biological activities of endotoxin (1,2).
Based on detailed structural investigations of lipid A, it was proposed
that the common and ubiquitous structural element in lipid A is the
C-1,4' diphosphorylated β 1,6-linked D-glucosamine disaccharide with
ester- and amide-bound fatty acids (3). Homma and coworkers have been
carrying out comparative studies on biological activities and structures of chemically degraded or modified natural lipid A preparations
from Pseudomonas aeruginosa and revealed the essential regions responsible for some biological activities (4-6).

Recently, attempts to chemically synthesize the proposed structure
of lipid A have been made (7,8), and some synthetic lipid A analogues
are now available. Since the structures of these compounds are finely
defined, these compounds are very good tools to elucidate the relationship between chemical structures and biological activities of lipid A.

In the present paper, mitogenic, polyclonal B cell activation (PBA),
interferon-inducing, pyrogenic, proclotting enzyme of horseshoe crab
activation and macrophage activation activities of synthetic lipid A
analogues solubilized with triethylamine and complexed with bovine
serum albumin are described.

2 Structures of synthetic lipid A analogues

Structures of the synthetic lipid A analogues (7,9-11) used in this
experiment are shown in Fig. 1. These are derivatives of β 1,6-linked
glucosamine disaccharide. Ester- and amide-linked substituents are
represented as R_1 and R_2, respectively. According to variation of R_1
and R_2, the analogues were divided into 4 groups. Analogues in group
A possess only tetradecanoyl groups (C_{14}) as both R_1 and R_2 substituents, those in group B possess C_{14} as R_1 and (R)-3-hydroxytetra-
decanoyl groups (C_{14}-OH) as R_2 substituents, those in group C possess
C_{14}-OH as both R_1 and R_2 substituents and those in group D possess

C_{14}-OH as R_1 and (R)-3-tetradecanoyloxytetradecanoyl groups (C_{14}-O-(C_{14})) as R_2 substituents.

The other substituents at C-4' and C-1 positions are represented as R_3 and R_4, respectively. In every group, nonphosphorylated compounds (R_3=H and R_4=H) such as 301, 302, 311 and 314, and C-4' monophosphorylated compounds (R_3=P and R_4=H) such as 304, 321, 312 and 315 were available. Since C-1 monophosphorylated (R_3=H and R_4=P) and C-1,4' diphosphorylated (R_3=P and R_4=P) compounds in groups C and D had not been chemically synthesized, only those compounds in groups A (303 and 305) and B (316 and 317) were available.

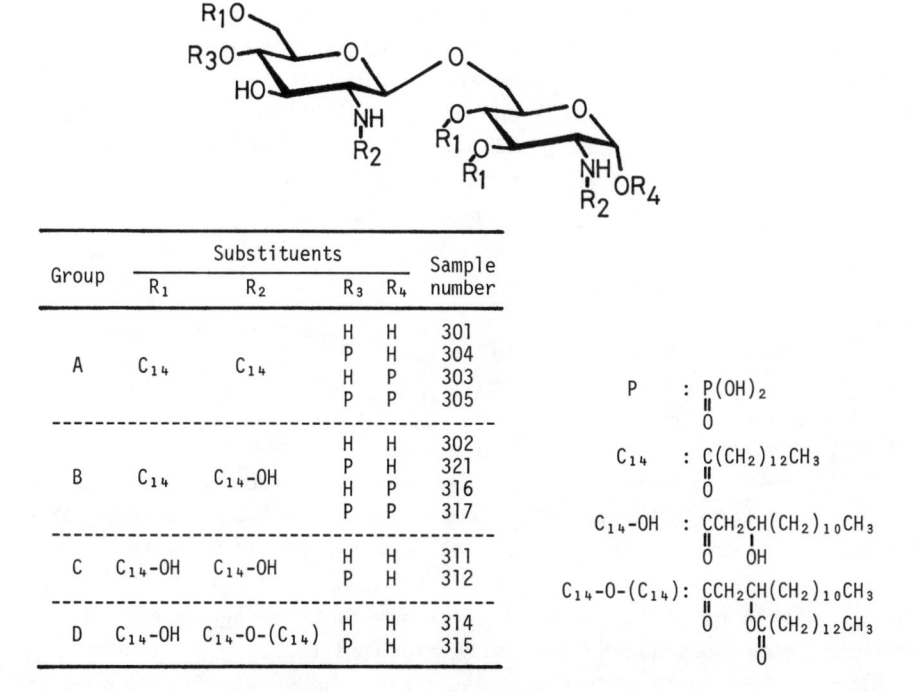

Group	Substituents				Sample number
	R_1	R_2	R_3	R_4	
A	C_{14}	C_{14}	H	H	301
			P	H	304
			H	P	303
			P	P	305
B	C_{14}	C_{14}-OH	H	H	302
			P	H	321
			H	P	316
			P	P	317
C	C_{14}-OH	C_{14}-OH	H	H	311
			P	H	312
D	C_{14}-OH	C_{14}-O-(C_{14})	H	H	314
			P	H	315

P : $\underset{\overset{\parallel}{O}}{P}(OH)_2$

C_{14} : $\underset{\overset{\parallel}{O}}{C}(CH_2)_{12}CH_3$

C_{14}-OH : $\underset{\overset{\parallel}{O}}{C}CH_2\underset{\overset{|}{OH}}{C}H(CH_2)_{10}CH_3$

C_{14}-O-(C_{14}): $\underset{\overset{\parallel}{O}}{C}CH_2\underset{\overset{|}{O\underset{\overset{\parallel}{O}}{C}(CH_2)_{12}CH_3}}{C}H(CH_2)_{10}CH_3$

Fig. 1. Structures of Synthetic Lipid A Analogues

3 Solubilization of synthetic lipid A analogues

Synthetic lipid A analogues are insoluble in water. To investigate their biological activities, the analogues and a natural lipid A were solubilized with triethylamine and complexed with bovine serum albumin; such pretreatment is frequently used to detect the biological activities of natural lipid A (12). The procedure used in this study was as follows. A sample (0.5 mg) was dissolved well with chloroform-methanol solution (3 to 5 ml) in a 10-ml short necked Kjeldahl flask. The solution was evaporated to dryness and a thin layer of sample formed on the inside wall of the flask. Distilled water (about 3 ml) was poured into the flask and sonicated to disperse the sample into water. Triethylamine (5 µl) and bovine serum albumin (0.5 mg in 0.5 ml water) were added to the solution, which was then sonicated. The resulting mixture was rotary evaporated to remove excess triethylamine. Finally, the volume of the solution was made to 2.5 ml by adding distilled water. This solution was used for biological testing as the starting material (200 µg of sample per ml).

To estimate the recovery, the sample was refluxed with methanolic 2N HCl for 5 hr and the fatty acid (C_{14} and/or C_{14}-OH) methyl esters obtained were then separated and determined with the column of 10% UCW-982 using 5840A Gas Chromatograph (Hewlett-Packard). Dodecanoic acid was used as an internal standard. Turbidities of the samples were also measured by reading the optical density at 660 nm. Recoveries of all the analogues were in the range from 35% to 94% and turbidities of the samples varied from 0.086 to 0.234.

4 Biological activities

4.1. Mitogenicity

To determine mitogenic activity, spleen cells from C3H/HeN and C3H/HeJ mice were cultivated with synthetic lipid A analogues at 37°C for 48 hr at a cell density of 4×10^5 cells per well in a total volume of 0.2 ml. One µCi of [^{3}H]thymidine was added to the culture for the last 4 hr of cultivation. Cells were harvested and radioactivity incorporated into the cells measured (13).

Mitogenic activity of the analogues assayed by use of spleen cells derived from C3H/HeN mice are shown in Table 1. Marked activity (+++, maximal stimulation index (SI_{max}) >10) was exhibited by a C-1 monophosphorylated analogue in group A (303), (SI_{max}=10.4). Significant activity (++, SI_{max}=5.1∼10) was exhibited by a C-1 monophosphorylated analogue in group B (316), nonphosphorylated and C-4' monophosphorylated analogues in group D (314 and 315). Positive activity (+, SI_{max}= 3.1∼5.0) was exhibited by C-4' monophosphorylated analogues in groups B and C (321 and 312) and a C-1,4' diphosphorylated analogue in group B (317). Mitogenic activity of the other analogues was obscure (±, SI_{max}=2.1∼3.0) or negative (-, SI_{max}≦2.0).

Table 1. Mitogenicity of Synthetic Lipid A Analogues

Group	Sample number	Substituent C-4'	Substituent C-1	Stimulation index[a] 1 µg/well Exp.1	1 µg/well Exp.2	3 µg/well Exp.1	3 µg/well Exp.2	10 µg/well Exp.1	10 µg/well Exp.2
A	301	H	H	0.6	1.1	1.0	1.1	1.6	1.3
	304	P	H	1.6	0.9	1.4	1.3	1.4	1.6
	303	H	P	2.4	4.6	7.9	10.4	7.8	8.8
	305	P	P	0.9	1.8	1.6	1.7	2.2	1.0
Control	Lipid A None			27.1 (1,885)*	10.8 (7,875)	21.6 (2,521)	20.9 (6,380)	15.1 (3,257)	16.1 (8,990)
B	302	H	H	1.4	NT**	1.4	NT	1.2	NT
	321	P	H	1.2	1.6	1.5	2.4	2.7	3.6
	316	H	P	2.3	1.5	3.2	2.6	3.4	7.1
	317	P	P	1.8	1.9	1.9	3.7	2.2	4.7
Control	Lipid A None			15.0 (4,817)	10.8 (7,875)	8.3 (7,624)	20.9 (6,380)	9.0 (11,682)	16.1 (8,990)
C	311	H	H	1.4	NT	1.3	NT	2.0	NT
	312	P	H	1.1	NT	2.3	NT	3.1	NT
Control	Lipid A None			21.4 (6,089)	NT NT	17.8 (5,512)	NT NT	18.2 (7,247)	NT NT
D	314	H	H	2.4	NT	5.2	NT	5.6	NT
	315	P	H	1.9	NT	3.3	NT	5.7	NT
Control	Lipid A None			18.6 (3,943)	NT NT	23.0 (4,021)	NT NT	22.3 (4,500)	NT NT

*tritiated thymidine uptake (cpm) of a control; **NT, not tested.

a) Stimulation index=(cpm in a test sample)/(cpm in a control)

As for mitogenic activity of all the analogues assayed by spleen cells from C3H/HeJ mice, only one, a nonphosphorylated analogue in group D (314), demonstrated significant activity. All of the other analogues were found to be negative.

4.2. Polyclonal B cell activation (PBA) activity

For determination of PBA activity, spleen cells from C3H/HeN and C3H/HeJ mice were cultivated with synthetic lipid A analogues at 37°C for 48 hr in a total volume of 1 ml at a cell density of 5×10^6 cells per ml. Numbers of plaque forming cells (PFC) per culture was determined by the hemolytic plaque technique of Cunningham and Szenber (14). Trinitrophenylated horse erythrocytes (TNP-HRBC) were used as indicator cells in the hemolytic plaque technique.

PBA activity of the analogues assayed using spleen cells derived from C3H/HeN mice are shown in Table 2. Significant activity (++, SI_{max}=5.1~10) was exhibited by a C-1 monophosphorylated analogue in group A (303, SI_{max}=9.1). Positive activity (+, SI_{max}=2.1~5.0) was exhibited by a C-1 monophosphorylated analogue in group B (316) and C-4' monophosphorylated analogues in groups C and D (312 and 315). PBA activity of the other analogues was obscure (±, SI_{max}=1.6~2.0) or negative (-, SI_{max}≦1.5).

When spleen cells of C3H/HeJ mice were employed, none of the analogues showed PBA activity.

Table 2. Polyclonal B Cell Activation by Synthetic Lipid A Analogues

Group	Sample number	Substituent C-4'	Substituent C-1	Stimulation index[a] Exp.1	Stimulation index[a] Exp.2
A	301	H	H	1.1	1.2
	304	P	H	1.0	0.7
	303	H	P	9.1	5.7
	305	P	P	1.9	1.6
Control	Lipid A None			21.6 (139±74)*	19.8 (231±92)
B	302	H	H	0.9	NT**
	321	P	H	1.2	NT
	316	H	P	2.1	2.5
	317	P	P	1.1	1.8
Control	Lipid A None			5.4 (473±79)	16.3 (121±114)
C	311	H	H	1.0	NT
	312	P	H	1.8	2.2
Control	Lipid A None			6.5 (809±99)	16.2 (136±51)
D	314	H	H	1.4	0.7
	315	P	H	2.4	NT
Control	Lipid A None			5.0 (442±43)	5.3 (437±79)

*[(anti-TNP-HRBC PFC/culture) ± standard deviation] of a control;
**NT, not tested.

a) Stimulation index = $\dfrac{\text{(anti-TNP-HRBC PFC/culture) of a test sample}}{\text{(anti-TNP-HRBC PFC/culture) of a control}}$

4.3. Interferon-inducing activity

For the assay of interferon-inducing activity (15), cells of spleen, bone marrow and mesentric lymph nodes of Japanese white rabbits weighing about 1 kg were mixed. The cells (1∿2 x 10^7 cells per ml) were cultivated at 25°C for 24 hr in an interferon-inducing medium consisting of Eagle's minimum essential medium supplemented with 10% heat inactivated calf serum. Interferon assay of the culture fluids was performed by the plaque reduction method using RK 13 line cells of rabbit kidney and vesicular stomatitis virus. The interferon titer was expressed as the reciprocal of the dilution showing 50% plaque reduction.

As shown in Table 3, interferon-inducing activity was exhibited by the analogues 303, 316, 312 and 315. Analogues 303 and 316 are C-1 monophosphorylated analogues in groups A and B, respectively, and analogues 312 and 315 are C-4' monophosphorylated analogues in groups C and D, respectively. Nonphosphorylated analogues in all groups (301, 302, 311 and 314) and diphosphorylated analogues in groups A and B (305 and 317) showed no activity.

Table 3. Interferon-inducing Activity of Synthetic Lipid A Analogues

Group	Sample number	Substituent C-4'	Substituent C-1	10	1.0	10^{-1}	10^{-2}	10^{-3}	10^{-4}
A	301	H	H	<10	<10	<10			
	304	P	H	<10	<10	<10			
	303	H	P	124	<10	<10			
	305	P	P	<10	<10	<10			
Control	Lipid A None				73	22	<10	<10	
				<10	<10	<10			
B	302	H	H	<10	<10	<10			
	321	P	H	<10	<10	<10			
	316	H	P	32	<10	<10			
	317	P	P	<10	<10	<10			
Control	Lipid A None				100	23	<10	<10	
				<10	<10	<10			
C	311	H	H	<10	<10	<10			
	312	P	H	146	94	33	<10		
Control	Lipid A None				427	352	109	<10	
				<10	<10	<10			
D	314	H	H	<10	<10	<10			
	315	P	H	175	33	<10			
Control	Lipid A None				400	240	15	<10	
				<10	<10	<10			

4.4. Pyrogenic, proclotting enzyme of horseshoe crab activation and macrophage activation activities

Pyrogenic activity was assayed by measuring temperature of rabbits for 3 hr after intravenous injection. Three rabbits were used for each dose of sample. The minimal dose giving a mean rise of higher than 1°F in temperature at 3 hr after the injection was determined to be the minimal effective dose (16). None of the analogues induced significant temperature rise even at such a high dose as 30 μg per kg.

Activation of the proclotting enzyme of horseshoe crab was assayed by gelation reaction of Limulus amebocyte lysate using a reagent kit. Most of the phosphorylated analogues induced gelation only at the highest concentration of 10 μg per ml. The other analogues and the control without a sample induced turbidity (±) in the reaction mixture at the same concentration although gelation was not completed. Consequently, none of the analogues were judged to have significant activity of proclotting enzyme of horseshoe crab activation.

Macrophage activation activity (17) was estimated by calculating percent cytostasis of target EL-4 leukemic cells. Inhibition of [^{3}H]-thymidine uptake into target EL-4 cells by the action of C57BL/6 mice peritoneal macrophages, which were treated with a sample after

Table 4. Pyrogenic, Proclotting Enzyme of Horseshoe Crab Activation and Macrophage Activation Activities of Synthetic Lipid A Analogues

Group	Sample number	Substituent C-4'	C-1	Pyrogenicity M.E.D.[a] (μg/kg)	LAL[b] gelation at the concentration (μg/ml) of 10	1.0	10^{-1}	10^{-2}	10^{-3}	10^{-4}	Mφ activation (% cytostasis)
A	301	H	H	>30	±	−	−				2.0
	304	P	H	>30	±	−	−				-6.7
	303	H	P	>30	+	±	−				-4.3
	305	P	P	>30	+	±	−				1.3
B	302	H	H	>30	±	−	−				-4.5
	321	P	H	>30	+	±	−				-7.0
	316	H	P	>30	+	±	−				-9.8
	317	P	P	>30	+	±	−				-4.3
C	311	H	H	>30	±	−	−				-3.2
	312	P	H	>30	+	±	−				6.0
D	314	H	H	>30	±	−	−				8.1
	315	P	H	>30	+	±	−				2.2
Control	Lipid A			0.03	+	+	+	+	+	±	72.5
	None			>30	±	−	−				7.7

[a]M.E.D.: minimal effective dose, [b]LAL: Limulus amebocyte lysate.

pretreatment with low concentration of macrophage activating factor, was measured. None of the analogues demonstrated macrophage activation activity.

These data are shown in Table 4.

5 Summary and discussion

It is proved in this study that some of the synthetic lipid A analogues can exhibit mitogenic, PBA and interferon-inducing activities. The synthetic lipid A analogues are classified according to their substituents and biological activities of each analogue summarized in Table 5. Mitogenic, PBA and interferon-inducing activities were simultaneously exhibited by both C-1 monophosphorylated analogues examined (303 and 316) regardless of their fatty acid substituents. As for C-4' monophosphorylated analogues, these three activities were simultaneously exhibited only by analogues in groups C and D (312 and 315). Only mitogenic activity among the three biological activities was shown by an analogue in group B (321), and none of the activities was detected in an analogue in group A (304). As for nonphosphorylated analogues, only mitogenic activity was shown by an analogue in group D (314) and none of the other analogues exhibited any of the activities. As for C-1,4' diphosphorylated analogues, only positive mitogenic activity was shown by an analogue in group B (317), but no positive activity was shown by an analogue in group A (305).

These results suggest that phosphate groups, especially the phosphate group at the C-1 position, play an important role for manifesting some of the biological activities of lipid A. It is also suggested that ester- and amide-bound fatty acid substituents take some part in manifestation of the activities and that C_{14}-OH or C_{14}-O-(C_{14}) is a more effective substituent than C_{14}.

It is interesting to note that a nonphosphorylated analogue in group D (314) was found to exhibit mitogenic activity not only in spleen cells of C3H/HeN mice but also those of C3H/HeJ mice which are known to be nonresponders to lipopolysaccharide. It is also interesting that the biological activities of natural lipid A are separately expressed by some of the synthetic lipid A analogues tested.

Table 5. Biological Activities of Synthetic Lipid A Analogues

	Phosphorylated position				Substituent		Group
	None	C-4'	C-1	C-1 & C-4'	Ester bound (R_1)	Amide bound (R_2)	
Sample number	301	304	303	305	C_{14}	C_{14}	A
Mitogenicity	−	−	+++	±			
PBA activity	−	−	++	±			
IFN-induction	−	−	+	−			
Limulus test	−	−	±	±			
Pyrogenicity	−	−	−	−			
Mφ activation	−	−	−	−			
Sample number	302	321	316	317	C_{14}	C_{14}-OH	B
Mitogenicity	−	+	++	+			
PBA activity	−	−	+	±			
IFN-induction	−	−	+	−			
Limulus test	−	±	±	±			
Pyrogenicity	−	−	−	−			
Mφ activation	−	−	−	−			
Sample number	311	312			C_{14}-OH	C_{14}-OH	C
Mitogenicity	−	+					
PBA activity	−	+					
IFN-induction	−	++					
Limulus test	−	±					
Pyrogenicity	−	−					
Mφ activation	−	−					
Sample number	314	315			C_{14}-OH	C_{14}-O-(C_{14})	D
Mitogenicity	++	++					
PBA activity	−	+					
IFN-induction	−	+					
Limulus test	−	±					
Pyrogenicity	−	−					
Mφ activation	−	−					

6 Acknowledgement

The authors wish to thank Dr. S. Kanegasaki, Institute of Medical Science, The University of Tokyo, for his valuable advice in carrying out this study.

7 References

(1) C. Galanos, O. Lüderitz, E. T. Rietschel, O. Westphal, Int. Rev. Biochem. 14 (1977) 239-335.

(2) O. Lüderitz, C. Galanos, V. Lehmann, H. Mayer, E. T. Rietschel, J. Weckesser, Naturwissensch. 65 (1978) 578-585.

(3) E. T. Rietschel, C. Galanos, O. Lüderitz, O. Westphal, Immunopharmacology and the regulation of leukocyte function, D. R. Webb (Ed.), Marcel Dekker, New York 1982, pp. 183-229.

(4) K. Tanamoto, C. Abe, J. Y. Homma, Y. Kojima, Eur. J. Biochem. 97, (1979) 623-629.

(5) Y. Cho, K. Tanamoto, Y. Oh, J. Y. Homma, FEBS Lett. 105 (1979) 120-122.

(6) K. Tanamoto, J. Y. Homma, J. Biochem. 91 (1982) 741-746.

(7) M. Inage, H. Chaki, S. Kusumoto, T. Shiba, Tetrahedron Lett. 21 (1980) 3889-3892.

(8) M. Kiso, H. Nishiguchi, S. Murase, A. Hasegawa, Carbohydr. Res. 88 (1981) C5-C9.

(9) M. Inage, H. Chaki, S. Kusumoto, T. Shiba, Chem. Lett. (1980) 1373-1376.

(10) M. Inage, H. Chaki, S. Kusumoto, T. Shiba, Tetrahedron Lett. 22 (1981) 2281-2284.

(11) M. Inage, H. Chaki, M. Imoto, T. Shimamoto, S. Kusumoto, T. Shiba, Tetrahedron Lett. 24 (1983) 2011-2014.

(12) C. Galanos, E. T. Rietschel, O. Lüderitz, O. Westphal, Y. B. Kim, D. W. Watson, Eur. J. Biochem. 31 (1972) 230-233.

(13) W. H. Alder, T. Takiguchi, B. Marsh, R. T. Smith, J. Exp. Med. 131 (1970) 1049-1078.

(14) A. J. Cunningham, A. Szenberg, Immunology 14 (1968) 599-601.

(15) Y. Kojima, J. Y. Homma, C. Abe, Jpn. J. Exp. Med. 41 (1971) 493-496.

(16) D. W. Watson, Y. B. Kim, J. Exp. Med. 118 (1963) 425-446.

(17) L. P. Ruco, M. S. Meltzer, J. Immunol. 120 (1978) 329-334.

Biological Activities of Synthetic Lipid A Analogues

Ken-ichi Tanamoto, Gery R. McKenzie, Chris Galanos, Otto Lüderitz, Otto Westphal, Ulrich Zähringer[1], Ulrich Schade[1], Ernst Th. Rietschel[1], Shoichi Kusumoto[2], and Tetsuo Shiba[2]

Max-Planck-Institut für Immunbiologie, D-7800 Freiburg, FRG
1) Forschungsinstitut Borstel, D-2061 Borstel, FRG
2) Faculty of Science, Osaka University, Osaka 560, Japan

1 INTRODUCTION

In the last decade, details of the chemical structure of Salmonella lipid A have been elucidated, and recently, its chemical synthesis has been approached. In a collaborative effort a number of synthetic preparations, representing lipid A analogues and intermediates of the Salmonella lipid A chemical synthesis were investigated for biological activities characteristic for endotoxins. These activities include serological cross reactivity with Salmonella lipid A antiserum, lethal toxicity, pyrogenicity, mitogenicity and prostaglandin release from macrophages. Some of the preparations were insoluble in water and could not be tested directly. Since it had previously been shown that succinylation of Re form lipopolysaccharide yields highly water-soluble dervatives without appreciable loss of endotoxic activity, some synthetic preparations were converted into succinyl esters and tested in this form. Results of these investigations are described in this paper.

2 MATERIALS AND METHODS

The isolation of Salmonella Re lipopolysaccharide (1), the preparation of lipid A (2), and the synthesis of the lipid A analogues (3) has been described previously.

Some of the synthetic (phosphate-containing) preparations could be solublized directly by suspending in 1N HCl (0°C) for a few minutes followed by centrifugation and neutralization with triethylamine. Succinylation of lipid A and the synthetic preparations was performed according to (4) with some modifications, and under steril conditions in order to avoid contamination with exogenous endotoxin (U. Zähringer, unpublished results).

For the demonstration of serological cross reactions with anti-Salmonella lipid A antiserum (1) two test systems were used (1): In the direct passive hemolysis test, erythrocytes were coated with the synthetic preparations (after treatment with NaOH) and hemolysis occurring on incubation with lipid A antiserum in the presence of complement was recorded. In the passive hemolysis inhibition test, the synthetic preparations were tested as inhibitors of the lytic system containing erythrocytes coated with lipid A, anti-lipid A antiserum, and complement.

Lethality tests were performed in C57BL/6 mice (over 10 weeks old) by injecting simultaneously the preparation (i.v.) and galactosamine (10 mg, i.p.). Galactosamine had been shown to increase the susceptibility of mice towards lipid A endotoxin by a factor of about 10^4 (5).

Pyrogen assays were performed in rabbits and the fever responses measured rectally by an recording temperature-measuring device (Hartmann and Braun, Frankfurt), as described previously (6).

Mitogenic activity of the preparations was determined by ^{3}H-thymidine incorporation into mouse spleen cells according to (7). The capacity of the preparations to inhibit the mitogenic response of Salmonella lipid A was also determined. Tests for anticomplementary (8) and Limulus gelation (9) activities were performed as described previously.

Some of the preparations were expected to exhibit hemolytic activity. This was tested by incubating different doses of the preparations with erythrocytes for different lengths of time; lysis was recorded by counting the remaining cells.

Mouse peritoneal and rabbit alveolar macrophages were used for the induction of prostaglandin release by synthetic preparations. Cells were incubated with the test sample (24 h, 37°C) and prostaglandins E_2 and $F_2\alpha$ released into the medium were determined by a radioimmunoassay (10).

3 RESULTS

The results of biological tests performed with the lipid A analogues are summarized in Table 1. Only preparation 317 could be solubilized in water, the other preparations were used either as fine suspensions or in the form of their succinylated derivatives. For serological tests, the synthetic preparations were treated with alkali in order to remove ester-linked fatty acids and to render the preparations soluble. It has been shown previously (1) that ester-linked fatty acids do not participate in the immune determinant structure of lipid A.

In order to test serological cross reactivity with Salmonella lipid A antiserum, the alkali treated synthetic preparations were coated to erythrocytes and incubated with dilutions of the antiserum and complement. Cross reactivity could also be demonstrated by using the synthetic preparations as inhibitors of the lytic system containing lipid A-coated erythrocytes, antiserum to lipid A, and complement. As shown in Table 1, most of the preparations exhibit serological cross reactivity, some in the same order of magnitude as lipid A (serum titer, 1024; minimal amount of inhibitor, 0.3 µg). This indicates that these preparations contain the structure which determines lipid A antigenic specificity and which had been identified previously as comprizing the linkage region of the amide-bound 3-hydroxytetradecanoic acid and parts of glucosamine.

Two glucosamine monosaccharide derivatives, the α- and ß-anomers of N-(3-hydroxymyristoyl-)glucosaminyl-1-phosphate, were also tested in this system and found to exhibit strong cross reactivity towards lipid A serum (titer, 128).

Lethal toxicity in galactosamine-sensitized mice was exhibited by the original preparations 316 and 317 and by succinylated preparations 314 and 315. In the case of lipid A, 0.1 µg per mouse resulted in 100% lethality. In order to obtain the same effect, 50 µg of the synthetic preparations had to be injected. No effect was seen in mice without galactosamine at this dose.

Pyrogenicity was tested in rabbits. Strong pyrogenic responses were obtained with 50 µg/kg doses of preparation 317, and of the succinylated preparations 302 and 316. The original preparation 316 exhibited moderate pyrogenicity. With the preparations of strong pyrogenic activity, the fever curves were biphasic with an increase in body temperature of about 2°C. A similar biphasic fever curve as that obtained with 50 µg/kg of the synthetic preparations is obtained with 0.1 µg/kg of lipid A. In the case of moderate pyrogenicity, the fever curves with 50 µg/kg of the preparation were monophasic and the temperature increase was about 0.6°C.

We have tested three different batches of preparation 302, two of which were synthesized in different ways. After succinylation, the three preparations were found to exhibit comparable pyrogenicity. This demonstrates reproducibility of the activation by succinylation of this preparation. On the other hand other succinylated preparations (i.e. 304, 312) were found to be devoid of pyrogenicity, indicating that the activity of succinylated preparation 302 was not due to contaminations introduced into the preparations as a result of their handling during and following the succinylation step.

None of the synthetic preparations was found to exhibit <u>Limulus</u> gelation activity when tested in doses up to 0.4 µg, while LPS is active in doses of about 2 pg.

Mitogenic activity could be demonstrate with preparations 305, 316, 317, and 321, and with the succinylated preparations 302, 311, 314, and 315. In the case of preparation 316, for instance, maximum thymidine incorporation was seen with 100 µg; for comparison, lipid A gave an about 4 times higher maximal response with 10 µg. Spleen cells from the lipopolysaccharide-nonresponder mouse strain C3H/HeJ, were stimulated neither by lipid A nor by preparation 316.

As seen from Table 1, some of the preparations were capable of inhibiting the mitogenicitiy of natural lipid A and some were found to cause lysis of erythrocytes. One preparation, the monosaccharide preparation 318 (N-(3-hydroxymyristoyl-)glucosaminly-α1-phosphate), was found to be non-hemolytic and non-mitogenic, but to strongly inhibit lipid A mitogenicity. Some preparations exhibited anticomplementary activity.

A number of the synthetic preparations stimulate macrophages to release prostaglandins E_2 and $F_{2\alpha}$ (not shown in Table 1). Some of the preparations exhibited an activity comparable to lipid A (15) with regard to the amounts of prostaglandin released into the medium. The

Endotoxic Activities

Table 1 Biological Activities of Synthetic Lipid A Analogues. -, o, 0: no, low, high activity; ( ): refers to succinylated preparations; 14, myristic acid; 14-OH, 3-hydroxymyristic acid; 14-O-14, 3-myristoyloxymyristic acid; P, phosphate.

Preparation Number	Substituents Ester E	Amide A	P4	P1	Antigenicity Hemo-lysis	Inhibition	Lethality	Pyrogenicity	Mitogenicity	Inhib. Mitogenicity	Hemolysis	Anti-C' Activity
302	14	14-OH	H	H	–	o	–	(0)	(o)	(0)	(0)	(o)
311	14-OH	14-OH	H	H	–	–	–	(o)	(o)	(0)	(0)	(o)
314	14-OH	14-O-14	H	H	0		(0)	–	(0)	–	(0)	–
304	14	14	P	H	0	0	–	–	–			(o)
321	14	14-OH	P	H	0	0	–	o	o	o	o	–
312	14-OH	14-OH	P	H	0	0	–	–	–		(o)	–
315	14-OH	14-O-14	P	H			(0)	(o)	(0)		–	–
316	14	14-OH	H	P			0	(0)	0	(o)	o	–
305	14	14	P	P	0	0	0	–	0	0	0	0
317	14	14-OH	P	P	0	0	0	0	0	0	0	0

monosaccharide preparations, N-(3-hydroxymyristoyl-) and N-(3-myristoyloxymyristoyl-) glucosamine, were also found to show strong activity.

4 DISCUSSION

A number of synthetic lipid A analogues have been tested for biological activities characteristic of enterobacterial lipid A. These most prominent activity of the preparations is their antigenic cross reactivity with anti-lipid A antiserum. Obviously, these preparations contain the immunodominant structure of the lipid A antigenic specificity, which has previously been determined to be represented by a discrete structure of the linkage region of an amide-linked fatty acid and parts of the glucosamine residue (11). Ester-linked fatty acids do not seem to participate in this immunodeterminant structure.

The present results with synthetic lipid A analogues show that (a) as with natural lipid A, deesterification by alkali treatment enhances cross reactivity to lipid A antiserum; (b) as lipid A, the synthetic preparations are able to coat erythrocytes to render them sensitive to lysis with lipid A antiserum and complement; (c) phosphate substitution of the glucosamine disaccharide is not a prerequisite for expression of antigenicity; (d) the amide-bound 3-hydroxymyristic acid can be replaced by myristic acid without loss of cross reactivity with anti-lipid A antiserum; and (e) N-acylglucosamine-1-phosphates (the α- and ß-anomers) exhibit strong cross reactivity with the antiserum. These results are in agreement with our previous findings (11) and those of Polish and Russian groups who showed that 3-hydroxymyristoyl-hydroxamate (12), and N-myristoyl-and N-D,L-3-hydroxymyristoyl-D-glucosa mine-6-phosphate (13) would exhibit lipid A cross reactivity.

Some preparations exhibit typical endotoxic effects, such as lethal toxicity and pyrogenicity, but much higher doses are required compared to lipid A. The results show that these two activities are separable. Thus, the succinylated preparation 302 was found to be pyrogenic, but not lethal, while the succinylated preparations 314 and 315 are lethal, but of no or low pyrogenicity. Furthermore, it is interesting to note that the original form of 316 is toxic and of low pyrogenicity, its succinylated form, however, is pyrogenic and of low toxicity.

None of the preparations was active in the Limulus gelation test. Even preparation 317, which is toxic and pyrogenic, does not exhibit gelation activity. We have found previously a reverse situation: The lipopolysaccharide of R. palustris, which lacks endotoxicity (pyrogenicity and lethal toxicity) is active in the Limulus gelation test. In contrast, the equally endotoxically inactive lipopolysaccharide of R. viridis does not express gelation activity (14). Such exceptions have to be kept in mind.

We have shown previously that the lipid A precursor I molecule, which contains the lipid A backbone substituted by four 3-hydroxymyristic acid residues, is toxic but of weak pyrogenicity (11). It could be imagined that preparation 316 resembles lipid A precursor I, and that its succinylated form, which probably contains acyloxyacyl residues (i.e., succinyl groups esterifying the 3-hydroxymyristoyl units) resembles lipid A in this respect.

Except for structural requirements of antigenicity, the present results do not yet allow to draw conclusions regarding relationships between chemical structure and endotoxic activities. As shown in this paper with the synthetic preparations as well as previously with natural lipid A's (11), individual biological effects, such as antigenicity, toxicity, pyrogenicity and others are separable. It seems probable that a number of factors are necessary for providing the structural and conformational prerequisites for these activities, solubility being very important. There is evidence that phosphate group(s), especially in position 1 (16), the presence of amide-bound, but also of ester-bound 3-hydroxyacyl and acyloxyacyl residues increase activity when different preparations containing or lacking one of these substituents are compared. However, none of the presently available preparations has a structure identical to that proposed for _Salmonella_ lipid A or lipid A precursor which may be the explanation for their relatively weak activity in biological assays other than antigenicity. We have to await further synthetic work, which is in progress, before being able to draw final conclusions about the specific structures of lipid A responsible for biological activities.

5 REFERENCES

(1) C. Galanos, O. Lüderitz, O. Westphal, Eur. J. Biochem. 9 (1969) 245-249.

(2) C. Galanos, O. Lüderitz, O. Westphal, Eur. J. Biochem. 24 (1971) 116-122.

(3) See contribution of Shiba et al., this series.

(4) E. Th. Rietschel, C. Galanos, A. Tanaka, E. Ruschmann, O. Lüderitz, O. Westphal, Eur. J. Biochem. 22 (1971) 218-224.

(5) C. Galanos, M. A. Freudenberg, W. Reutter, Proc. Natl. Acad. Sci. USA 76 (1979) 5939-5943.

(6) D. W. Watson, Y.B. Kim, J. Exp. Med. 118 (1963) 425-446.

(7) J. Anderssen, F. Melchers, C. Galanos, O. Lüderitz, J. Exp. Med. 137 (1973) 943-953.

(8) C. Galanos, E. Th. Rietschel, O. Lüderitz, O. Westphal, Eur. J. Biochem. 19 (1971) 143-152.

(9) E. T. Yin, C. Galanos, S. Kinsky, R. A. Bradshaw, S. Wessler, O. Lüderitz, M. E. Sarmiento, Biochim. Biophys. Acta 261 (1972) 284-289.

(10) A. Jobke, B.A. Peskar, and B.M. Peskar (1973) FEBS letters, 37, 192-196.

(11) O. Lüderitz, C. Galanos, V. Lehmann, H. Mayer, E. Th. Rietschel, J. Weckesser, Naturwissensch. 65 (1978) 578-585.

(12) C. Lugowski, E. Romanowska, Eur. J. Biochem. 48 (1974) 81-87.

(13) V. I. Gorbach, I. N. Krasikova, P. A. Lukyanov, O. Y. Razmakhnina, T.F. Solov'eva, Y. S. Ovodov, Eur. J. Biochem. 98 (1979) 83-86.

(14) O. Lüderitz, M. A. Freudenberg, C. Galanos, V. Lehmann, E. Th. Rietschel, D. H. Shaw, in: Membrane Lipids of Prokaryotes (S. Razin and S. Rottem, eds.) Current Topics in Membranes and Transport. Academic Press, Inc., New York, 17 (1982) 79-151.

(15) U. Schade, E.Th. Rietschel, Klin. Wochenschr. 60 (1982) 743-745.

(16) T. Yasuda, S. Kanegasaki, T. Tsumita, T. Tadakuma, J.Y. Homma, M. Inage, S. Kusumoto, T. Shiba, Eur. J. Biochem. 124 (1982) 405-407.

Shwartzman Activities of Synthetic Lipid A Analogues and Their Effects on Hepatic Enzyme Activities

Nobuhiko Kasai, Kiyoshi Egawa, Junichi Mashimo

Department of Microbial Chemistry, School of Pharmaceutical Sciences, Showa University, Hatanodai, Shinagawa-ku, Tokyo, 142

Tetsuo Shiba, Shoichi Kusumoto

Department of Chemistry, Faculty of Science, Osaka University. Toyonaka, Osaka, 560, Japan

Abstract

The synthetic lipid A analogs and related compounds were tested on their ability to prepare the local Shwartzman reaction in rabbits. One of the nineteen preparations tested, 1-α-monophosphate of β-1,6-glucosamine disaccharide which carries two amide-bound (R)-3-hydroxy-tetradecanoyl and three ester-bound tetradecanoyl residues was found to exhibit a weak hemorrhagic necrosis when Salmonella minnesota R595 endotoxic glycolipid was used as a provocative endotoxin.

The depression of hepatic aminopyrine demethylase activity in mice, seen on endotoxin administration, was not shown by any analog tested. However, some of lipid A analogs were found to show significant alterations in the levels of hepatic cytochrome P-450, δ-aminolevulinic acid synthetase and/or heme oxygenase activity.

The results suggested that the nature of linkage of fatty acids and phosphates in lipid A might play an important role in the elicitation of these biological activities.

Abbreviations: LPS: Lipopolysaccharide; ALA: δ-Aminolevulinic acid.

1 Introduction

Since the direct evidence that lipid A moiety of endotoxic lipopolysaccharide (LPS) is essential for the manifestation of most of biological activities was shown by Lüderitz et al. (1), a number of studies have been focussed on the elucidation of structure-activity relationships of lipid A. However, free lipid A derived from LPS was a mixture of analogs as clearly demonstrated with thin-layer chromatography (2, 3). Furthermore, there are species variations and minor heterogeneity even within one species in the chemical structure of natural lipid A in LPS (4, 5).

Recently, Shiba, Kusumoto and their coworkers (6-9) have synthesized a number of Salmonella type lipid A analogs and related compounds. Biological study using synthetic pure compounds may lead us to the elucidation of structures responsible to the toxic and beneficial activities of lipid A.

In this study, we examined some toxic activities of synthetic lipid A analogs and related compounds in comparison with those of the lipid A and LPS preparations.

2 Shwartzman activity

Our early investigations (2, 10) on the Shwartzman activity of endotoxins have revealed following points: 1) the purified lipid A_2 or A_3 isolated from E. coli LPS and the free lipid A (HCl) derived from Salmonella minnesota R595 glycolipid after hydrolysis with 1N HCl at 100°C for 30 minutes were practically inactive. 2) some of preparations derived from R595 glycolipid by short-time hydrolysis with 1% acetic acid were much more active than the free lipid A (AcOH) which was prepared by the hydrolysis at 100°C for 2 hours. 3) the mild hydrolysis of R595 glycolipid in ammonia-saturated anhydrous methanol at 37°C resulted in the liberation of mainly ester-bound 3-hydroxytetradecanoic and tetradecanoic acids with marked loss of the Shwartzman activity. 4) the phosphate-free glycolipid (11) derived from Selenomonas ruminantium was moderately active at the dose of 20 µg, which the intensity was comparable to that of 10 µg of Salmonella free lipid A.

Thus, our previous results seemed to show that the nature of linkage of fatty acids and phosphate residues in lipid A region may be important to elicit the Shwartzman activity. Therefore, comparison of the Shwartzman activity of synthetic lipid A analogs and natural lipid A or free lipid A preparations was made.

2.1 Shwartzman assay

The Shwartzman assay of test materials was carried out by usual procedure; the solutions to be determine were injected intradermally into the shaved abdomen of rabbits weighing between 2 and 3 kg, and after 24 hours the animals were challenged with 200 ug of standard endotoxin given intravenously. The purified glycolipid from <u>Salmonella minnesota</u> R595 was used as the standard endotoxin in this experiment. Usually, five rabbits to each preparation were used. The abnormally reacting rabbits were excluded from data reported here.

2.2 Shwartzman activity of synthetic lipid A analogs

The synthetic lipid A analogs and related compounds were solubilized with chloroform-methanol and/or triethylamine and complexed to bovine serum albumin (BSA), if needed, followed by brief sonication. The ratio of lipid and BSA was 2:1. The results of the Shwartzman assay were summarized in Table 1 and 2.

Table 1. Shwartzman activity of synthetic lipid A analogs

Sample number	R1	R2	R3	R4	Preparative dose (μg)	Shwartzman activity (a) Hemorrhage	Necrosis
301	C_{14}	C_{14}	H	H	160	-	-
302	C_{14}	C_{14}-OH	H	H	80	±	-
307	H	C_{14}-OH	H	H	160	-	-
311	C_{14}-OH	C_{14}-OH	H	H	160	-	-
314	C_{14}-OH	C_{14}-O-(C_{14})	H	H	160	+	-
304	C_{14}	C_{14}	P	H	80	-	-
312	C_{14}-OH	C_{14}-OH	P	H	160	-	-
315	C_{14}-OH	C_{14}-O-(C_{14})	P	H	160	+	-
303	C_{14}	C_{14}	H	P	80	-	-
316	C_{14}	C_{14}-OH	H	P	160	+	+
305	C_{14}	C_{14}	P	P	160	-	-
317	C_{14}	C_{14}-OH	P	P	160	+	-

C_{14} : tetradecanoyl
C_{14}-OH : (R)-3-hydroxytetradecanoyl
C_{14}-O-(C_{14}) : (R)-3-tetradecanoyloxytetradecanoyl
P : P(OH)$_2$ ‖ O

(a) Shwartzman activity was arbitrarily graded from negative(-) to positive(+) on the basis of number of rabbits which hemorrhagic or necrotic lesion was seen, and (±) denotes that only one or two out of five rabbits showed hemorrhagic lesion.

Among the synthetic lipid A analogs tested, compounds (301, 302, 304 and 305), which both residues of R^1 and R^2 in the structure are substituted by only tetradecanoyl residues, did not showed any hemorrhagic or necrotic lesion. Compound (307) carrying amide-linked 3-hydroxytetradecanoyl residue alone was also inactive.

On the other hand, compounds (314 and 315) which carry 3-hydroxytetradecanoyl residues at R^1 and 3-tetradecanoyloxytetradecanoyl residues at R^2 and compounds (302, 316 and 317) which carry tetradecanoyl at R^1 and 3-hydroxytetradecanoyl residues at R^2 showed a weak hemorrhage, but the necrotic lesion was only observed with the compound 316. The necrosis-producing activity of the compound 316 was much (two order of magnitude) lower than Salmonella lipid A (AcOH) in the intensity.

The D-glucosamine derivatives carrying various substituents were inactive, as shown in Table 2, though compound 313 carrying an amide-linked acyloxy group showed a weak hemorrhage.

Table 2. Shwartzman activity of synthetic compounds

Sample number	R^2	R^4	Preparative dose (μg)	Shwartzman activity Hemorrhage	Necrosis
308	C_{14}-OH	H	160	-	-
309	C_{14}	H	80	-	-
310	(S)-C_{14}-OH	H	160	-	-
313	C_{14}-O-(C_{14}-OH)	H	80	±	-
318	C_{14}-OH	P	160	-	-
319	C_{14}-OH	β-P	160	-	-
320	C_{14}	P	160	-	-

Thus, our previous inference that the nature of linkage of fatty acids and phosphate residues may play an important role was also suggested in this study.

2.3 Shwartzman activity of natural lipid A

We have also tested the Shwartzman activity of various LPS and some free lipid A preparations. Bacterial strains from which LPS were isolated, were Salmonella typhimurium LT-2, SL1034, TV149, SL1069, SL 1102, Salmonella minnesota R595, Pseudomonas aeruginosa ATCC7700,

Chromobacterium violaceum IFO 12614 and Caulobacter crescentus CB13. LPS and free lipid A from Klebsiella pneumoniae O3K⁻ were donated by Dr. N. Kato.

In the Shwartzman assay of these preparations (data not shown), LPS from C.violaceum was found to be much more active than those of other strains. Also, LPS which showed the least activity was the preparation from Caulobacter crescentus. At the present time, it is not clear whether or not such the differences in the Shwartzman activity due to the microheterogeneity of lipid A region.

3 Effect on hepatic enzyme activities

In previuos studies (12, 13) we found that the administration of R595 glycolipid or free lipid A (AcOH) results in the significant depression of hepatic drug-metabolizing enzyme activities, cytochrome P-450 levels and ALA synthetase activity and the marked increase in the levels of heme oxygenase activity in mice.

In order to investigate the structure of lipid A responsible for such a metabolic perturbation by endotoxin, the effect of some synthetic lipid A analogs on these enzyme activities were studied.

3.1 Assays of enzymes

Male ddy mice (Shizuoka Jikkendobutsu Co., Shizuoka) weighing 23-27 g were, maintained on a laboratory diet and given water ad lib. throughout the experiment. Mice, except in the case of the assay of aminopyrine demethylase activity, were fasted overnight and injected i.p. with synthetic lipid-BSA complex prepared as described previously. The R595 glycolipid was used as a reference standard endotoxin, and control mice were injected with the vehicle only.

Assay of enzyme activity was carried out as described in previous paper (13); mice were injected with each preparation (80-160 ug/mouse) and killed 18 hours after administration. Aminopyrine demethylase activity was assayed by the method of Cochin and Axelrod (14), cytochrome P-450 levels by the method of Omura and Sato (15), ALA synthetase by the method of Marver et al. (16) and heme oxygenase activity by the method of Tenhunen et al. (17). Protein content was determined by the method of Miller (18) using bovine serum albumin as a standard.

3.2 Effects of synthetic lipid A analogs

Table 3 shows the effects of some of synthetic lipid A analogs on the hepatic cytochrome P-450 levels and related enzyme activities.

Table 3. Effects of synthetic lipid A analogs on hepatic enzyme activities in mice

Sample number (a)	R^1	R^2	R^3	R^4	aminopyrine demethylase	cyto-chrome P-450	ALA synthetase	heme oxygenase
301	C_{14}	C_{14}	H	H	−	−	−	
303	C_{14}	C_{14}	H	P	−	−	−	−
304	C_{14}	C_{14}	P	H	−	−	−	−
305	C_{14}	C_{14}	P	P	−	−	−	
307	H	C_{14}-OH	H	H	−	±	−	+
316	C_{14}	C_{14}-OH	H	P	−	±	−	+
317	C_{14}	C_{14}-OH	P	P	−	±		
311	C_{14}-OH	C_{14}-OH	H	H	−	±	+	
312	C_{14}-OH	C_{14}-OH	P	H	−			+
314	C_{14}-OH	C_{14}-O-(C_{14})	H	H	−	+	+	+
315	C_{14}-OH	C_{14}-O-(C_{14})	P	H	−	+		+

(a) Dose (i.p.): 80-160 µg/mouse
(b) Alteration in the levels of each enzyme activity were arbitrarily graded from negative (−) to positive (+) on the basis of the statistical significance of the effect and the reproducibility, and (±) indicates that was significant in only one or two repeated experiments.

All the compounds tested did not show the significant depression of hepatic microsomal aminopyrine demethylase activity, seen on endotoxin administration. However, all the analogs carrying 3-hydroxytetradecanoyl residues at the R^1 and/or R^2, whether the compound carry phosphate group or not, showed a significant increase of the heme oxygenase activity. A similar effect was also observed in the assays of cytochrome P-450 levels, though in this case the analogs which are substituted by 3-hydroxytetradecanoyl residues at the R^1 and 3-tetradecanoyloxytetradecanoyl residues at R^2 seemed to be more active than other analogs. In the assay on ALA synthetase activity it was suggested that the substitution of 3-hydroxytetradecanoyl residues at the position R^1 as well as R^2 is required to exhibit the activity as the depressor. In addition, doses of the synthetic analogs which needed to exhibit the activity were much higher than Salmonella lipid A.

4 Summary

In this study we have demonstrated that some synthetic lipid A analogs exhibit the Shwartzman reaction and the alterations in the levels of hepatic cytochrome P-450 , ALA synthetase and/or heme oxygenase activities.

It seems to be premature to draw definite conclusion on structure-activity relationships, but the results suggested that the nature of linkage of fatty acids including 3-hydroxyfatty acid in β-1,6-glucosamine disaccharide is of most importance in the elicitation of the observed activities, and that the phosphate residues also might concern to some of the activities.

However, the facts that the observed activities were far less than those of natural lipid A in the intensity and that no synthetic compound which depresses the aminopyrine demethylase activity, seen on natural lipid A, was found, suggest that there are some differences in the fine structures between the synthetic analogs and natural lipid A.

5 References

(1) O. Lüderitz, C. Galanos, V. Lehmann, M. Nurminen, E. T. Rietschel, G. Rosenfelder, M. Simon, O. Westphal, J. Infect. Dis. $\underline{128}$ (1973) S17-S29.

(2) N. Kasai, Ann. N. Y. Acad. Sci. $\underline{133}$ (1966) 486-507.

(3) N. Kasai in J. Y. Homma, K. Saito, N. Kasai, M. Niwa (Eds.), Bacterial Endotoxin (in Japanese), Kodansha Academic Press, Tokyo 1973, pp. 134-151.

(4) C. Galanos, O. Lüderitz, E. T. Rietschel, O. Westphal, Int. Rev. Biochem. $\underline{14}$ (1977) 239-335.

(5) C. H. Chen, A. G. Johnson, N. Kasai, B. A. Key, J. Levin, A. Nowotny, J. Infect. Dis. $\underline{128}$ Suppl. (1973) S43- S51.

(6) M. Inage, H. Chaki, S. Kusumoto, T. Shiba, Tetrahedron Lett. $\underline{21}$ (1980) 3889-3892.

(7) M. Inage, H. Chaki, S. Kusumoto, T. Shiba, Chem. Lett. (1980) 1373-1376.

(8) M. Inage, H. Chaki, S. Kusumoto, T. Shiba, Tetrahedron Lett. 22 (1981) 2281-2284.

(9) M. Inage, H. Chaki, S. Kusumoto, T. Shiba, Chem. Lett. (1982) 1281-1284.

(10) N. Kasai, R. Matsushita, N. Komatsu, Japan. J. Bacteriol. 28 (1973) 96.

(11) Y. Kamio, K. C. Kim, H. Takahashi, Agr. Biol. Chem. 36 (1972) 2195-2201.

(12) K. Egawa, N. Kasai, Microbiol. Immunol. 23 (1979) 87-94.

(13) M. Yoshida, K. Egawa, N. Kasai, Toxicol. Lett. 12 (1982) 185-190.

(14) J. Cochin, J Axelrod, J. Pharmacol. Exp. Ther. 125 (1959) 105-110.

(15) T. Omura, R. Sato, J. Biol. Chem. 239 (1964) 2370-2378.

(16) H. S. Marver, D. P. Tschudy, M. G. Perlroth, A. Collins, J. Biol. Chem. 241 (1966) 2803-2890.

(17) R. Tenhunen, H. S. Marver, R. Schmid, J. Lab. Clin. Med. 75 (1970) 410-420.

(18) G. L. Miller, Anal. Chem. 31 (1959) 964.

Bone Marrow Reactions Produced with Synthetic Lipid A Analogues

Masao Yoshida, Michimasa Hirata, Nobuko Tsunoda and Katsuya Inada

Department of Bacteriology, School of Medicine, Iwate Medical University, Uchimaru, Morioka, 020 Japan

Tetsuo Shiba and Shoichi Kusumoto

Department of Chemistry, Faculty of Science, Osaka University, Toyonaka, Osaka, 560 Japan

Abstract

Synthetic lipid A analogs were studied on four bone marrow reactions reported as endotoxic reactions Analogs (#307, 311, 314, and 312) with 3-hydroxymyristic acid, except for #315, showed positive leukocyte migration. All analogs with 3-hydroxymyristic acid showed positive hemorrhage reactions. Only #314 and 312 showed slight cytotoxicity reaction. No analog showed a positive procoagulant reaction. #305, which did not have 3-hydroxymyristic acid but phosphate, did not show any bone marrow reaction. Phosphate residue does not seem to be involved in the manifestation of these reactions. Compared to natural lipid A, the analogs showed weaker reactions which were reproducible with some degree of variance.

1 Introduction

Host responses to bacterial endotoxin are very multiple, and contain many kinds of biological responses, for example, hemodynamic injuries accompanied by inflammatory damages and beneficial responses to stimulate immune mechanisms. In addition, many kinds of cells and blood

Endotoxic Activities

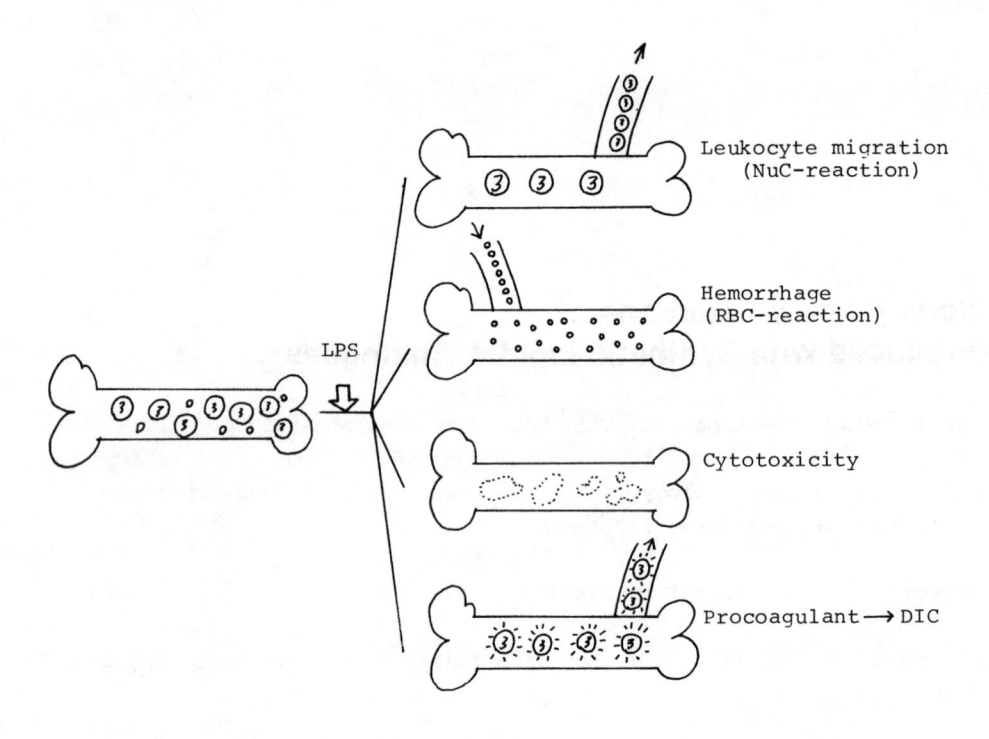

Fig. 1. Bone marrow reactions

Table 1. Structures of synthetic lipid A analogs

Sample number	R1	R2	R3	R4
307	H	C_{14}-OH	H	H
311	C_{14}-OH	C_{14}-OH	H	H
314	C_{14}-OH	C_{14}-O-(C_{14})	H	H
312	C_{14}-OH	C_{14}-OH	P	H
315	C_{14}-OH	C_{14}-O-(C_{14})	P	H
305	C_{14}	C_{14}	P	P

fluid components are involved in these endotoxic responses.

To investigate the biological activities of synthetic lipid A analogs, several different biological responses should be checked, if possible, simultaneously using the same experimental animal.

Yoshida et al.(1-4), Hirata et al.(5-6) have studied the bone marrow reactions to endotoxin in mouse. As shown in Fig. 1, the bone marrow reactions to endotoxin are as follows; (1) a decrease in number of nucleated cells resulting from leukocyte migration into the peripheral circulation (NuC-reaction or leukocyte migration), (2) an increase in number of erythrocytes due to hemorrhage in the marrow (RBC-reaction or hemorrhage), (3) the cytotoxic damage checked by trypan blue exclusion test (cytotoxicity) and (4) procoagulant activity of marrow cells attributable to thromboplastin (P-reaction). In these four reactions at least three of them are able to be examined simultaneously in a mouse without much difficulty. Since the difference of individual sensitivity to endotoxin is not small, it is noteworthy that several parameters are examined in the same experimental animal.

2 Materials and Methods

2.1. Synthetic lipid A analog and the solubilization. The synthetic lipid A analogs used in the present study and their structures are shown in Table 1. These water-insoluble lipid A analogs were injected into mice with the solution prepared as follows: Two mg of natural or synthetic lipid A was solubilized with both 40 µl of triethylamine and 60 µl of methanol, followed by adding 1.9 ml of saline. The solution was then diluted with saline five times, and finally contained 0.4% triethylamine, 0.6% methanol and 200 µg/ ml of lipid A. Natural lipid A and the analog #315 became thereby a clear solution. The solubilities of #311, 312 and 314 were also rather good, but the solutions of #305 and #307 were turbid even after sonication. Therefore, the solubilities of the latter would not be good. The control mice were given only the solvents mentioned above.

2.2 Natural lipid A and other preparations. S-LPS (Westphal type) and lipid A of S. minnesota were kindly supplied by Prof. O. Lüderitz, Max-Planck. Institut., Freiburg, Germany, and used as reference and standard preparations, respectively. K-antigenic polysaccharide of E. coli. K 1 supplied by Drs. Jann, Max-planck Institut, Freiburg, Germany was used as a carrier to solubilize #307 and #311 analogs. Instead of saline, saline containing 20 µg/ml K-antigen preparation was used to dilute lipid A analog solution of triethylamine and methanol, as described above.

2.3. Bone marrow reactions. Mice; 8-10 weeks-old ddY mice (Shizuoka Nokyo Farm) were injected intraperitoneally with a solution of the test sample, and 18-20 hr later were sacrificed for the assays.

Estimation of absolute counts of both nucleated and red blood cells (2-4). Two femora of the mouse was put into 2 ml ice-cold PBS (phosphate buffer saline) in a weighing glass vessel, and cut fine with a pair of small scissors as used in ophthalmological operations. The cell suspension was filtered with nylon-wool, and pipetted twice to make a homogenous suspension. These procedures were carried out in an ice bath. The cells were then washed twice by 1,000 rpm 5 min centrifugation at 4° C, the final cell pack was suspended into 1.0 ml PBS and diluted 500 times with Cell-Kit-7 (TOA Medical Electric Co., LTD), thereafter, whole cell (nucleated and red blood) count was estimated by microcell counter (CC-1002, BL-2, TOA Medical Electric Co., LTD, Kobe, Japan). The nucleated cell count was likewise estimated after adding of a drop of saponin solution. The difference between whole cell and nucleated cell counts was regarded as the count of red blood cells.

Cytotoxicity (3-5). Humeri from the mouse, as mentioned above, were used for the cytotoxicity test. 4.25% NaCl and 0.2% trypan blue solutions were mixed at 1:4 ratio just before use. A drop of the mixture was put on a slide, then a piece of humerus cut longitudinally was immersed, followed by gentle moving the piece of bone with a pincette to prepare cell suspension. Eventually, dead cells were stained with a final 0.16% trypan blue, the cell suspension was microscoped in a hemocytometer, and the cells stained per 400 cells were counted. The cytotoxicity was then expressed by percentage.

Estimation of procoagulant activity. This was performed according to Hirata et al.(5-6). Briefly, the bone marrow cells obtained from the mouse mentioned above were suspended in saline or in RPMI 1640, and washed 3 times by 1,000 rpm 5 min centrifugation at 4°C. Cell suspension (0.1 ml) was preincubated with 0.1 ml substrate plasma (pooled citrate-plasma of mice). Then trostin-$CaCl_2$ was added to the mixture and the clotting time was measured using a fibrometer (Bioquest Division, Becton, Dickinson and Co., Cokeysville, USA).

3 Results

3.1. Leukocyte migration reaction (NuC-reaction). Combined results are shown in Table 2 with dose-response relationship as to leukocyte migration. Analog #315 and 305 did not induce this reaction, although the experiments were done with only one dose of as much as 100 µg. In natural lipid A very good dose-response relationship was observed, while in synthetic lipid A analogs, dose-response relationships were not good. In only #314, slight dose-response relationship was observed. Nevertheless, positive reactions as to leukocyte migration i.e. decrease of nucleated cell numbers in the marrow were observed at certain dose levels of #307, 311, 314 and 312, compared to the control.

The regression lines of #307, 311, 314 and 312 were statistically investigated on parallelism (7) for the standard regression line of natural lipid A. Parallelisms were not then denied in all regression lines of analogs, although dose-response relationships were not so good. Eventually relative potencies were calculated, as stated below.

Table 2. Dose response relationships of nucleated cell counts ($\times 10^7$/two femora)

Dose (μg)	Natural lipid A	Synthetic lipid A's			
		307	311	314	312
100	1.98±0.16 (10)	2.00±0.2 (10)	1.84±0.18 (15)	2.38±0.15 (15)	2.73±0.19 (10)
50	2.39±0.13 (15)	2.17±0.24 (5)	2.71±0.21 (5)	2.62±0.24 (5)	2.29±0.33 (5)
25	3.19±0.08 (4)	1.93±0.16 (5)	2.41±0.17 (5)	2.71±0.24 (5)	2.59±0.13 (5)

All figures are nucleated cell counts of two femora of a mouse. Control showed 2.78±0.13 (30), 100 μg of either 315 or 305 showed 2.62±0.21 (5) or 2.73±0.4 (5). All figures in parentheses show numbers of mice used.

3.2. Hemorrhage reaction (RBC reaction). Combined results as to hemorrhage reaction are also shown in Table 3 with dose-response relationship. In the hemorrhage reaction i.e. increase or RBC numbers in the marrow, natural lipid A showed good dose-response relationship. Analogs #307, 311 and 314 showed positive reactions, but dose-response relationships were not good. Parallelisms were not denied in #311 and 314 for the standard regression line of natural lipid A. In #315 a strong positive reaction was observed with one dose (100 μg) experiment. Analog #305 did not induce a positive reaction, although the experiment was so far done with only one dose of as much as 100 μg. In analog #312, no mean value of any dose was over the control in Table 3. In addition to these experiments, a further experiment was performed as to RBC-reaction, and a positive reaction was obtained, in #312 although relative potency was not calculated.

Table 3. Dose-response relationship of RBC counts
($\times 10^6$/two femora)

Dose (µg)	Natural lipid A	307	311	314	312
100	16.64±2.67 (10)	11.07±1.88 (10)	7.09±1.07 (15)	8.98±0.99 (15)	6.09±0.69 (10)
50	12.85±1.88 (15)	10.06±0.68 (5)	8.88±0.88 (5)	5.48±0.76 (15)	4.22±0.43 (5)
25	9.32±1.12 (25)	10.44±1.58 (5)	6.86±0.91 (5)	6.88±0.85 (5)	7.22±0.84 (5)

All figures are RBC counts of two femora of a mouse. Control showed 7.3±0.61, 100 µg of either 315 or 305 showed 16.20±3.30 (5) or 5.36±1.23 (5). All figures in parentheses show numbers of mice used.

3.3 Reproducibility of reactions due to synthetic lipid A.

Sensitivities to endotoxin of mice are rather various even in inbred strains, thus the mice reponses always show certain variances (4). To investigate the variance of reactions due to the synthetic lipid A, two active analogs #307 and 311 were chosen and studied as to the leukocyte migrations and hemorrhage. The results are shown in Table 4. These results show that there are variances in some degree also in the experiments of synthetic lipid A analogs. It is thus likely that the variances in the analog experiment were not as much as compared to the variances in the experiments of endotoxin or natural lipid A.

3.4. Analog solubilization with K-antigenic polysaccharide.

Macrosocopically, the solubilities of lipid A analogs are not necessarily good even after solubilization with organic solvents as stated in Materials and Methods. Therefore, K 1 preparation of K-antigenic acidic polysaccharides which could solubilize endotoxin or lipid A in a natural situation was added into the saline to dilute the analog-organic solvent solution. Thus, increase of solubility of the analog was expected. Increased solubility of #307 was easily observed by the naked eye as diminution of turbidity, while that of #311 was not able to be observed because of the clear solubility of the first solution. In Table 5, the results of #311 and K 1 are presented. As shown in this table, addition of K 1 preparation did not necessarily augment the activities of #311. The results of #307 and K 1 are also presented in Table 5. Analog #307 + K 1 showed negative activity in spite of slight but reliable activity of the analog only.

Table 4. Reproducibilities of reactions of analogs

		Leukocyte migration	Hemorrhage
#307	Exp. 1	±	+
	Exp. 2	+	−
	Exp. 3	++	−
#311	Exp. 1	+	±
	Exp. 2	±	++
	Exp. 3	−	+

In Exp. 1-3 #307 was used, and in Exp. 4-6 #311 was used.
++ : Difference from control was statistically significant.
+ : Difference from control was not significant, but a reasonable difference was observed.
± : A little difference was observed.
− : No reasonable difference was observed.

Table 5. Solubilization of analog with K-antigen

	Leukocyte migration		Hemorrhage	
	Exp. 1	Exp. 2	Exp. 1	Exp. 2
K1	+	−	−	−
#311	+	+	+	++
#311 + K1	−	++	+	++
K1	−		−	
#307	+		−	
#307 + K1	−		−	

−∼++ : See the legend of Table 4.

3.5. Cytotoxicity and procoagulant activity. In Table 6, two experimental results as to the cytotoxicity and procoagulant activity are shown; the latter was expressed as thromboplastin unit (8) converted from the clotting times of mouse plasma. Analogs #314 and 312 had slight cytotoxic activity, but #311 and 315 did not. No analog examined showed procoagulant activity.

Table 6. Cytotoxic and procoagulant activity

	Cytotoxicity (%)	Thromboplastin unit
Control (Exp. 1)	13.7±1.7	1447±141
Natural lipid A 50 μg	25.8±2.3*	3711±340*
Lipid A analogs 100 μg 305	14.8±1.6	1707±247
312	18.1±0.7**	1536±161
Control (Exp. 2)	15.7±1.0	3620±320
Natural lipid A 100 μg	34.9±4.5*	5321±591*
Lipid A analogs 100 μg 311	17.0±1.9	3477±339
314	21.2±2.2	2655±115

*P<0.05 compared with control.
**0.1>P>0.05 compared with control.

3.6. Relative potencies of both leukocyte migration and hemorrhage.
From the dose-response regressions in Table 2-3, relative potencies were calculated in spite of poor regressions. The values are presented in Table 7, where the results as to both cytotoxicity and procoagulant are also summarized. Relative potencies were calculated putting the activities of natural lipid A as 1.00. In leukocyte migration, analog #307 had very high activity, and #311, 314 and 312 had similar level activity to natural lipid A as to this reaction.

In hemorrhage reaction, the relative potency of #307 may be inaccurate, because the parallelism was denied. Analog #315 showed high activity which would be equivalent to 1.00, although relative potency was not able to be calculated. All relative potencies except for that of #307 were 15-20% of that of natural lipid A.

Table 7. Relative potencies

	Leukocyte migration (NuC-reaction)	Hemorrhage (RBC-reaction)	Cytotoxicity	Procoagulant
Natural lipid A	1.00	1.00		
S-LPS	4.41	1.76		
307	3.23	0.43*		(−)
311	1.36	0.14	(−)	(−)
314	0.91	0.18	(+)	(−)
312	0.96	0.04	(+)	(−)
315	(−)	(+)		(−)
305	(−)	(−)	(−)	(−)

*Parallelism was not recognized.

4 Discussion

The bone marrow reactions are inflammatory responses shown in mouse to endotoxin; the hemorrhage, cytotoxicity and procoagulant are injurious or toxic changes, while leukocyte migration is an essentially beneficial one for the host animal.

In the present study, at least two and at maximun four of these parameters were checked simultaneously with synthetic lipid A analogs. Thus, this elaboration contemplates the relationship between structure and function of lipid A.

Except for #305, other analogs examined had 3-hydroxymyristic acid at R1 and/or R2 positions, and these analogs had activities to induce one or all of the bone marrow reactions, except for procoagulant. On the other hand, #305 had phosphates at R3 and R4, but not 3-hydroxymyristic acid. This #305 did not induce any bone marrow reactions. Among analogs containing 3-hydroxymyrisitc acid, #312 and 315 had phosphate at R3 position. These analogs had neither higher nor broader activities as to bone marrow reactions, compared to other analogs which did not have phosphate, but 3-hydroxymyristic acid. Analogs #314 and 315 had double acylated hydroxymyristic acid at R2 position. These analogs did not necessarily have higher or broader activities compared to others. These results show that 3-hydroxy or double acyl-myristic acid is important for the exhibition of bone marrow reactions, whereas phosphate is not important.

In natural lipid A, it has been reported that the phosphoryl group is important to manifest endotoxicity, and in synthetic analogs also, most investigators reported necessity of phosphate at R3 or R4. The present study shows, however, that phosphoryl residue is not always necessary to manifest the bone marrow reactions. Kasai (9) reports in this book that 3-hydroxy fatty acid in synthetic lipid A is essential for alteration of activities of hepatic heme enzymes. Kasai (10) also reported that lipid A of Selmonas ruminatum did not have phosphoryl residue, but had the activities of both pyrogenicity and Shwartzmann reaction. This contradiction remains for further study.

In only one experiment, 100 µg of #315 was injected. Hemorrhage reaction was strongly positive, while leukocyte migration was negative. This reason is unexplainable at present.

As shown in Table 7, leukocyte migration was the most sensitive reaction among four bone marrow reactions in this study. A few mechanisms (11) such as endothelium injury of blood vessels, increased adhesion of granulocytes and release of leukocytosis inducing factor are involved in the leukocyte migration from the marrow.

Hemorrhage reaction was positive in all analogs containing 3-hydroxymyristic acid. Damage of endothelial cells of marrow sinus, in addition to histamine and serotonin (12), are involved in this reaction.

It is of interest that cytotoxicity and procoagulant are insensitive reactions and different from the former two reactions i.e. leukocyte migration and hemorrhage. This difference would result from that of mechanisms. We have reported that lysosomal enzyme and biomembrane are involved in cytotoxicity and a change of cytoplasmic membrane is essential for procoagulant production (3), (5), furthermore, that these two reactions correlate intimately with each other.

It is noteworthy, however, that the leukocyte migration is the strongest reaction among the bone marrow reactions induced by lipid A analogs, and that this reaction is an essentially beneficial one of the bone marrow reactions.

When the migrating leukocytes have procoagulant activity, it can cause intravascular blood coagulation in periphery (5.6.8.).

Fortunately, the analogs did not have the activity to produce procoagulant, thus leukocyte migration would be more beneficial for the host in the case of synthetic analogs.

We reported a similar finding about endotoxins of E. coli (4), compared to those of Salmonella. This might suggest a possibility of clinical use of synthetic lipid A analogs.

Dose-response regressions of synthetic analogs were not good even as to leukocyte migration in which analogs showed comparable activities to natural lipid A.

The reproducibilities of leukocyte migration and hemorrhage seem to be comparable to that of lipid A or endotoxin. However, it is conceivable that activities of the analogs are much lower than that of natural lipid A. Nevertheless, it is unclear that the difference between biological properties of synthetic and natural lipid A's depend upon the difference of chemical structure or that of physical properties.

Since solubilities of analogs were generally worse than that of natural lipid A, promotion of the solubility was performed in this study by adding K-antigenic polysaccharide. No better result was obtained by this procedure. Especially, although promoted solubility of #307 was confirmed with the naked eye, the increase of reactions were never observed, contrarily all reactions examined became negative.

5 References

(1) M. Yoshida, M. Hirata, Y, Hatano, K. Inada, Jap. J. Exp. Med. 38 (1968) 335-346.

(2) M. Yoshida, F. Parant, M. Hirata, L. Chedid, Jap. J. Med. Sci. Biol. 25 (1972) 243-247.

(3) M. Yoshida, M. Hirata, M.K. Agarwal, In Animal, Plant and Microbial Toxins. A. Ohsaka, K. Hayashi and Y. Sawai ed. Plenum New York (1976), pp. 509-520.

(4) M. Yoshida, J. Iwate Med. Ass. 30 (1978) 491-505.

(5) M. Hirata, K. Inada, N. Tsunoda, M. Yoshida, In Bacterial Endotoxin and Host Response. M.K. Agarwal ed. Elsevier/North-Holland Biomedical Press, (1980), pp. 255-272.

(6) M. Hirata, N. Tsunoda, K. Inada, M. Yoshida, Jap. J. Med. Sci. Biol. 33 (1980) 30-31.

(7) D.J. Finney. in Statistical Method in Biological Assay, Griffin London 1964, pp. 99-117.

(8) M. Hirata, M. Yoshida, N. Tsunoda, K. Inada. a paper in this book.

(9) N. Kasai, K. Egawa, J. Mashimo, M. Yoshida, M. Sasaki, T. Shiba, S. Kusumoto. a paper in this book.

(10) N. Kasai, R. Matsushita, N. Komatsu, Japan. J. Bacteriol. 28 (1973) 96 in Japanese.

(11) M. Yoshida, M. Hirata, K. Inada, N. Tsunoda, H. Mohri, S. Matsui, in Bacterial Endotoxin and Host Response. M.K. Agarwal ed. (1980), pp. 231-254.

(12) M. Hirata, J. Infect. Dis. 132 (1975) 611-616.

Biological Activities of Synthetic Lipid A Analogues in Artificial Membrane Vesicles

Shiro Kanegasaki, Tatsuji Yasuda, Toru Tsumita
The Institute of Medical Science, The University of Tokyo, Shirokanedai, Minato-ku, Tokyo 108, Japan

Takushi Tadakuma
Keio University School of Medicine, Shinjuku-ku, Tokyo 160, Japan

J. Yuzuru Homma
The Kitasato Institute, Shirokane, Minato-ku, Tokyo 108, Japan

Shoichi Kusumoto, Tetsuo Shiba
Faculty of Science, Osaka University, Toyonaka, Osaka 560, Japan

Abstract

To compare biological activities of chemically-synthesized lipid A analoues under the condition without being affected by their physico-chemical difference as happens in aqueous solution, we incorporated the analogues into liposomes and investigated adjuvant and mitogenic activities. Among the synthetic analogues tested, those which showed significant adjuvant activity were acylated β-1,6 linked D-glucosamine disaccharides 1) carrying either phosphate in position 1 of the reducing sugar or 2) amide bound 3-myristoxymyristic acid, or 3) carrying both phosphate in position 4 of non reducing sugar and amide bound 3 hydroxymyristic acid. The disacchrides without such substitution(s) were not active. All tested glucosomine monosaccharide-type analogues with various substiturions did not show the activity as adjuvant. Significant mitogenic activity was found in liposomes containing the analogues which showed adjuvant activity.

Endotoxic Activities

1. Introduction

The bound lipid moiety of lipopolysaccharide, called lipid A is known to be essential for various endotoxic activities of the molecule and, in fact, free lipid A introduced into suitable systems expresses most of the activities. To confirm the proposed chemical structure (1) and to establish the structure(s) involved in the biological activities, various analogues of lipid A have recently been synthesized (2,3,4). Some of these analogues, however, are quite hydrophobic and it is difficult to handle them quantitatively as aqueous suspension. Furthermore, different physicochemical properties of micells derived from each analogue may influence biological activities.

To overcome such difficulties, we incorporated the analogues into a hapten-sensitized liposomal system and compared the adjuvant activity. We showed previously that two out of five analogues tested were effective as adjuvant (5). Since we last reported, many similar analogues have been synthesized chemically making it possible to investigate in more detail the structure and function of these analogues. In the present paper, we report that certain new analogues showed better adjuvanticity when incorporated into the liposomal system than the analogues reported previously (5). We also compare mitogenic activity of each analogues in liposomes.

2. Adjuvant activity of synthetic analogues in a liposomal system

We used multilamellar liposomes as immunogens (Fig. 1). Liposomes were actively sensitized by the addition of 10% trinitrophenyl-aminocaproylphosphatidylethanolamine (TNP-Cap-PE) to a mixture containing dipalmitoylphosphatidylcholine (DPPC), cholesterol, and phosphatidic acid (PA) in molar ratios of 1:1:0.1, respectively. Lipid A or lipid A analogues in a chloroform-methanol mixture were added to the other lipids (5). The lipids were dried in a pea-shaped flask on a rotary evaporator, followed by standing for 1 hr under high vacuum in a desiccator. The dried lipid film was dispersed in phosphate buffered saline.

As adjuvant activity, numbers of IgM plaque forming cells were determined 4 days after intraperitoneal injection to C57BL/6 mice of liposomes composed of DPPC (100 nmol), cholesterol (100 nmol) and PA (10 nmol) containing 10 nmol of TNP-Cap-PE and 0.2 to 20 µg of lipid A analogue per mouse (5).

The structures of the synthetic analogues are summarized in Table 1.

The adjuvant activity of synthetic lipid A analogues and natural lipid A are shown in Fig. 2. In this experiment 10 µg of each lipid A

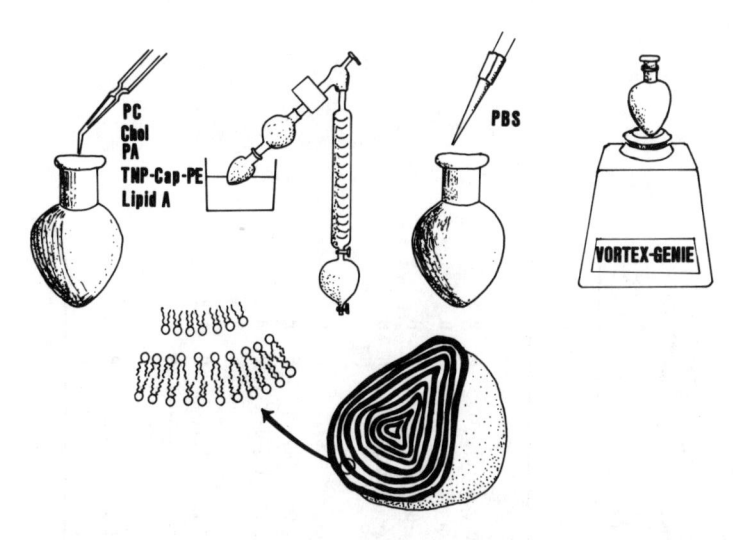

Fig. 1. Method of preparation of multilamellar liposomes.

analogue was incorporated with TNP-Cap-PE sensitized DPPC liposomes. Arrow shows the control level of anti-TNP PFC response immunized with TNP-Cap-PE liposomes.

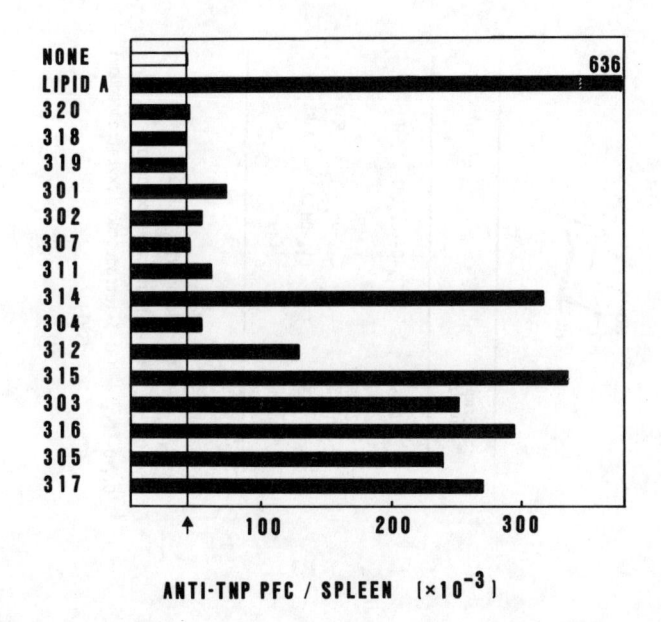

Fig. 2. Adjuvant activity of lipid A and analogues in liposomes.

Endotoxic Activities

Table 1. Structures of chemically synthesized analogues of lipid A used in the present investigations.

Sample number	R¹	R²	R³	R⁴
301	C_{14}	C_{14}	H	H
302	C_{14}	C_{14}-OH	H	H
307	H	C_{14}-OH	H	H
311	C_{14}-OH	C_{14}-OH	H	H
314	C_{14}-OH	C_{14}-O-(C_{14})	H	H
304	C_{14}	C_{14}	P	H
312	C_{14}-OH	C_{14}-OH	P	H
315	C_{14}-OH	C_{14}-O-(C_{14})	P	H
303	C_{14}	C_{14}	H	P
316	C_{14}	C_{14}-OH	H	P
305	C_{14}	C_{14}	P	P
317	C_{14}	C_{14}-OH	P	P

Sample number	R¹	R²	R³	R⁴
313		C_{14}-O-(C_{14}OH)		H
318		C_{14}-OH		P
319		C_{14}-OH		β-P
320		C_{14}		P

C_{14} : tetradecanoyl
C_{14}-OH : (R)-3-hydroxytetradecanoyl
C_{14}-O-(C_{14}) : (R)-3-tetradecanoyloxytetradecanoyl
P : $\overset{O}{\underset{\parallel}{P}}(OH)_2$

104

Among the synthetic analogues tested in our system, one group which showed significant adjuvant activity is glucosamine disaccharides carrying the phosphate in position 1 of the reducing sugar. These are analogues 303, 316, 305, and 317. A second group with adjuvant activity is the analogues having amide- bound 3-myristoxymyristic acid and probably esterified hydroxymyristic acids as well. These are analogues 314 and 315. A third group is the analogues carrying both phosphate in position 4 of the nonreducing sugar and amide- (or ester-) bound 3-hydroxymyristic acid (312). In contrast, the disaccharides without such substitution(s) were not active (301, 302, 307, 311 and 304). All tested monosaccharide type analogues with various substitutions did not show the activity as adjuvant (313, 320, 318 and 319). (The result for 313 is not shown in Fig. 2).

3. Structure and function relationship of synthetic analogues : detailed study

To go into more detail about the structure-function relationship of lipid A analogues, we focused attention on certain substitutions and compared the dose response of the analogues.

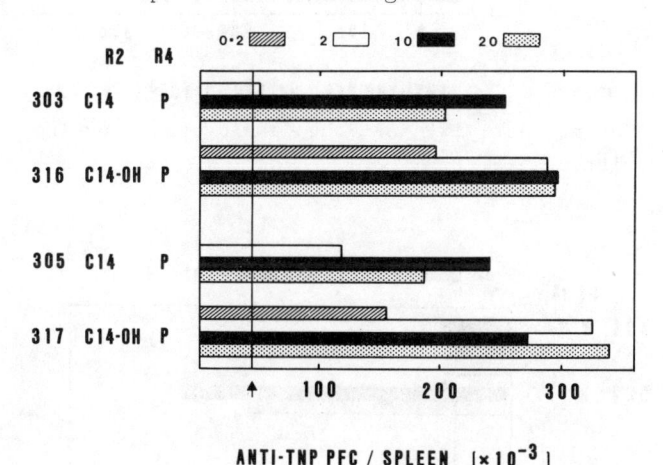

Fig. 3. Composition of the adjuvant activity of the analogues carrying phosphate in position 1.

Figure 3 shows that the activity of the analogues which carry the phosphate group solely in position 1 (303) was enhanced by the coexistence of either a position 4' phosphate of nonreducing glucosamine (305) or/and an amide-bound 3-hydroxymyristic acid (316 and 317). The differences between 303 and 316 or 305 and 317 are amide-bound fatty acid residue. The analogues which have 3-hydroxymyristic acid showed stronger adjuvanticity.

Regardless of whether the analogues carry a position 4' phosphate or not, those in which the disaccharide was substituted by an amide-bound 3-myrystoxymyristic acid (315 or 314) instead of a 3-hydroxymyristic acid (312 or 311, respectively) showed stronger adjuvant activity (Fig. 4). Branched fatty acids at position 2 of glucosamine contribute greatly in this system.

As shown in Fig. 5, the analogues carrying a position 1 phosphate showed stronger adjuvant activity than the analogue which carries a position 4' phosphate (312). The differences between analogues 304 and

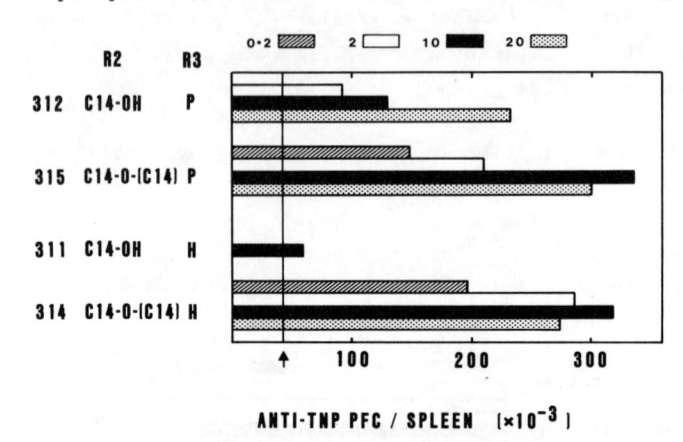

Fig. 4. Adjuvant activity of the analogues carrying amide-bound 3-myrystoxymyristic acid.

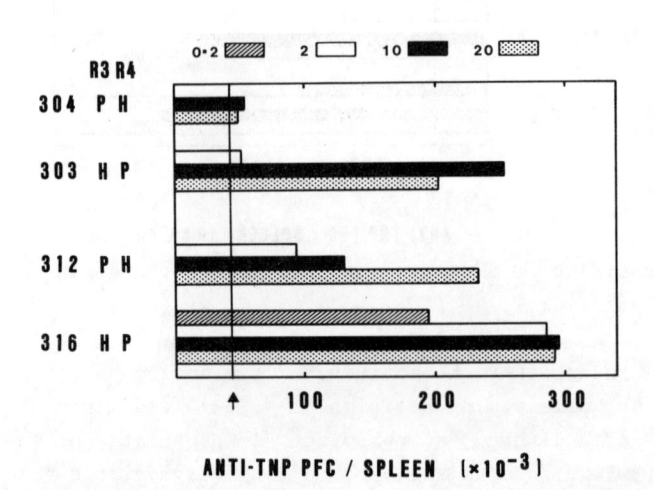

Fig. 5. Comparison of the activity of analogues carrying position 1 phosphate and position 4' phosphate.

303, or 312 and 316 are the phosphate position. Analogues 303 and 316 which have a position 1 phosphate showed stronger adjuvant activity than the analogues with a position 4' phosphate.

4. Adjuvant activity of synthetic analogues in nonresponder mice

C3H/HeJ mice are known to be defective in B cell activity and are unable to respond to most lipopolysaccharides or lipid A's. We further tested to see whether the stimulating effect of lipid A analogues on antibody formation was only observed in lipopolysaccharide responder mice and not in nonresponder mice. All the analogues which showed adjuvant activity were tested. Nonresponder C3H/HeJ mice produced a comparable number of anti-TNP PFC whether or not the antigenic liposomes contained synthetic analogues as was the case of liposomes which contained natural lipid A. (Result not shown).

5. Mitogenic activity of synthetic analogues

To test whether the analogues which showed adjuvant activity also have mitogenic activity, liposomes composed of DPPC (100 µM) and cholesterol (100 µM) containing 8 µg per well (200 µl) of the synthetic analogue were incubated for 3 days with spleen cells derived from C57BL/6 mice. One µCi/ml (final concentration) of [3H]-thymidine

^{3}H-TdR INCORPORATION (cpm/CULTURE)

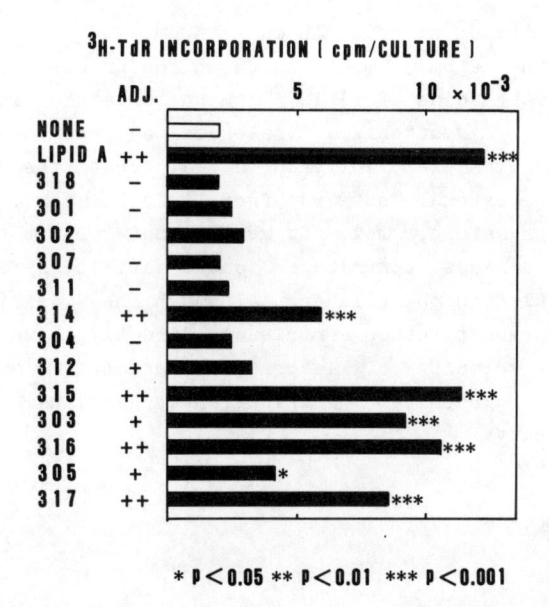

	ADJ.	
NONE	–	
LIPID A	++	***
318	–	
301	–	
302	–	
307	–	
311	–	
314	++	***
304	–	
312	+	
315	++	***
303	+	***
316	++	***
305	+	*
317	++	***

* $p < 0.05$ ** $p < 0.01$ *** $p < 0.001$

Fig. 6. Mitogenic activity of lipid A analogues.

was added to the culture 24 hours before harvest and radioactive incorporation into the cells was assayed.

As shown in Fig. 6, significant activity was found in liposomes containing 303, 305, 314, 315, 316 and 317 which also showed adjuvant activity as described.

6. Concluding remarks

In the present investigation, we employed liposomal model membrane systems to investigate the biological activities of chemically synthesized analogues of lipid A. By using such systems, we were able to compare the activities of each analogue without their being affected too much by their physicochemical difference as happens when they are suspended in water. We showed that certain analogues, glucosamine disaccharides carring amide- and ester- bound fatty acids, acted as adjuvant in the hapten-sensitized liposomal system and as mitogen in the liposomal system, if the synthetic analogues fulfilled at least one of the following requirements: 1) The synthetic analogue carries a glycosidic bound phosphate residue in position 1 of the reducing glucosamine. 2) The amide-bound fatty acids are 3-myristoxymyristic acid (and ester-bound fatty acids are 3-hydroxymyristic acid). 3) The synthetic analogue carries ester-bound phosphate in position 4 of the nonreducing glucosamine in addition to amide-bound 3-hydroxymyristic acids. The disaccharides without such substitution(s) were not active as adjuvant or as mitogen at least in our system.

According to the atomic model based on the proposed structure on lipid A (1), all fatty acids of lipid A can be oriented in the same direction. Chemically synthesized analogues may thus localize in the artificial membrane vesicles in such a way that the hydrophilic saccharide portion extends outwards from lipid bilayers. The results presented here, therefore, may suggest that substitutions of glucosamine disaccharides contribute to the activity by changing the depth of the molecules in the bilayers, although the possibility still remains that each substitution contributes directly to the biological activity. Different velocity of lateral diffusion of analogues in membrane and/or topographical distribution of each analogue may also influence the activity.

7. References

(1) E. T. Rietschel, C. Galanos, O. Luderitz, O. Westphal, Immunopharmacology and the regulation of leukocyte function, D. R. Webb (Ed.), Marcel Dekker, New York 1982 pp. 183-229.

(2) M. Inage, H. Chaki, S. Kusumoto, T. Shiba, Tetrahedron Lett. 21 (1980) 3889-3892.

(3) M. Inage, H. Chaki, S. Kusumoto, T. Shiba, Tetrahedron Lett. 22 (1981) 2281-2284.

(4) M. Inage, H. Chaki, S. Kusumoto, T. Shiba, Chemistry Lett. (1980) 1373-1376.

(5) T. Yasuda, S. Kanegasaki, T. Tsumita, T. Tadakuma, J.Y. Homma, M. Inage, S. Kusumoto, T. Shiba, Eur. J. Biochem. 124 (1982) 405-407.

Comparative Studies on the Immunobiological Activities of Synthetic Lipid A Analogues and Lipophilic Muramyl Peptides

Shozo Kotani[1], Haruhiko Takada[1], Masachika Tsujimoto[1], Tomohiko Ogawa[1], Yoshihide Mori[2], Tetsuo Shiba[3], Shoichi Kusumoto[3], Masaru Inage[3], Nobuhiko Kasai[4]

[1]Department of Microbiology and Oral Microbiology, [2]The Second Department of Oral and Maxillofacial Surgery, Osaka University Dental School, Kita-Ku, Osaka 530; [3]Department of Chemistry, Faculty of Science, Osaka University, Toyonaka, Osaka 560; [4]Department of Microbial Chemistry, School of Pharmaceutical Sciences, Showa University, Tokyo 142; Japan

Abstract

Studies have been made on immunobiological activities of representative lipophilic derivatives of muramyl peptides, variously acylated and phosphorylated β(1'-6)glucosamine disaccharides which were prepared to mimic a so far proposed structure of Salmonella type lipid A, and compounds related to each of them, in comparison with those of bacterial lipopolysaccharide, Re-glycolipid and lipid A of natural sources. Characteristics of immunobiological activities of each type of compounds were discussed.

1 Introduction

Bacteria parasitic or indigenous to animal hold a variety of cellular constituents having mainfold, biological response modifying (BRM) activities, especially those to modulate host defence mechanisms. These are cell wall peptidoglycans (PG) (1), endotoxic lipopolysaccharides (LPS) of gram-negative bacteria (2), lipoteichoic acids of gram-positives (3), trehalose-6,6'-dimycolates and related compounds in mycobacteria (4), and pertussigen (pertussis toxin) of Bordetella pertussis (5) and others. It is worthy of note that most of bacterial BRM principles are localized in cell surface layers, while cellular constituents encased by the cytoplasmic membrane are scarcely immunobiologically active. It may be not unreasonable to speculate that biologically active surface layer components of bacterial cells are products of mutation and selection during an extremely long history of interactions between hosts and parasitic or indigenous bacteria through cell surface layers of the latter, and BRM activities have survival values on both sides of host and bacteria.

Among the above-stated immunomodulators, LPS and PG have so far been most extensively studied on their BRM activities from various points of view. There has been rapid progress of studies on chemical entities responsible for BRM activities of PG and LPS for the past several years. It has been disclosed by synthesis that N-acetylmuramyl-L-alanyl-D-isoglutamine (MDP for muramyl dipeptide) is a minimal essential structure responsible for majority, if not all, of BRM activities of bacterial cell wall PG. A number of analogs and derivatives of MDP have then been prepared with a view to get compounds with the maximum beneficial BRM activities and the minimum undesirable effects which can be useful in clinical or preventive medicine (6-9). As regards bacterial LPS, on the other hand, a lipid A moiety of the molecule has been shown to carry most, if not all, of BRM or endotoxic properties of it (10-12). In recent studies, Shiba, Kusumoto and their colleagues (13) have established a synthetic route for the preparation of O,N-polyacyl glucosamine β(1'-6) disaccharide 1,4'-diphosphate (10) which corresponds to the proposed structure of Salmonella type lipid A.

In this article, we will summarize our recent studies examining immunobiological activities of lipid A analogs and related compounds which have been synthesized to mimic lipid A, and muramyl peptides and their analogs or derivatives which are a mimicry of structural units of PG, in comparing with those of lipid A and LPS preparations of natural sources.

2 Test materials

2.1 Lipopolysaccharides

Lipopolysaccharides prepared by Boivin method (14), LPS-B, or those by Westphal method (15), LPS-W, from Salmonella enteritidis and Escherichia coli 0127:B8 were purchased from Difco Laboratories (Detroit,

Mich.). Use was also made of an LPS-W specimen which was prepared from Pseudomonas aeruginosa ATCC 7700 in Kasai's laboratory (16).

2.2 Re-glycolipid

Two reference specimens of Salmonella minnesota R595 were used. One (Re-glycolipid 1) of them was prepared as described previously (17), and the other (Re-glycolipid 2) (18) was a gift of Dr. Dennis W. Watson (University of Minnesota, Minneapolis, Minn.).

2.3 Lipid A

A lipid A specimen of S. minnesota R595 was prepared by acid hydrolysis of Re-glycolipid with 1% acetic acid, and that of P. aeruginosa was isolated by hydrolysis of LPS-W with 5% acetic acid (17).

2.4 Synthetic lipid A analogs and related compounds

Preparation methods were described previously (13, 19-24). Figure 1 represents compounds used in this study, with their chemical structures, literatures describing synthetic methods, and compound numbers. In some experiments, D-glucosamine hydrochloride, GlcN was used as a reference.

2.5 Muramyl peptides and related compounds

The following compounds, whose structures and preparation methods in terms of literatures were shown in Figure 2, were submitted to in vivo and in vitro assays in the present study. The first group of compounds, totally synthetic ones, N-acetylmuramyl-L-alanyl-D-isoglutamine (MDP), 6-O-stearoyl-N-acetylmuramyl-L-alanyl-D-isoglutamine (L18-MDP), 6-O-(2-tetradecylhexadecanoyl)-MDP (B30-MDP), N^{α}-MDP-N^{ε}-stearoyllysine [MDP-Lys(L18)], an L-valine analog of MDP-Lys(L18) whose L-alanine was replaced with L-valine, N-acetylmuramyl-L-alanyl-D-glutaminyl-n-butyl ester [MDP(Glu)-OnBu, murabutide], N-[N^2-(heptanoyl-γ-D-glutamyl)-meso-2-(L),2'(D)-diaminopimel-1-yl]-D-alanine (FK-565), and inactive analogs, D-isoglutamine residue of whose peptide moiety was substituted by L-isoglutamine with MDP, L18-MDP and MDP-Lys(L18) or D-isoasparagine with B30-MDP, were generously supplied by Daiichi Seiyaku Co. Ltd., Tokyo, Japan. The second group of compounds, semisynthetic ones, sodium β-N-acetylglucosaminyl-(1-4)-N-acetylmuramyl-L-alanyl-D-isoglutaminyl-(L)-stearoyl-(D)-meso-2,6-diaminopimelic acid-(D)-amide-(D)-alanine (GM-53) and O-methyl or O-butyl ester of GM-53 [GM-53(OMe) and GM-53(OBu)], and a parent compound of the above three, β-N-acetylglucosaminyl-(1-4)-N-

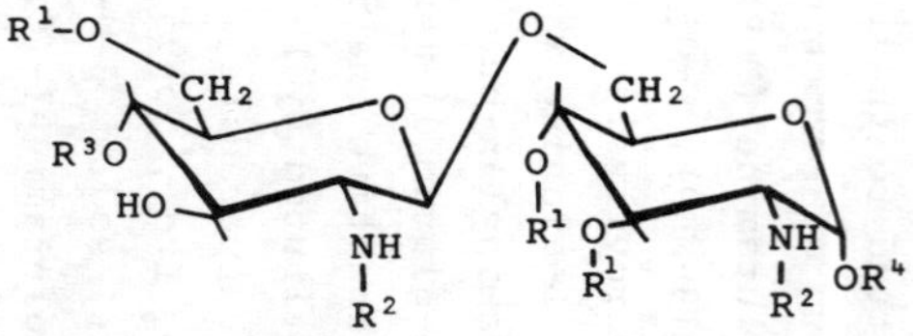

Endotoxic Activities

Compound number	R¹	R²	R³	R⁴	Reference number for synthesis
301	C₁₄	C₁₄	H	H	19
302	C₁₄	C₁₄-OH	H	H	20
307	H	C₁₄-OH	H	H	13
311	C₁₄-OH	C₁₄-OH	H	H	23
314	C₁₄-OH	C₁₄-O-(C₁₄)	H	H	23
304	C₁₄	C₁₄	P	H	21
321	C₁₄	C₁₄-OH	P	H	13
312	C₁₄-OH	C₁₄-OH	P	H	23
315	C₁₄-OH	C₁₄-O-(C₁₄)	P	H	23
303	C₁₄	C₁₄	H	P	21
316	C₁₄	C₁₄-OH	H	P	23
305	C₁₄	C₁₄	P	P	21
317	C₁₄	C₁₄-OH	P	P	23

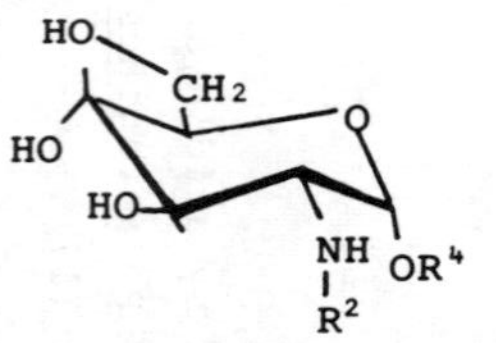

Compound number	R¹	R²	R³	R⁴	Reference number for synthesis
308		C₁₄-OH		H	24
309		C₁₄		H	24
310	(S)-C₁₄-OH			H	24
313	C₁₄-O-(C₁₄-OH)			H	24
318		C₁₄-OH		P	22
319		C₁₄-OH		β-P	22
320		C₁₄		P	22

C₁₄ : tetradecanoyl
C₁₄-OH : (R)-3-hydroxytetradecanoyl
C₁₄-O-(C₁₄) : (R)-3-tetradecanoyloxytetradecanoyl
C₁₄-O-(C₁₄-OH) : (R)-3-[(R)-3-hydroxytetradecanoyl-oxytetradecanoyl]

P : P-(OH)₂
    ‖
    O

Fig. 1. Compound number, structures and literatures for synthesis of lipid A analogs (glucosamine disaccharide derivatives) and related compounds (glucosamine monosaccharide derivatives)

Immunobiological Activities of Lipid A Analogues

MDP and its 6-O-acyl derivatives (25-27)

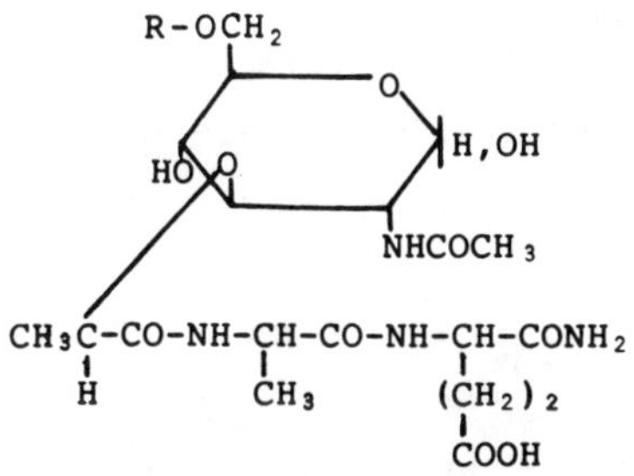

Compound	R
MDP	H
L18-MDP	$CH_3(CH_2)_{16}CO-$
B30-MDP	$\begin{array}{l}CH_3(CH_2)_{13}\\CH_3(CH_2)_{13}\end{array}\!\!\!>\!\!CHCO-$

MDP-Lys(L18) (28)

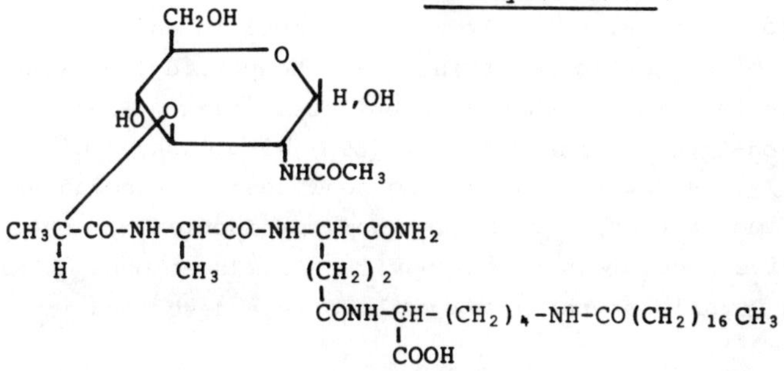

Murabutide (29)

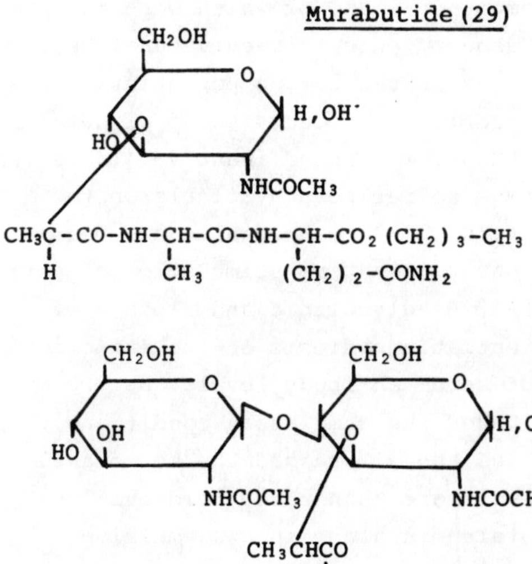

FK-565 (30)

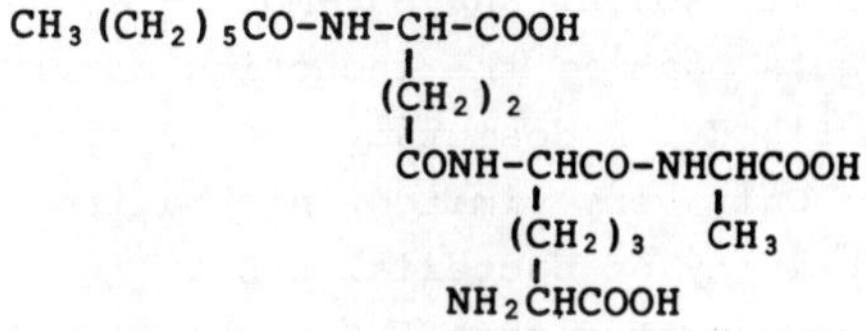

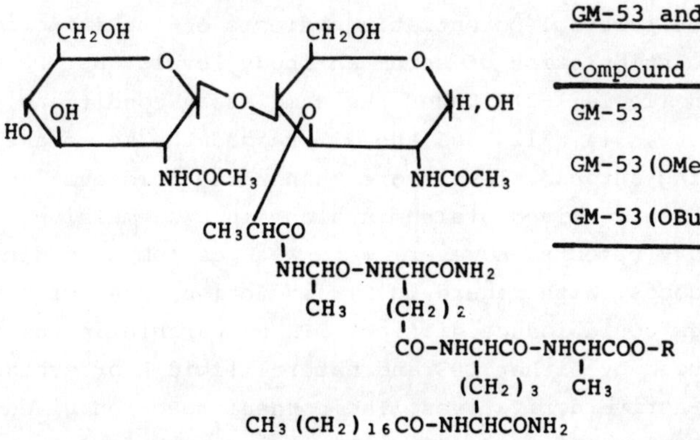

GM-53 and its derivatives (31)

Compound	R
GM-53	Na
GM-53 (OMe)	CH_3
GM-53 (OBu)	$(CH_2)_3CH_3$

Fig. 2. Chemical structures of MDP and related compounds

acetylmuramyl-L-alanyl-D-isoglutaminyl-(L)-meso-2,6-diaminopimelic acid-(D)-amide-(L)-D-alanine (GMP4)—an enzymatic degradation product of Lactobacillus plantarum (ATCC 8014) cell walls (32)—were a gift of Dainippon Pharmaceutical Co. Ltd., Osaka.

3 Immunostimulating (adjuvant) activity

Groups of 5 albino guinea pigs were immunized by intra-footpad injection of an indicated dose of test materials with 1 (or 2) mg of ovalbumin as a test antigen. The vehicle of administration was water-in-oil emulsion (w/o emulsion, Difco Freund's incomplete adjuvant), liposomes consisting of synthetic lecithin (DL-α-phosphatidyl choline, dipalmitoyl) and cholesterol, or phosphate buffered saline (PBS). The induction of delayed-type hypersensitivity (DTH), a prototype of cell-mediated immunity, was determined by a corneal test on the 2nd and 3rd week after the immunization, and the antibody production was estimated by the quantitative precipitin reaction with serum specimens obtained on the 4th week. Details of the assay methods were described in previous papers (33-35).

Results are summarized in Table 1. When w/o emulsion was used as a vehicle, all of test MDP-related compounds, many of which were made lipophilic by introduction of acyl groups, except inactive analogs, i.e. MDP(L-isoGln) and B30-MDP(D-isoAsn), exhibited strong potentiating effects both on the induction of DTH and on the production of circulating antibody. A desmuramyl derivative, FK-565 was found inactive (Table 1-1).

Only very limited informations have so far been available on the activity of bacterial LPS to induce DTH (36-38), though it has been demonstrated that LPS works as a potent adjuvant on stimulation of antibody production. All specimens of LPS, Re-glycolipid and lipid A of natural sources showed powerful potentiating effects on the induction of DTH as well as the stimulation of serum antibody levels, as MDP and related compounds did (Table 1-2). Under the same assay conditions, synthetic compounds 302, 314, 312, 305 and 317 raised the serum antiovalbumin precipitating antibody level more than three fold over that of control animals which received ovalbumin alone in w/o emulsion. Their immunostimulatory effects, however, were weak as compared with those of natural products. With regard to DTH induction, none of synthetic lipid A analogs could induce distinct DTH comparable in the intensity to those induced by either LPS and natural lipid A or synthetic MDP and its adjuvant-active derivatives. The corneal reaction given by the animals receiving some compounds (for example, compounds 305 and 317) lied on the borderline between the positive and the negative

Table 1-1. Adjuvant activity to induce delayed-type hypersensitivity and elevate serum antibody level against chicken ovalbumin (1 mg/animal) in guinea pigs (MDP and related compounds)

Test material	w/o emulsion Dose[a]	C.R.[b]	Ab[c]	Liposome Dose[a]	C.R.[b]	Ab[c]	PBS Dose[a]	C.R.[b]	Ab[c]
MDP	100	2.8	461± 7**	131	0.4	220±60**	260	0.8	132±28
MDP(L-isoGln)	100	0	124± 8	131	ND[d]		260	ND[d]	149±23
L18-MDP	100	2.8	568±68**	131	1.7	426±62**	260	0.6	
L18-MDP(L-isoGln)	100	ND[d]		131	0.7	88±13		ND[d]	
B30-MDP	100	2.8	553±107**	100	2.1	571±157**	100	2.6	439±67**
				33	1.7	337±61**	33	2.1	391±52**
B30-MDP(D-isoAsn)	100	0.4	46±11		ND[d]		260	0.3	95±45
MDP-Lys(L18)	56	2.6	278±79**		ND[d]		56	0.4	130±43
MDP(Val)-Lys(L18)		ND[d]		100	1.9	354±74**		ND[d]	
Murabutide	100	2.8	475±87**	33	0.9	162±25**	33	0.6	87±27
FK-565	100	0.8	83± 6		ND[d]			ND[d]	
GMP4	51	3.0	407±83**		ND[d]			ND[d]	
GM-53	40	2.5	225±51*	33	1.0	126±24*	33	0.9	246±62*
GM-53(OMe)		ND[d]			ND[d]		33	0.8	184±40
GM-53(OBu)		ND[d]			ND[d]		33	1.6	233±31*
None		0.5	96±18		0.3	49±14		0.4	99±23

[a] μg(MDP equivalent).
[b] Corneal response at the 3rd week. Maximum value is 3.0.
[c] μgN/ml of serum taken at the 4th week.
[d] ND, Not done.
* Differs significantly from control ($P < 0.05$)
** Differs significantly from control ($P < 0.01$)

Table 1-2. Adjuvant activity to induce delayed-type hypersensitivity and elevate serum antibody level against chicken ovalbumin (1 mg/animal) in guinea pigs (LPS, Re-glycolipid and lipid A)

Test material	w/o emulsion Dose[a]	C.R.[b]	Ab[c]
LPS-W (*S. enteritidis*)	100	2.3	426±62**
LPS-W (*E. coli*)	100	1.8	533±77**
LPS-W (*P. aeruginosa*)	100	2.0	653±94**
Re-glycolipid-1 (*S. minnesota*)	100	2.4	628±42**
Re-glycolipid-2 (*S. minnesota*)	100	2.7	365±102**
Lipid A (*S. minnesota*)	100	1.7	188±29**
Lipid A (*P. aeruginosa*)	100	2.3	810±125**
None		0.5	71± 8

[a] μg
[b] Corneal response at the 3rd week. Maximum value is 3.0.
[c] μgN/ml of serum taken at the 4th week.
** Differs significantly from control ($P < 0.01$)

judgement. With regard to monosaccharide derivatives, the assay was so far made on compound 313 which was proved to be distinctly active in most in vitro assays (describe later), but this compound showed only marginal effects at most in stimulation of immune responses in vivo (Table 1-3).

Freund's type water-in-mineral oil emulsion is a very potent and useful vehicle for stimulation of immune responses of laboratory animals to protein antigens, but its high irritative property makes it impracticable to apply it in human beings. So, attempts were made to use liposomes or PBS in place of w/o emulsion as vehicles with MDP derivatives. Among compounds tested, L18-MDP, B30-MDP and MDP-Lys(L18) (L-Val analog) were distinctly immunostimulatory with liposomes, while murabutide and GM-53 as well as MDP were only marginally active (Table 1-1). When PBS was used as a vehicle, B30-MDP and GM-53(OBu) were found to be only compounds which could significantly stimulate both humoral and cell-mediated immune responses to ovalbumin, and all of other test compounds including murabutide were scarcely active in the induction of DTH, but GM-53 and GM-53(OMe) were weakly active in the stimulation of antibody production (Table 1-1).

Table 1-3. Adjuvant activity to induce delayed-type hypersensitivity and elevate serum antibody level against chicken ovalbumin (1 mg/animal) in guinea pigs (Synthetic lipid A and related compounds)

Test material	Dose[a]	w/o emulsion C.R.[b]	Ab[c]
301	100	0.4	137±42
302	100	1.0	298±30**
307	100	0.4	159±29**
311	100	1.1	188±40**
314	100	1.0	221±25**
304	100	1.0	52±16
321	100	1.1	41± 6
312	100	0.5	242±45**
315	100	0.6	192±35**
303	100	0.6	85±19
316	100	0.5	149±29**
305	100	1.3	305±25**
317	100	1.0	323±53**
313	100	0.8	140± 5
None		0.5	71± 8

[a] µg
[b] Corneal response at the 3rd week. Maximum value is 3.0.
[c] µgN/ml of serum taken at the 4th week.
** Differs significantly from control (P < 0.01).

4 Mitogenicity

Mitogenic activities of test materials were determined in terms of the increase of [^{3}H]thymidine incorporation by splenocytes of athymic BALB/c nu/nu mice, namely B lymphocytes, by a conventional methods. Briefly, splenocytes (5×10^6 / ml) were suspended in RPMI medium supplemented with 20% fetal bovine serum and antibiotics. The cell suspensions (0.1 ml each) in a flat-bottomed micro tissue culture tray were added with test specimens (none in control) either suspended by ultrasonication or dissolved in 0.1 ml of RPMI medium. Cultivation was made at 37°C for 48 h in 5% CO_2-95% air in triplicat or in quadruplicate. Tritiated thymidine (0.5 µCi) was added to each culture 24 h before the harvest. After completion of cultivation, thymidine incorporation into the cells was measured by the liquid scintillation method. The increase of thymidine uptake by a test specimen was expressed as a stimulation index, a ratio of dpm of a test culture to dpm of the respective

Table 2-1. Mitogenic effect on splenocytes from athymic BALB/c nu/nu mice (synthetic lipid A analogs and related compounds)

Test material	Dose: µg / culture			
	0.1	1	10	100
301[a]	0.61±0.02	0.70±0.07	0.66±0.03	0.59±0.05
302[b]	1.12±0.10	1.23±0.05*	2.11±0.23**	3.77±0.34**
307[b]	1.00±0.14	1.06±0.14	1.19±0.05	1.18±0.06
311[b]	1.09±0.09	1.06±0.05	1.21±0.06	0.99±0.16
314[b]	1.17±0.09	1.28±0.08*	1.55±0.04**	2.66±0.13**
304[a]	0.84±0.11	0.68±0.06	0.75±0.06	0.51±0.01
321[b]	0.94±0.08	0.94±0.05	0.99±0.04	1.50±0.10
312[b]	1.31±0.12	1.06±0.05	1.04±0.06	1.09±0.10
315[c]	0.95±0.06	0.85±0.03	0.84±0.04	0.92±0.05
303[b]	0.93±0.07	1.10±0.09	1.09±0.04	0.88±0.09
316[a]	0.92±0.08	1.14±0.08	1.15±0.07	2.06±0.15**
305[b]	0.77±0.08	0.82±0.05	0.64±0.05	0.52±0.01
317[b]	1.14±0.04	1.51±0.04**	1.76±0.03**	1.11±0.09
308[a]	0.80±0.02	0.63±0.03	0.50±0.09	0.34±0.04
309[a]	0.88±0.09	0.81±0.07	0.93±0.06	0.79±0.08
310[a]	0.72±0.06	0.63±0.05	0.69±0.06	0.53±0.05
313[b]	1.25±0.08	2.60±0.19**	4.37±0.09**	0.08±0.02
318[b]	1.69±0.10**	1.37±0.06**	0.58±0.01	0.08±0.02
319[b]	0.92±0.06	0.87±0.09	0.92±0.08	0.13±0.06
320[b]	0.94±0.13	1.02±0.13	0.64±0.04	0.07±0.02
LPS-W[b,d]	11.87±0.32**	14.72±0.45**	15.52±0.30**	14.50±0.35**
Re-glycolipid 1[b,e]	13.09±0.37**	15.45±0.30**	14.55±0.29**	4.36±3.13
Lipid A[b,e]	5.54±0.29**	9.49±0.33**	12.70±0.21**	0.91±0.21
MDP[b]	1.39±0.05**	1.56±0.12**	1.88±0.09**	1.62±0.20*
PHA[b,f]	1.07±0.03	1.23±0.04	1.22±0.16	ND

a,b,c The counts in control cultures were; 6,267±382 dpm (a), 8,851±242 dpm (b), and 9,410±363 dpm (c), respectively.

d Prepared from S. enteritidis.

e Prepared from S. minnesota R595.

f Phytohemagglutinin (purified Phytohemagglitinin HA16; Wellcome Reagents).

* Differs significantly from control (P<0.05).

** Differs significantly from control (P<0.01).

control culture.

Results of representative assays were summarized in Table 2. Among 13 glucosamine disaccharide derivatives, compounds 302, 314 and 316 showed a distinct mitogenic activity on murine B cells, and compounds 321 and 317 caused a slight, but significant increase of thymidine uptake at appropriate dosages. Concerning glucosamine derivatives, compound 313 exerted a strong mitogenic effect. The peak response was obtained at dosage of 10 µg (3.3 µg in another experiment, data not shown). Compound 318 gave a slight stimulation at dosages of 0.1 - 1.0 µg. Overdosage (100 µg) of some monosaccharide derivatives was toxic to cells. The peak response induced by monosaccharide compound 313 was significantly higher than those by either active glucosamine disaccharide derivatives or MDP, but far less than that by natural LPS, Re-glycolipid and lipid A (Table 2-1). Some of lipophilic derivatives of

Table 2-2. Mitogenic effect on splenocytes from athymic BALB/c nu/nu mice (MDP and related compounds)

Test material	Dose: µg / culture			
	0.01	0.1	1	10
MDP[a]	1.17±0.07	1.28±0.03*	1.41±0.12*	1.56±0.03**
MDP(L-isoGln)[a]	0.86±0.02	0.85±0.08	0.92±0.07	0.81±0.06
L18-MDP[a]	2.08±0.40*	1.96±0.21*	1.67±0.08**	0.31±0.11
L18-MDP(L-isoGln)[a]	1.05±0.04	1.00±0.05	1.06±0.09	0.42±0.16
B30-MDP[a]	0.98±0.06	1.10±0.05	1.21±0.03	1.65±0.07**
B30-MDP(D-isoAsn)[a]	1.15±0.21	0.88±0.09	1.01±0.09	1.00±0.47
MDP-Lys(L18)[a]	1.95±0.09**	1.82±0.02**	1.77±0.07**	0.90±0.08
MDP(L-isoGln)-Lys(L18)[a]	0.88±0.04	0.82±0.03	0.96±0.05	0.66±0.07
Murabutide[a]	1.14±0.03	0.98±0.06	1.21±0.08	1.16±0.08
GMP$_4$[b]	1.06±0.10	1.08±0.08	1.06±0.13	1.44±0.12*
GM-53[b]	1.24±0.16	1.32±0.13	1.49±0.12*	1.08±0.18
GM-53(OMe)[b]	1.26±0.01*	1.31±0.08*	1.47±0.09*	1.63±0.01**
GM-53(OBu)[b]	1.12±0.06	1.34±0.03*	1.71±0.10**	1.78±0.20*
FK-565[a]	1.96±0.11**	1.70±0.08**	1.51±0.04**	1.65±0.17*
LPS-W (*S. enteritidis*)[a]	5.11±0.35**	8.30±0.11**	13.88±0.15**	16.64±0.20**
PHA[a,c]	ND	1.05±0.10	1.25±0.10	ND

 a,b The counts in control cultures were; 4,390±368 dpm (a) and 8,161±509 dpm (b), respectively.

 c Phytohemagglutinin (purified Phytohemagglutinin HA16; Wellcome Reagents).

 * Differs significantly from control (P<0.05).

 ** Differs significantly from control (P<0.01).

MDP, L18-MDP and MDP-Lys(L18) were found to stimulate murine B lymphocytes more effectively than MDP in terms of an effective dose and an extent of stimulation. Stimulation by these compounds was small in the extent, but was significant in view of the fact that their inactive analogs lacked the activity at all. The activity of these MDP derivatives, though structure-dependent, specific in other words, was far less than those of LPS and related natural compounds. Noticeable findings are that a desmuramyl compound (FK-565) which lacked the antigen-specific immunostimulatory activity in vivo was definitely mitogenic on murine B lymphocytes, while murabutide which was definitely active in the antigen-specific immunostimulation was hardly active. These findings suggest that antigen-specific immunostimulating activity in vivo and B cell mitogenicity in vitro are dissociable. Semisynthetic compounds, GM-53(OMe) and GM-53(OBu), especially the latter, exhibited somewhat stronger mitogenic effect than parent compounds (GM-53 and GMP_4).

5 Polyclonal B cell activation (PBA)

This activity was determined in terms of the increase of backgroud anti-sheep red blood cell (SRBC) plaque forming cell (PFC) number in murine splenocyte cultures treated with test specimens. Briefly, splenocytes (6 x 10^6) of BALB/c mice were suspended in 1.6 ml of RPMI medium supplemented with 0.2 ml of fetal bovine serum and antibiotics. The cell suspensions in a 24-wells flat-bottomed tissue culture tray were added with test specimens (none in control) in 0.2 ml of RPMI medium, and were cultured triplicately at 37°C for 5 days in 5% CO_2 - 95% air. The number of anti-SRBC antibody-secreting cells (per 5 x 10^6 cells) was counted by the conventional method with a Cunningham's chamber, and the mean PFC number and SE of the mean were calculated with each dose of each specimen. PBA activity was expressed as a ratio of PFC number in a test culture to that in the respective control, namely as a stimulation index.

The results of representative experiments were shown in Table 3. Among 13 test glucosamine disaccharide derivatives, compounds 314, 312, 315, 316, 305 and 317 exhibited a weak, but significant PBA activity. The activity of compound 315 was comparable to that of MDP, but far less than those of natural lipid A, Re-glycolipid and LPS specimens. Concerning monosaccharide derivatives, compound 309, 313, 318, 319 and 320 exhibited a distinct PBA effect, stronger than those of the disaccharide derivatives. Glucosamine itself was found to be quite inactive (Table 3-1).

Table 3-1. PBA activities on splenocytes from BALB/c mice
(synthetic lipid A analogs and related compounds)

Test material	Dose: μg / ml		
	1	10	100
301[a]	0.92±0.10	0.83±0.13	0.90±0.19
302[b]	1.18±0.11	1.08±0.08	1.38±0.12
307[c]	0.77±0.07	1.12±0.06	1.05±0.15
311[c]	0.88±0.07	0.88±0.15	1.17±0.17
314[d]	1.30±0.10	1.60±0.09**	1.91±0.16**
304[b]	0.72±0.09	1.21±0.14	1.41±0.07
321[c]	0.62±0.06	1.04±0.17	1.16±0.04
312[c]	0.88±0.22	0.68±0.33	1.50±0.02*
315[d]	1.40±0.12*	2.44±0.11**	2.10
303[b]	0.79±0.20	0.79±0.11	0.92±0.12
316[a]	1.55±0.24	1.63±0.07**	1.16±0.04
305[b]	1.69±0.08*	2.02±0.14**	1.87±0.14*
317[d]	1.05±0.08	1.37±0.22	1.60±0.13*
308[e]	0.90±0.08	1.04±0.12	1.05±0.18
309[e]	1.49±0.25	1.92±0.26*	2.12±0.38*
310[e]	0.82±0.12	0.97±0.32	1.14±0.30
313[f]	1.98±0.18*	3.54±0.36**	2.96±0.15**
318[f]	2.35±0.53	2.32±0.54	3.53±0.46**
319[f]	1.19±0.17	1.58±0.26	2.47±0.30*
320[f]	1.16±0.03	1.47±0.24	2.21±0.27*
GlcN[a]	0.87±0.02	0.96±0.08	0.87±0.09
LPS-W(*S. enteritidis*)[c]	19.11±1.67**	27.00±3.19**	43.53±4.32**
LPS-W(*P. aeruginosa*)[g]	25.11±4.20**	35.11±2.16**	42.73±3.36**
Re-Glycolipid 1(*S. minnesota*)[h]	9.73±0.61**	10.00±1.89**	8.79±1.99**
Lipid A(*S. enteritidis*)[g]	9.60±0.65**	4.60±0.16**	46.60±3.76**
Lipid A(*P. aeruginosa*)[g]	30.40±4.09**	44.93±6.17**	55.51±1.66**
MDP[d]	2.04±0.07**	2.22±0.09**	1.54±0.04**

a,b,c,d,e,f,g,h The number of PFC in control assays were: 15.3±0.88 (a), 13.0±2.08 (b), 19.0±2.08 (c), 20.0±1.58 (d), 20.0±1.15 (e), 19.0±3.60 (f), 22.7±1.76 (g), and 11.0±1.73 (h), respectively.

* Differs significantly from control (P<0.05)
** Differs significantly from control (P<0.01)

Regarding synthetic MDP derivatives, the PBA activity of MDP was significantly increased by the introduction of acyl groups. Such is the case with L18-MDP, B30-MDP and MDP-Lys(L18). But the fact that inactive analogs of L18-MDP and MDP-Lys(L18) caused some increase of PFC number suggests that a part of the PBA activity of lipophilic derivatives of MDP may be due to the amphipathic property of the molecules. The activity of murabutide and FK-565 was weaker than that of MDP. The PBA activity of semisynthetic GM-53, on the other hand, was comparable to that of MDP, and the ability of its O-methyl or O-butyl ester was somewhat lower than that of the parent molecule. At any rate, the activity of MDP and related compounds to cause polyclonal B cell activation was far less than that of LPS (Table 3-2).

Table 3-2. PBA activities on splenocytes from BALB/c mice
(MDP and related compounds)

Test material	Dose: µg / ml		
	1	10	100
MDP [a]	1.37± 0.24	2.34±0.46*	3.86±0.67*
L18-MDP [b]	3.51±0.22**	4.54±0.33**	6.24±0.79**
L18-MDP(L-isoGln) [c]	1.04±0.11	1.66±0.30	1.48±0.15*
B30-MDP [a]	1.13± 0.22	2.49±0.18**	9.06±0.45**
B30-MDP(L-isoAsn) [c]	1.08±0.13	1.58±0.30	1.08±0.13
MDP-Lys(L18) [b]	1.64±0.14	1.86±0.14	3.94±0.59
MDP(L-isoGln)-Lys(L18) [c]	1.51±0.06**	1.37±0.12*	1.02±0.09
Murabutide [a]	1.25± 0.10	1.66±0.17*	1.32±0.08*
GMP$_4$ [d]	1.06± 0.10	1.39±0.27	0.86±0.22
GM-53 [d]	1.49±0.16*	1.47±0.07**	4.42±0.46**
GM-53(OMe) [d]	1.33±0.05**	1.41±0.12*	2.35±0.65
GM-53(OBu) [d]	2.10± 0.25**	1.58±0.25	1.33±0.11
FK-565 [a]	1.42±0.16*	1.81±0.15**	1.93±0.23*
LPS-W(S. enteritidis) [d]	26.20± 3.58**	41.90±3.90**	43.37±2.61**

a,b,c,d The numbers of PFC in control assays were; 19.7±1.33 (a), 21.3±1.86 (b), 25.7±0.88 (c), and 16.3±0.67 (d), respectively.
* Differs significantly from control (P<0.05).
** Differs significantly from control (P<0.01).

6 Enhancing effects on the migration of human blood monocytes and polymorphonuclear leukocytes (PMNL)

In the course of studies (39,40) on migration-enhancing effects of bacterial cell walls and muramyl peptides on human peripheral monocytes, control experiments showed that, though LPS was inactive in this respect, lipid A specimens enhanced the monocyte migration. Thus we attempted to compare the monocyte-migration enhancing effects of various LPS, Re-glycolipid, lipid A and synthetic lipid A analogs, MDP and its related compounds. On the other hand, there are a number of studies on chemotaxigenic effect of LPS on PMNL in the presence of fresh serum as a source of chemotactic mediators, but there are no studies on possible, direct (without involvement of complement) migration-enhancing activity of LPS on PMNL, except a recent study of Adamu and Sperry (41). Se we examined the migration-enhancing effects of the above test materials on PMNL.

Assay for the migration-enhancement of monocytes and PMNL was performed by use of a multiwell chemotaxis assembly (Neuro Probe) and a filter sheet of 5-μm (for monocytes) or that of 3-μm (for PMNL). Determination was made on the number (per oil immersion field) of monocytes or PMNL which migrated to the surface of the membrane sheet adjacent to the lower well containing a test specimen, to obtain the mean ± S.E.. Stimulation of the migration by test specimens was expressed as a ratio of the value in a test to that in the respective control. In all experiments, two reference specimens served as positive controls: one of them is a fresh serum specimen which was activated by LPS-B of S. enteritidis (Difco) according to the conventional method and diluted 1:10, and

Table 4-1. Stimulation of the migration of human monocytes (LPS, Re-glycolipid and lipid A)

Test material	Dose: μg/ml			
	0.0001	0.001	0.01	0.1
LPS-W(S. enteritidis)[a]	1.27±0.09	1.18±0.09	1.18±0.09	1.18±0.09
LPS-B(S. enteritidis)[a]	1.18±0.09	1.27±0.09	1.09±0.18	1.00±0.09
LPS-W(E. coli)[a]	1.18±0.09	1.18±0.09	1.27±0.09	1.27±0.09
LPS-B(E. coli)[a]	1.18±0.09	1.15±0.05	1.09±0.09	1.18±0.09
LPS-W(P. aeruginosa)[a]	0.82±0.18	1.10±0.08	1.09±0.09	0.72±0.09
Re-glycolipid(S. minnesota)[b]	1.08±0.08	2.41±0.08*	2.92±0.25*	2.58±0.08*
Lipid A(S. minnesota)[c]	1.18±0.09	2.55±0.18*	1.27±0.18	1.27±0.09
Lipid A(P. aeruginosa)[c]	1.73±0.03*	2.18±0.09*	1.36±0.09	0.82±0.09

[a,b,c] The stimulation indexes of positive controls were: 6.45±0.36(a), 7.86±0.29(b) and 6.50±0.42(c) for FMLP(10^{-8} M), and 8.27±0.18(a), 9.43±0.29(b) and 8.67±0.58(c) for LPS-activated serum(1:10); those of negative controls were: 1.00±0.08(a), 1.00±0.05(b) and 1.00±0.03(c).
 * Differs significantly from control(P<0.05).
 ** Differs significantly from control(P<0.01).

Endotoxic Activities

the other is N-formylmethionyl-leucyl-phenylalanine (FMLP), 10^{-8} M for monocytes and 10^{-6} M for PMNL. Detailed method was described previously (24,39,40).

Representative experimental results with monocytes were shown in Table 4. A control study with LPS and related compounds of natural sources showed that none of test LPS preparations exerted any significant migration-enhancing effects on monocytes, but Re-glycolipid and lipid A specimens caused a distinct augmentation of monocyte migration (Table 4-1). The effect was comparable to that of MDP in the extent, but the peak response was attained by the former at lower dosages than by the latter; namely 1-10 ng / ml as compared with 100 ng / ml of MDP. All of lipophilic derivatives (either synthetic or semisynthetic) of muramyl peptides, except GM-53(OMe), and a desmuramyl compound, FK-565 significantly enhanced the monocyte migration in dose-response patterns

Table 4-2. Stimulation of the migration of human monocytes (MDP and related compounds)

Test material	Dose:μg(MDP equiv.)/ml			
	0.001	0.01	0.1	1.0
MDP[a]	1.13±0.04	1.70±0.02**	1.98±0.05**	1.68±0.04**
MDP(L-isoGln)[a]	1.06±0.07	1.02±0.08	1.11±0.08	0.85±0.05
L18-MDP[a]	1.41±0.07*	2.09±0.10**	1.44±0.06*	1.46±0.02**
L18-MDP(L-isoGln)[a]	1.00±0.07	1.04±0.07	0.91±0.02	0.92±0.02
B30-MDP[a]	1.04±0.02	1.71±0.02**	1.63±0.08**	1.54±0.13*
B30-MDP(D-isoAsn)[a]	1.00±0.07	1.20±0.05	1.13±0.01	1.02±0.07
MDP-Lys(L18)[a]	1.02±0.02	1.37±0.02*	1.07±0.09	1.21±0.04
MDP(L-isoGln)-Lys-L18[a]	1.00±0.03	1.09±0.07	1.09±0.05	1.04±0.05
Murabutide[b]	1.49±0.15	1.82±0.06**	1.96±0.11**	1.69±0.14*
GMP4[b]	1.37±0.13	1.59±0.06*	1.77±0.15*	1.65±0.07*
GM-53[b]	1.51±0.11*	1.78±0.09**	1.51±0.04*	1.72±0.04**
GM-53(OMe)[b]	1.22±0.02	1.12±0.18	1.24±0.04	0.94±0.07
GM-53(OBu)[b]	1.63±0.11*	1.65±0.09*	1.49±0.11*	1.72±0.04**
FK-565[b]	1.69±0.11*	1.71±0.10*	1.92±0.14**	1.25±0.07

[a,b] The stimulation indexes of positive controls were: 3.39±0.08(a) and 4.02±0.11(b) for FMLP(10^{-8} M), and 4.72±0.06(a) and 3.59±0.21(b) for LPS-activated serum(1:10); those of negative controls were: 1.00±0.09(a) and 1.00±0.14(b).

* Differs significantly from control(P<0.05).
** Differs significantly from control(P<0.01).

similar to that of MDP. The peak responses were obtained with these compounds in lower dosage (10 ng / ml) as compared with MDP (100 ng / ml) (Table 4-2).

Synthetic lipid A analogs, except compounds 301, 302 and 304, exerted a moderate to strong migration-enhancing activity. The activity of the most active compounds were comparable in the intensity to those of

Table 4-3. Stimulation of the migration of human monocytes
(synthetic lipid A analogs and related compounds)

Test material	Dose: µg/ml 0.0001	0.001	0.01	0.1
301[b]	1.15±0.23	1.54±0.08*	1.00±0.15	1.00±0.15
302[b]	1.15±0.08	1.54±0.08*	1.38±0.15	0.77±0.15
307[c]	3.00±0.08**	3.00±0.08**	3.58±0.33**	2.67±0.17**
311[c]	2.92±0.03**	2.92±.033**	2.08±0.17*	2.08±0.25*
314[c]	2.58±0.33*	3.25±0.17**	3.92±0.25**	2.17±0.33*
304[b]	1.23±0.23	1.08±0.15	1.00±0.15	0.38±0.08
321[c]	1.67±0.08**	1.67±0.08**	2.00±0.03**	1.17±0.08
312[c]	2.83±0.17**	2.92±0.08**	3.08±0.17**	2.41±0.17**
315[f]	0.75±0.13	2.25±0.13*	3.00±0.13**	0.75±0.05
303[b]	1.31±0.08	2.00±0.08**	1.23±0.00	1.08±0.04
316[a]	2.14±0.29**	2.43±0.42*	2.71±0.29**	2.29±0.43*
305[a]	2.00±0.29*	3.00±0.29**	2.71±0.29**	2.57±0.43*
317[e]	1.92±0.25*	3.17±0.08**	2.17±0.08**	1.08±0.08
308[d]	2.27±0.09**	2.36±0.09**	1.55±0.09*	1.27±0.09
309[d]	2.55±0.18**	2.64±0.09**	2.18±0.18**	2.18±0.18**
310[d]	2.18±0.09**	1.82±0.36	1.82±0.27*	1.27±0.18
313[c]	1.50±0.08	3.08±0.08**	2.08±0.08**	1.33±0.08
318[c]	2.25±0.25*	1.50±0.03*	1.58±0.25	1.17±0.08
319[c]	3.58±0.03**	2.17±0.17**	2.08±0.08**	1.42±0.17
320[c]	1.50±0.08	1.42±0.17	1.50±0.08	1.17±0.08

[a,b,c,d,e,f] The stimulation indexes of positive controls were: 7.86±0.29 (a), 5.15±0.15 (b), 5.67±0.17 (c), 6.64±0.09 (d), 6.50±0.42 (e) and 7.67±0.17 (f) for FMLP (10^{-8} M), and 9.43±0.29 (a), 4.92±0.08 (b), 7.67±0.08 (c), 7.09±0.27 (d), 8.67±0.03 (e) and 4.92±0.17 (f) for LPS-activated serum (1:10); those of negative controls were: 1.00±0.05 (a), 1.00±0.08 (b), 1.00±0.17 (c), 1.00±0.09 (d), 1.00±0.03 (e) and 1.00±0.08 (f).
* Differs significantly from control ($P<0.05$).
** Differs significantly from control ($P<0.01$).

Endotoxic Activities

natural lipid A, Re-glycolipid and MDP. All of glucosamine monosaccharide derivatives except compound 320 were more or less active in this assay (table 4-3).

Table 5-1. Checkerboard analysis of monocyte migration enhancement (lipid A)[a]

Concn (μg / ml) of lipid A in lower well	Cells / O.I.F. ± SE[b] at concn (μg / ml) of lipid A in upper well			
	0	0.001	0.005	0.01
0	14 ± 2[c]	27 ± 2	29 ± 4	36 ± 2
0.001	23 ± 3	26 ± 3	28 ± 2	24 ± 0
0.005	25 ± 2	26 ± 1	20 ± 1	28 ± 1
0.01	29 ± 1	24 ± 0	26 ± 5	29 ± 2

[a] Test material was added to both wells or either of the lower and upper wells of the multiwell chemotaxis chamber at indicated concentrations. Monocytes were added to the upper well.

[b] The extent of monocyte migration was determined as described in the text. The number of monocytes per oil immersion field (O.I.F.) on the lower surface of a polycarbonate filter was triplicately counted on 20 microscopic fields to obtain the mean ± standard error (SE).

[c] The values between diagonal lines indicate those obtained when the concentration of specimen on both sides of the filter is identical.

Table 5-2. Checkerboard analysis of monocyte migration enhancement (compound 314)[a]

Concn (μg / ml) of compound 314 in lower well	Cells / O.I.F. ± SE[b] at concn (μg / ml) of compound 314 in upper well			
	0	0.00001	0.0001	0.001
0	19 ± 2[c]	19 ± 1	45 ± 2	53 ± 2
0.00001	27 ± 1	24 ± 0	24 ± 1	25 ± 1
0.0001	29 ± 1	26 ± 1	32 ± 1	27 ± 1
0.001	39 ± 1	25 ± 2	29 ± 1	38 ± 2

[a,b,c] See footnotes to Table 5-1.

We carried out checkerboard analysis with representative compounds to see whether the enhanced migration of monocytes was due to increased randam migration (chemokinesis) or to chemotaxis directed by the concentration gradient of an active agent. It turned out that the enhanced monocyte migration caused by a lipid A specimen of S. minnesota R595 (Table 5-1) and a synthetic lipid A analog, compound 314 (Table 5-2), was mainly due to chemokinesis, since the intensity of migration enhancement was not significantly affected by the concentration gradient of active principles between the upper and lower wells. On the other

Table 5-3. Checkerboard analysis of monocyte migration enhancement (MDP)[a]

Concn (µg / ml) of MDP in lower well	\multicolumn{4}{c}{Cells / O.I.F. ± SE[b] at concn (µg / ml) of MDP in upper well}
	0	0.001	0.01	0.1
0	4 ± 1[c]	6 ± 1	4 ± 1	4 ± 1
0.001	5 ± 1	2 ± 1	3 ± 1	5 ± 1
0.01	24 ± 1	13 ± 1	5 ± 1	2 ± 1
0.1	33 ± 1	27 ± 1	5 ± 1	7 ± 1

[a,b,c] See footnotes to Table 5-1.

hand, the enhanced monocyte migration caused by MDP was found to be directed by the gradient of MDP concentration, that is to say, due to chemotaxis as already reported (40) (Table 5-3). The mechanism of migration enhancement by monosaccharide derivatives dose not seem to

Table 5-4. Checkerboard analysis of monocyte migration enhancement (compound 313)[a]

Concn (µg / ml) of compound 313 in lower well	\multicolumn{4}{c}{Cells / O.I.F. ± SE[b] at concn (µg / ml) of compound 313 in upper well}
	0	0.0125	0.025	0.05
0	18 ± 1[c]	29 ± 2	32 ± 3	30 ± 1
0.0125	16 ± 1	19 ± 1	18 ± 1	26 ± 2
0.025	21 ± 1	23 ± 2	21 ± 0	15 ± 1
0.05	26 ± 1	25 ± 1	21 ± 0	15 ± 1

[a,b,c] See footnotes to Table 5-1.

be identical with those by natural lipid A and synthetic compound 314, since the response of monocytes to the concentration gradient of compound 313 was different from those of either lipid A or compound 314:

Table 6-1. Stimulation of the migration of human polymorphonuclear leukocytes (LPS, Re-glycolipid and lipid A)

Test material	Dose:µg/ml			
	0.0001	0.001	0.01	0.1
LPS-W(*S. enteritidis*)[c]	1.10±0.10	1.60±0.10*	1.60±0.20	1.00±0.00
LPS-B(*S. enteritidis*)[c]	1.30±0.20	1.60±0.20	2.80±0.00**	1.10±0.10
LPS-W(*E. coli*)[b]	1.78±0.33	2.00±0.11**	1.41±0.15	1.26±0.07
LPS-B(*E. coli*)[b]	1.85±0.19*	2.37±0.25**	1.89±0.03**	1.30±0.14
LPS-W(*P. aeruginosa*)[a]	1.83±0.17**	3.42±0.33**	1.50±0.17*	1.08±0.08
Re-glycolipid(*S. minnesota*)[e]	0.89±0.22	1.00±0.11	2.00±0.06**	0.61±0.06
Lipid A(*S. minnesota*)[e]	1.44±0.22	1.44±0.22	2.00±0.11**	1.27±0.17
Lipid A(*P. aeruginosa*)[a]	1.75±0.25*	3.00±0.25**	2.00±0.17**	1.17±0.08

[a,b,c,d] The stimulation indexes of positive controls were: 10.33±0.58(a) 5.33±0.15(b), 11.80±0.50(c) and 7.67±0.33(d) for FMLP(10^{-6} M), and 7.83±0.08(a), 5.30±0.19(b), 18.90±0.90(c) and 5.11±0.50(d) for LPS-activated serum(1:10); those of negative controls were: 1.00±0.08(a), 1.00±0.04(b), 1.00±0.10(c) and 1.00±0.06(d).
* Differs significantly from control(P<0.05).
** Differs significantly from control(P<0.01).

the number of monocyte migrated towards the lower well was almost constant when the concentrations of 313 were the same on both sides of the filter, but the monocyte number on the lower surface of the filter increased as the concentration of 313 increased in upper wells, in the absence of the compound in the lower well (Table 5-4). This finding implies that compound 313 holds both chemotactic and chemokinetic effects on human monocytes.

Representative assay result with PMNL (Table 6) shows that all of natural products including LPS preparations (except LPS-W of S. enteritidis) definitely enhanced the migration of human PMNL (Table 6-1). Among synthetic lipid A analogs and related compounds, disaccharide derivatives 311, 314, 303 and 305 and monosaccharide derivatives 309, 310 and 313 caused a definite enhancement of PMNL migration. Limited amounts available did not permit us to examine the activity of compounds 312, 325, 316 and 317 towards PMNL (Table 6-2). Among MDP derivatives, on the other hand, only semisynthetic compounds, GM-53(OBu) and GM-53(OMe), significantly enhanced the migration of human PMNL, though a parent compound, GM-53 was inactive.

Table 6-2. Stimulation of the migration of human polymorphonuclear leukocytes (synthetic lipid A analogs and related compounds)

Test material	Dose: μg/ml			
	0.0001	0.001	0.01	0.1
301^g	1.42±0.08*	1.67±0.17*	0.25±0.08	0.08±0.03
302^d	0.82±0.09	0.82±0.09	0.64±0.18	0.36±0.09
307^d	0.55±0.09	0.91±0.09	0.82±0.09	0.64±0.02
311^e	1.33±0.17	1.00±0.17	2.50±0.33*	1.00±0.33
314^a	3.00±0.17**	4.00±0.25**	2.92±0.17**	1.25±0.08
304^g	1.08±0.17	1.50±0.17	0.41±0.08	0.33±0.08
321^c	0.44±0.06	0.89±0.02	0.61±0.02	0.33±0.02
312	ND^h	ND	ND	ND
315	ND	ND	ND	ND
303^f	0.75±0.05	2.38±0.25**	2.63±0.38*	1.75±0.13*
316^b	1.30±0.10	1.60±0.30	1.30±0.40	1.00±0.10
305^e	1.80±0.17**	1.50±0.17*	3.33±0.33**	1.33±0.03**
317^c	1.00±0.11	1.28±0.30	1.05±0.39	0.50±0.02
308^d	0.45+0.03	0.64+0.02	1.64+0.09**	0.64+0.09
309^e	1.33+0.07*	1.67+0.17*	3.00+0.07**	1.33+0.17*
310^f	0.88±0.13	1.25±0.13	1.50±0.13	3.50±0.25**
313^g	1.08±0.08	2.42±0.17**	1.25±0.08	0.41±0.17
318^d	0.73±0.09	1.45±0.09*	0.64±0.09	0.64±0.09
319^f	0.88±0.13	1.13±0.13	1.13±0.13	1.75±0.25
320^e	1.33±0.03**	1.33±0.03**	1.50±0.03**	1.33±0.03**

a,b,c,d,e,f,g The stimulation indexes of positive controls were: 10.33±0.58(a), 8.50±0.30(b), 7.67±0.33(c), 9.55±0.73(d), 10.67±1.00(e), 8.00±0.25(f) and 6.33±0.25(g) for FMLP(10^{-6} M), and 7.83±0.08(a), 18.90±0.90(b), 5.11±0.50(c), 5.36±0.36(d), 15.33±0.95(e), 10.50±0.75(f), and 12.08±0.75(g) for LPS-activated serum(1:10); those of negative controls were: 1.00±0.08(a), 1.00±0.10(b), 1.00±0.06(c), 1.00±0.09(d), 1.00±0.03(e), 1.00±0.13(f) and 1.00±0.08(g).
h ND, Not done.
* Differs significantly from control(P<0.05).
** Differs significantly from control(P<0.01).

Table 6-3. Stimulation of the migration of human polymorphonuclear leukocytes (MDP and related compounds)

Test material	Dose: μg(MDP equiv.)/ml			
	0.001	0.01	0.1	1.0
MDP[a]	1.00±0.05	1.09±0.08	1.00±0.05	0.91±0.05
MDP(L-isoGln)	ND[b]	ND	ND	ND
L18-MDP	1.04±0.08	1.04±0.05	1.02±0.02	1.08±0.02
L18-MDP(L-isoGln)	ND	ND	ND	ND
B30-MDP	1.05±0.08	1.08±0.05	1.16±0.08	1.09±0.03
B30-MDP(D-isoAsn)	ND	ND	ND	ND
MDP-Lys(L18)	1.00±0.10	1.02±0.05	1.02±0.05	0.85±0.07
MDP(L-isoGln)-Lys-L18	ND	ND	ND	ND
Murabutide	1.04±0.07	1.02±0.04	1.02±0.10	0.86±0.02
GMP$_4$	1.05±0.02	1.00±0.04	1.05±0.04	1.02±0.05
GM-53	1.05±0.07	1.11±0.02	1.11±0.07	1.09±0.03
GM-53(OMe)	1.26±0.07*	1.66±0.10**	1.77±0.77**	1.02±0.05
GM-53(OBu)	1.15±0.03	2.35±0.17**	1.57±0.10**	0.84±0.02
FK-565	0.95±0.02	1.13±0.07	1.02±0.07	1.05±0.07

[a] The stimulation indexes of positive controls were: 10.05±0.17 for FMLP(10^{-6} M), and 5.01±0.07 for LPS-activated serum(1:10); that of negative control was 1.00±0.07.

[b] ND, Not done.

* Differs significantly from control(P<0.05).
** Differs significantly from control(P<0.01).

7. Macrophage activation

Many studies have demonstrated that bacterial LPS activates macrophages in various ways. Among them, a study of Wilton et al (42) showed that LPS increased the glucosamine uptake by guinea pig peritoneal macrophages in the presence of B lymphocytes. Similar macrophage-stimulating activity, though far less in the extent, was found with MDP (43). So we compared the stimulatory activity of natural Re-glycolipid, lipid A, synthetic lipid A analogs and related glucosamine derivatives, on guinea pig peritoneal macrophages.

Stimulation of peritoneal macrophages was examined by determination of [^{14}C]glucosamine incorporation according to the method previously described (24,43). In brief, thioglycolate-induced peritoneal macrophages were cultured for 3 days in a 24 flat-bottomed well, plastic

Table 7. Stimulation of glucosamine incorporation of peritoneal macrophages from a guinea pig by test materials

Test material	Dose: µg / culture			
	0.1	1	10	100
LPS-W[a,g]	ND	1.60±0.40	1.77±0.02**	2.45±0.12**
Re-glycolipid 1[a,h]	ND	1.29±0.09	1.48±0.05**	1.82±0.14**
Lipid A[b,h]	ND	1.40±0.19*	1.51±0.18**	1.80±0.07**
Lipid A[a,g]	ND	2.02±0.16**	2.18±0.19**	2.99±0.11**
316[b]	ND	0.64±0.03	0.77±0.09	0.88±0.05
317[b]	ND	1.32±0.14*	1.02±0.07	1.34±0.04**
313[b]	ND	0.82±0.04	0.69±0.05	1.52±0.13**
MDP[c]	1.61±0.25*	1.75±0.45	1.56±0.26*	1.76±0.41*
MDP(L-isoGln)[c]	1.05±0.01	1.10±0.22	1.01±0.13	1.08±0.04
L18-MDP[d]	1.39±0.42	1.57±0.35	2.21±0.04*	ND
B30-MDP[e]	1.09±0.09	1.28±0.10*	1.52±0.23*	1.61±0.21*
MDP-Lys(L18)[e]	1.25±0.12	1.58±0.18*	1.33±0.11*	1.65±0.02**
Murabutide[f]	0.88±0.06	1.16±0.01	1.39±0.11*	ND
GM-53(OBu)[f]	2.06±0.05**	1.92±0.05**	2.27±0.63	ND
FK-565[f]	0.82±0.07	0.95±0.20	1.33±0.08*	ND

a,b,c,d,e,f The counts in control cultures were; 199±13.7 dpm (a), 81±3.5 dpm (b), 127±8.6 dpm (c), 128±18.1 dpm (d), 84±6.9 dpm (e) and 257±13.4 dpm (f), respectively.

g Prepared from P. aeruginosa ATCC 7700.
h Prepared from S. minnesota R595.
* Differs significantly from control ($P<0.05$).
** Differs significantly from control ($P<0.01$).

culture tray as a monolayer adherent on well surface. During the final 8 h of the culture, macrophages were pulsed with 0.25 µCi of [^{14}C]glucosamine. After the completion of culture, macrophage monolayers were washed three times with Hanks balanced salt solution and then were dissolved by addition of 0.5 ml of Lowry's alkaline copper solution. Dissolved cells were submitted to the estimation of protein content and glucosamine incorporation. Stimulation of glucosamine incorporation by macrophages exposed to test specimens over that by control macrophages was calculated by the following formula;

$$\frac{\text{mean (of 3 - 4 cultures) of dpm of } [^{14}C] / \text{µg of protein in test culture}}{\text{mean (of 3 - 4 cultures) of dpm of } [^{14}C] / \text{µg of protein in control culture}}$$

Table 7 shows that all of 4 reference natural specimens (LPS-W of P. aeruginosa, Re-glycolipid of S. minnesota and lipid A from S. minnesota and P. aeruginosa) stimulated thioglycolate-induced peritoneal macrophages to increase the uptake of glucosamine significantly, although there were considerable variations in the extent of stimulation. Among 3 synthetic lipid A analogs which were submitted to this assay, compounds 317 and 313 exerted a weak but significant stimulating activity, while compound 316 lacked the activity. All of MDP and related compounds tested, except an adjuvant-inactive MDP(L-isoGln), significantly increased the glucosamine uptake of macrophages. GM-53(OBu) exerted the strongest effect, while murabutide and FK-565 were only slightly active.

8 Activation of human serum complement

It is well-known that endotoxic LPS activates both the classical and alternative pathways of complement system. Morrison and Klein (44) demonstrated that lipid A region of the LPS was responsible for the activation of classical pathway, due to a mechanism independent of antigen-antibody reaction. On the other hand, a number of studies including ours (8,45) showed that PG as well as whole cell walls activated complement, and some of them affected almost exclusively the alternative pathway. We further demonstrated that B30-MDP and 6-O-(3-hydroxy-2-docosylhexacosanoyl)-MurNAc-L-Ala-D-isoGln-L-Lys-D-Ala, i.e. BH48-MDP-L-Lys-D-Ala, activated human complement system by mechanisms different with each other (see later), although MDP and L18-MDP were inactive. Comparisons were thus made on the complement activating ablity of synthetic lipid A analogs with that of LPS, Re-glycolipid and lipid A of natural sources and with those of some MDP derivatives in the following way.

Two tenth ml aliquots of fresh, pooled human serum were incubated with test materials dissolved or suspended in 0.2 ml of physiological saline, at 37°C for 60 min. The reaction mixtures were centrifuged at 9,000 x g for 5 min, the 50% hemolytic unit of complement (CH50) of the supernatant fluids was determined by the method of Mayer (46) and was compared with that of a control serum treated with saline alone. Blocking effect of the addition of EGTA (10 mM) and $MgCl_2$ (5 mM) to the reaction mixture was determined to see whether the activation of complement system by test or reference specimens occured through the classical or the alternative pathway.

Analysis by the addition of EGTA to block the classical pathway showed that activation of human complement by B30-MDP was almost completely suppressed by EGTA, while that by BH48-MDP-L-Lys-D-Ala was only

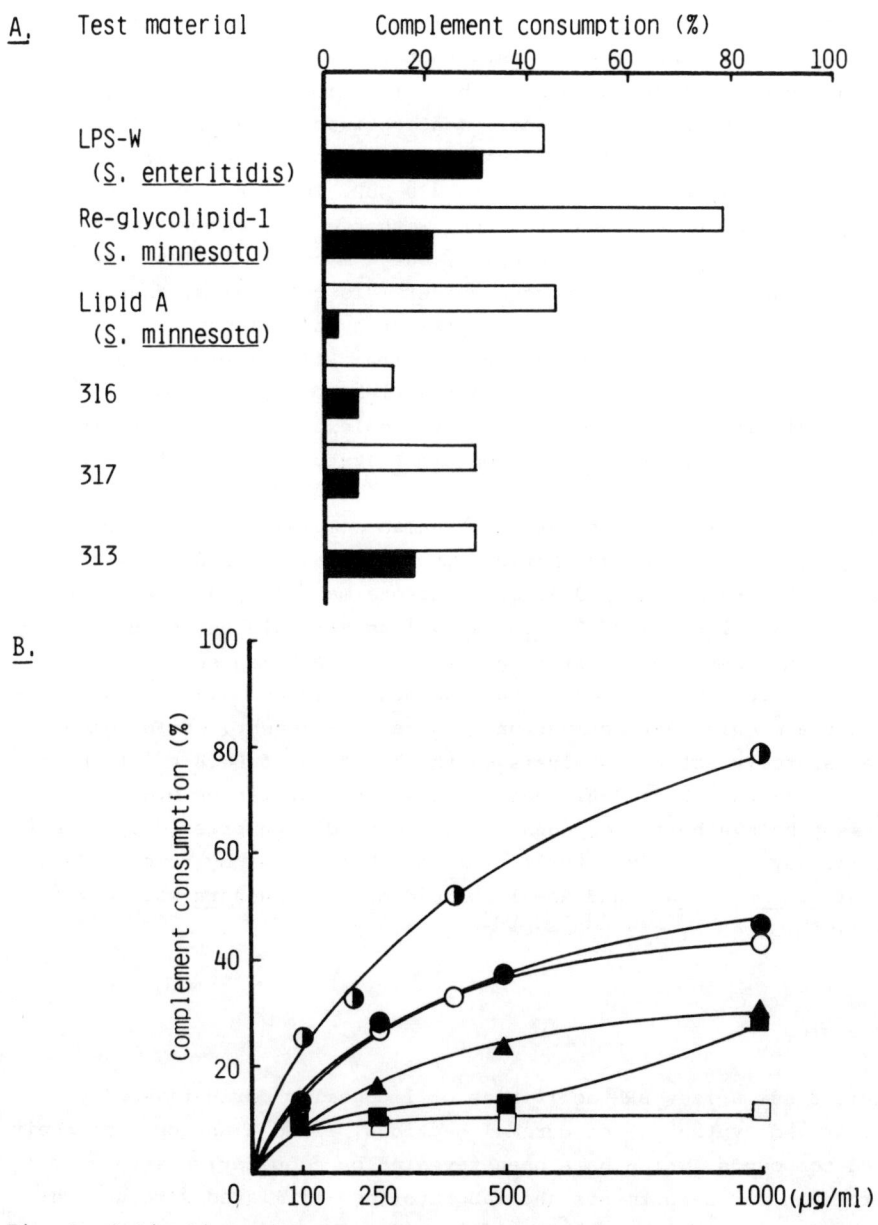

Fig. 3. Activation of human complement system by LPS, Re-glycolipid, lipid A and compound 316, 317 and 313.
A: Dose-response curve -○: LPS-W (S. enteritidis), ◐: Re-glycolipid (S. minnesota), ●: lipid A (S. minnesota), □: compound 316, ▲: compound 317 and ■: compound 313.
B: Blocking effects on the classical pathway by addition of EGTA -□: absence and ■: presence of EGTA.

partially blocked (8,45). An experiment using a partially purified preparation of C1 component suggests that complement activation by B30-MDP is not dependent on antigen-antibody reaction (45). It should be pointed out, however, that complement activation by B30-MDP is mainly due to its acyl moiety, because B30-MDP(D-isoAsn) activated complement system in a similar way to B30-MDP. The findings described in the proceeding paragraphs, together with the previous observation (8,45) that a polymer of PG subunits, but a monomer, isolated from an enzymatic digest of Staphylococcus epidermidis cell walls, exerted a strong ability to activate human complement system almost exclusively through the alternative pathway, suggest that MDP structure, which are known to carry the majority of BRM activities of PG, needs some additional physical properties or molecular complexities to exhibit its potential activity to stimulate complement system via the alternative pathway.

Figure 3-A shows the activation of human complement by LPS, Re-glycolipid, lipid A, and synthetic compounds 316, 317 and 313. It can be seen that a Re-glycolipid of S. minnesota markedly, and LPS-W of S. enteritidis and lipid A of S. minnesota less markedly but significantly activated the complement system in fresh, pooled human serum. Among three synthetic compounds examined, compounds 317 and 313 caused a weak, but a significant activation of human complement, while compound 316 was scarcely active. Analysis by the addition of EGTA to block the classical pathway (Fig. 3-B) revealed that the activation of human complement by synthetic compound 317 was markedly suppressed by EGTA in a similar way to that by natural lipid A, while those by compound 313, LPS-W of S. enteritidis and Re-glycolipid of S. minnesota was only partially blocked by the presence of EGTA.

9 Discussion

Table 8 summarizes BRM activities of lipophilic derivatives (synthetic or semisynthetic) of muramyl peptides, lipid A analogs and their related compounds, which have been revealed by cooperative studies performed in our departments and laboratories affiliated with us. This table indicates that muramyl peptide derivatives and some synthetic lipid A analogs share many BRM activities, as parental compounds, PG and LPS, do so.

Some of test lipid A analogs, especially compound 317 showed distinct activities in the majority of the present assays. The activities of them detected in vitro assays were comparable to those of muramyl peptide derivatives, but far less than those of lipid A, Re-glycolipid

Table 8. Summary of BRM activities of synthetic lipid A analogs, synthetic or semisynthetic lipophilic muramyl peptides and related compounds (selected)

Test material	Adjuvancy w/o DTH/Ab	Adjuvancy PBS DTH/Ab	Non-specific resistance [a]	Arth-rito-genicity [b]	Pyro-genicity [c]	Granuloma formation [d]	Hemorrhagic necrosis [e]
LPS	2+/3+		3+	(−)	+		−
Re-glycolipid	2+/3+				+		
Lipid A	2+/3+				+		
301	−/−						
302	±/2+						
307	/±						
311	±/±						
314	±/+						
304	±/−						
321	±/−						
312	−/+						
315	−/±						
303	−/−						
316	−/±						
305	−/2+						
317	−/2+						
313	−/±						
MDP	3+/3+	−/−	+(+)[1]	−(+)[5] [b']	+[8]	−(+)	+
L18-MDP	3+/3+	−/−	2+(+)[1]	+[5]	+[8]	−(+)	+
B30-MDP	3+/3+	3+/3+	−[1]	+[5]	−(+)[8]	+(+)	−/+ [e']
MDP-Lys(L18)	3+/3+	−/−	2+(+)[1]	+[6]	±[9]	−(+)	+
Murabutide	3+/3+	−/−	+[2]	−[7]	−(−)[10]	−[13]	−[14]
GM-53	3+/2+	±/2+	2+[3]		−[11]		±[14]
GM-53(OBu)		2+/2+	2+[3]				+[14]
FK-565	−/−		3+(2+)[4]		+[12]		

[a] Enhancement of nonspecific resistance of mice to microbial infections. s.c. route (p.o. route) ([1]9, [2]47, [3]48, [4]unpublished).

[b] Induction of polyarthritis in rats. Seven daily (2 mg / kg) i.v. injections as PBS solution or suspension ([5]49, [6]9, [7]50) ([b']single s.c. injection as w/o emulsion; 51, 52).

[c] Administration to rabbits by i.v. (i.c.v.) route ([8]8, [9]9, [10]47, [11]48, [12]unpublished).

[d] S. c. injection in guinea pigs as 1% Tween 80-PBS solution or suspension (w/o emulsion) (53, [13]54).

Table 8. Summary of BRM activities of synthetic lipid A analogs, synthetic or semisynthetic lipophilic muramyl peptides and related compounds (selected) -continued-

Test material	Ileal strip contraction [f]	Macrophage [g] GlcN uptake	Macrophage [g] DNA sup. pres.	B-cell mitogenicity	PBA	Enhanced migration monocyte	Enhanced migration PMNL	Complement activation
LPS	−	+(+)	(+)	3+	3+	−	+	2+
Re-glycolipid		+		3+	3+	2+	+	2+
Lipid A		+		3+	3+	2+	2+	2+
301		(−)	(−)	−	−	−	−	
302		(−)	(−)	+	−	−	−	
307		(−)	(−)	−	−	2+	−	
311		(−)	(−)	−	−	+	+	
314		(−)	(−)	+	+	2+	2+	
304		(−)	(−)	−	−	−	−	
321				±	−	+	−	
312		(−)	(−)	−	+	2+		
315		(−)	(−)	−	2+	2+		
303		(+)	(+)	−	−	+	+	
316		−		+	+	2+		±
305		(−)	(−)	−	2+	2+	2+	
317		+		+	+	2+		+
313		+		2+	2+	2+	+	+
MDP	+	+(+)	(+)	+	2+	2+	−	−
L18-MDP	+	+(+)	(+)	+	2+	2+	−	−
B30-MDP	+	+(+)	(+)	+	2+	+	−	2+
MDP-Lys(L18)		+(+)	(+)	+	2+	+	−	
Murabutide	±			±	+	2+	−	
GM-53		(+)	(+)	+	2+	+	−	
GM-53(OBu)		+(+)	(+)	+	+	+	+	
FK-565		±		+	+	2+	−	

[e] Induction of extensive hemorrhagic necrosis by i.v. or s.c. injection of PBS solution or suspension at the skin site prepared by s.c. injection of w/o emulsion added with heat-killed tubercle bacilli (9, 55, ^{14}56). ([e'] Suspensions in PBS / solution in Nikkol HCO-60).

[f] Refer to 66.

[g] Stimulation of [^{14}C]glucosamine uptake and suppression of [^{3}H]thymidine incorporation by guinea pig's peritoneal macrophages induced with thioglycolate (liquid paraffin; 56, 57).

or LPS of natural sources, in the intensity and effective dose.

A recent analytical study of Imoto et al (58) which was performed on a chemically pure specimen of Re-lipid A by means of a recently developed NMR technique (2D-NMR) strongly suggests that 1,4'-diphospho-β(1',6)-glucosamine disaccharide in this lipid A specimen was acylated at 3 and 3' hydroxyl groups as well as two amino groups, leaving free at 6'-hydroxyl position. This important finding is supported by a study of Strain et al (59) that the so far accepted 3-deoxy-D-manno-octulosonate (KDO) linkage site at C-3' of glucosamine disaccharide structure of E. coli lipid A appears to be not supported by the ^{13}C chemical shift data, and by a study of Takayama et al (60) that a novel glycolipid which accumulates in E. coli mutant defective in phsphatidyl-glycerol synthesis contains two β-hydroxymyristate groups, one attached as an amide at C-2 and the other as an ester at C-3 of the sugar. Therefore, there is a possibility that weak BRM activities of synthetic lipid A analogs as compared with those of natural lipid A, are attributable to the fact that compounds so far available have been synthesized to mimic the "incorrect" target structure in which the hydroxyl groups at C-3' position of diglucosamine is linked to KDO, so free from acyl groups. This possibility will be verified in the near future by examination of BRM activities of compounds synthesized according to a revised structural model of E. coli lipid A.

In view of the state of affairs described in the preceeding paragraphs and discrepancies among findings reported by various research groups on BRM activities of the same synthetic specimens (see 61 and other articles in this book), it would not to be advisable to draw any definite conclusions about structure-activity relationships among synthetic lipid A analogs. However, it is tempting to suggest that on the basis of available informations described in this paper that 3-hydroxylation of the tetradecanoyl residue on R^2 position (see Fig. 1) tends to increase biological activities, as illustrated by the fact that the potencies of compounds 302, 321, 316 and 317 are generally more obvious than those of compounds 301, 304, 303 and 305, respectively. This does not seem to be true with the acyl groups on R^1 position. For example, compounds 305 and 317 were the most active compounds in the majority of assays, though they are devoid of the hydroxyl groups on tetradecanoyl residues. Moreover, the introduction of double acyl residue on R^2 position also seems to increase various activities, namely compounds 314 and 315 was more active than compounds 311 and 312, respectively. Another general tendency may be that the presence of phosphoryl groups on R^3 or R^4 position more or less increases a variety of immunobiological activities. Thus, compound 317, β(1',6)-linked glucosamine disaccharide which carries (R)-3-hydroxytetradecanoyl on R^3 position and tetradecanoyl residues on R^1 position, and phosphoryl groups on R^3 and R^4 positions, was proved to be the most active one among glucosamine disaccharide derivatives so far examined. It may be added here that this compound was definitely active in almost all of in vivo and in vitro assays performed in other laboratories (see other articles in this book) as well as ours.

A remarkable finding revealed by present study is that some of acyl derivatives of glucosamine (monosaccharide), particularly compound 313, N-(R)-3-[(R)-3-hydroxytetradecanoyloxy]tetradecanoyl glucosamine, exhibited distinct immunobiological activities in all of the present in vitro assays, though this compound was found to be only marginally active as an immunoadjuvant in vivo. This finding would permit us a speculation that lipid A does not necessarily need a glucosamine disaccharide structure as a backbone to manifest many of its immunobiological activities in vitro. However, a possibility still remains that the mode of action of compound 313 is entirely different from those of natural lipid A or active synthetic lipid A analogs, for instance, due to properties as a detergent. This possibility might be verified by studies using low- and non-responders to LPS or lipid A as test animals.

Concerning MDP and related compounds, discussion will be mainly concerned in the following four compounds: murabutide - a compound which is shown by French investigators to be highly adjuvant-active, but devoid of several undesirable effects of MDP (47,62), B30-MDP - a candidate for a useful adjuvant to potentiate vaccines for human use, MDP-Lys(L18) - a compound of choice to enhance nonspecific resistance of hosts to microbial infections, GM-53(OBu) - a semisynthetic compound which was prepared by using GMP4 which was isolated from the enzymatic digest of L. plantarum cell wall as a starting material and shared with B30-MDP the potent adjuvant activity in PBS as a vehicle, and FK-565 - a compound which is similar to synthetic desmuramyl peptides described by Miglior-Samour et al (63) as useful immunomodulators to enhance the host resistance to microbial infections. Table 1-1 indicates that B30-MDP and GM-53(OBu) are characterized by distinctive immunoadjuvant activities in both DTH induction and antibody formation without help of an extremely irritative Freund's incomplete type w/o emulsion, while neither murabutide nor MDP-Lys(L18), when administered as PBS solution, were immunostimulatory. Inability of desmuramyl peptides and their lipophilic derivatives as an antigen-specific immunoadjuvant (64) was confirmed by demonstration that FK-565 was quite inactive to stimulate immune responses to ovalbumin even when administered as a w/o emulsion, though antigen-specific immunostimulation by some desmuramyl peptides was reported in other laboratories (63,65).

None of murabutide, B30-MDP and GM-53(OBu) was pyrogenic by intravenous injection into rabbits at usual dosages. While murabutide is reported to be nonpyrogenic even in administration through a very sensitive intracerebroventicular (i.c.v.) route, rabbits receiving i.c.v. injection of B30-MDP showed distinctive febrile responses. No data are available on the pyrogenicity of GM-53(OBu) administered by i.c.v. route.

The remarkable finding is that murabutide is devoid of the activities to induce polyarthritis by repeated i.v. injections into rats and to provoke extensive hemorrhagic necrosis at the skin site of guinea pigs prepared by s.c. injection of heat-killed tubecle bacilli in w/o emulsion. Under the same experimental conditions, B30-MDP and MDP-Lys(L18) exhibited both of the activities. In this connection, B30-MDP dissolved in Nikkol HCO-60, a detergent for clinical use, induced the hemorrhagic necrosis, but B30-MDP suspension in PBS did not. GM-53(OBu), on the other hand, showed the activity to induce the hemorrhagic necrosis, though GMP_4, a parent compound of GM-53(OBu), did not show the necrosis-provoking activity, unlike MDP, muramyl tripeptide (either L-Lys or meso-A_2pm type) (9). No assay has so far been done on the arthritogenicity of GM-53(OBu), and no informations have been available on the arthritogenicity and the necrosis-inducing activity of FK-565.

There does not seem to be essential differences among activities of the above four compounds in other assay systems, except that GM-53(OBu) was only a compound capable of enhancing human PMNL migration, and murabutide did not show any detectable, murine B cell mitogenicity.

10 References

(1) D. E. S. Stuwart-Tull, Ann. Rev. Microbiol. 34 (1980) 311-340.

(2) D. C. Morrison, J. L. Ryan, Adv. Immunol. 28 (1979) 294-450.

(3) A. J. Wicken, K. W. Knox in D. Schlessinger (Ed.), Microbiology 1977, American Society for Microbiology, Washington, D.C. 1977, pp. 360-365.

(4) M. B. Goren, P. J. Brennan in G. P. Youmans (Ed.), Tuberculosis, W. B. Saunders, Philadelphia 1979, pp. 63-193.

(5) A. C. Wardlaw, R. Parton, Pharmacol. Ther. 19 (1983) 1-53.

(6) L. Chedid, F. Audibert, A. G. Johnson, Prog. Allergy 25 (1978)

63-105.

(7) A. Adam, J.-F. Petit, P. Lefrancier, E. Lederer, Mol. Cell. Biochem. 41 (1981) 27-47.

(8) S. Kotani, H. Takada, M. Tsujimoto, T. Ogawa, K. Kato, T. Okunaga, Y. Ishihara, A. Kawasaki, I. Morisaki, N. Kono, T. Shimono, T. Shiba, S. Kusumoto, M. Inage, K. Harada, T. Kitaura, S. Kano, S. Inai, K. Nagaki, M. Matsumoto, T. Kubo, M. Kato, Z. Tada, K. Yokogawa, S. Kawata, A. Inoue in H. Friedman, T. W. Klein, A. Szentivanyl (Eds.), Immunomodulation by Bacteria and Their Products, Plenum Press, New York 1981, pp. 231-273.

(9) S. Kotani, H. Takada, M. Tsujimoto, T. Kubo, T. Ogawa, I. Azuma, H. Ogawa, K. Matsumoto, W. A. Siddiqui, A. Tanaka, S. Nagao, O. Kohashi, S. Kanoh, T. Shiba, S. Kusumoto in J. Leljaszewicz, G. Pulverer, W. Roszkowski (Eds.), Academic Press, London 1982, pp. 67-107.

(10) O. Lüderitz, C. Galanos, V. Lehmann, H. Mayer, E. T. Rietshel, J. Weckesser, Naturwissenschaft. 65 (1978) 578-585.

(11) E. T. Rietschel, H.-W. Wollenweber, U. Zähringer, O. Lüderitz, Klin. Wochenschr. 60 (1982) 705-709.

(12) O. Westphal, K. Jann, K. Himmelspach, Prog. Allergy 33 (1983) 9-39.

(13) S. Kusumoto, M. Inage, H. Chaki, M. Imoto, T. Shimamoto,c T. Shiba in L. Anderson, S. M. Unger (Eds.), Carbohydrates in Bacterial Lipopolysaccharides: Synthesis and Biomedical Significance, The American Chemical Society, (in press).

(14) A. Boivin, I. Mesrobeanu, L. Mesrobeanu, Comp. Rend. Soc. Biol. 114 (1933) 307-310.

(15) O. Westphal, O. Lüderitz, F. Bister, Z. Naturforsch. 7b (1952) 148-155.

(16) A. H. Fenson, G. W. Gray, Biochem. J. 114 (1969) 185-196.

(17) K. Egawa, N. Kasai, Microbiol. Immunol. 23 (1979) 87-94.

(18) Y. B. Kim, D. W. Watson, J. Bacteriol. 94 (1967) 1320-1326.

(19) M. Inage, H. Chaki, S. Kusumoto, T. Shiba, Tetrohedron Lett, 21 (1980) 3889-3892.

(20) M. Inage, H. Chaki, S. Kusumoto, T. Shiba, Chem. Lett. (1980) 1373-1376.

(21) M. Inage, H. Chaki, S. Kusumoto, T. Shiba, Tetrahedron Lett. 24 (1981) 2281-2284.

(22) M. Inage, H. Chaki, S. Kusumoto, T. Shiba, Chem. Lett. (1982) 1281-1284.

(23) M. Inage, H. Chaki, M. Imoto, T. Shimamoto, S. Kusumoto, T. Shiba, Tetrohedron Lett. 24 (1983) 2011-2014.

(24) S. Kotani, H. Takada, M. Tsujimoto, T. Ogawa, Y. Mori, M. Sakuta, A. Kawasaki, M. Inage, S. Kusumoto, T. Shiba, N. Kasai, Infect. Immun. 41 (1983) (in press).

(25) S. Kusumoto, Y. Tarumi, K. Ikenaka, T. Shiba, Bull. Chem. Soc. Jpn. 49 (1976) 533-539.

(26) S. Kusumoto, S. Okada, K. Yamamoto, T. Shiba, Bull. Chem. Soc. Jpn. 51 (1978) 2122-2126.

(27) S. Kusumoto, M. Inage, T. Shiba, I. Azuma, Y. Yamamura, Tetrahedron Lett. 49 (1978) 4899-4902.

(28) K. Matsumoto, T. Otani, T. Une, Y. Osada, H. Ogawa, I. Azuma, Infect. Immun. 39 (1983) 1029-1040.

(29) P. Lefrancier, M. Derrien, X. Jamet, J. Choay, E. Lederer, F. Audibert, M. Parant, F. Parant, L. Chedid, J. Med. Chem. 25 (1982) 87-90.

(30) Japanese patent, 56-45449 (1981).

(31) Japanese patent, 57-82178 (1982).

(32) S. Kawata, T. Takemura, Y. Takase, Y. Yokogawa, Agric. Biol. Chem. (submitted).

(33) S. Kotani, T. Narita, D. E. S. Stewart-Tull, T. Shimono, Y. Watanabe, K. Kato, S. Iwata, Biken J. 18 (1975) 77-92.

(34) S. Kotani, F. Kinoshita, I. Morisaki, T. Shimono, T. Okunaga, H. Takada, M. Tsujimoto, Y. Watanabe, K. Kato, T. Shiba, S. Kusumoto, S. Okada, Biken J. 20 (1977) 95-103.

(35) M. Tsujimoto, J. Osaka Univ. Dent. Soc. 26 (1981) 63-83. (in Japanese with English summary).

(36) D. Hagimoto, Bull. Res. Inst. Dis. Chest Kyushu Univ. 13 (1969) 19-34.

(37) S. Kuwajima, K. Kobayshi, K. Okawa, S. Oka, Y. Tanaka, A. Kitano, H. Nakajima, T. Tamura, M. Kato, M. Hori, S. Kotani, S. Yamamoto, M. Masui in Y. Yamamura, S. Kotani, I. Azuma, A. Koda, T. Shiba (Eds.), Immunomodulation by microbial products and related synthetic compounds, Excerpta Medica, Amsterdam 1982, pp. 343-346.

(38) M. Ohta, I. Nakashima, N. Kato, Cell. Immunol. 66 (1982) 111-120.

(39) T. Ogawa, S. Kotani, K. Fukuda, Y. Tsukamoto, M. Mori, S. Kusumoto, T. Shiba, Infect. Immun. 38 (1982) 817-824.

(40) T. Ogawa, S. Kotani, S. Kusumoto, T. Shiba, Infect. Immun. 39 (1983) 449-451.

(41) S. A. Adamu, J. F. Sperry, Infect. Immun. 33 (1981) 806-810.

(42) J. M. Wilton, D. L. Rosenstreich, J. J. Oppenheim, J. Immunol. 114 (1975) 388-393.

(43) H. Takada, M. Tsujimoto, K. Kato, S. Kotani, S. Kusumoto, M. Inage, T. Shiba, I. Yano, S. Kawata, K. Yokogawa, Infect. Immun. 25 (1979) 48-53.

(44) D. C. Morrison, L. F. Kline, J. Immunol. 118 (1977) 362-368.

(45) A. Kawasaki, J. Osaka Univ. Dent. Soc. 27 (1982) 46-61.

(46) M. M. Mayer in E. A. Kabat, M. M. Mayer (Eds.), Experimental Immunochemistry, 2nd Edit., Charles C. Thomas, Springlfield 1961, pp. 133-240.

(47) L. A. Chedid, M. A. Parant, F. M. Audibert, G. J. Riveau, F. J. Parant, E. Lederer, J. P. Choay, P. L. Lefrancier, Infect. Immun. 35 (1982) 417-424.

(48) S. Kawata, personal communication.

(49) O. Kohashi, Y. Kohashi, S. Kotani, A. Osawa, Ryumachi 21(Suppl.) (1981) 149-156.

(50) O. Kohashi, personal communication.

(51) S. Nagao, A. Tanaka, Infect. Immun. 28 (1980) 624-626.

(52) O. Kohashi, A. Tanaka, S. Kotani, T. Shiba, S. Kusumoto, K. Yokogawa, S. Kawata, A. Ozawa, Infect. Immun. 29 (1980) 70-75.

(53) A. Tanaka, S. Nagao, K. Emori, K. Imai, K. Kushima in Japan Medical Research Foundation (Ed.), Sarcoidosis, Univ. Tokyo Press, Tokyo 1981, pp. 131-146.

(54) A. Tanaka, S. Nagao, personal communication.

(55) S. Nagao, Y. Iwata, A. Tanaka in (37), pp. 189-192.

(56) S. Nagao, A. Tanaka, personal communication.

(57) S. Nagao, A. Tanaka, Microbiol. Immunol. 27 (1983) 377-387.

(58) M. Imoto, S. Kusumoto, T. Shiba, H. Naoki, T. Iwashita, E. T. Rietshel, H.-W. Wollenwever, C. Galanos, O. Lüderitz, Tetrahedron Lett. (submitted).

(59) S. M. Strain, S. W. Fesik, I. M. Armitage, J. Biol. Chem. 258 (1983) 2906-2910.

(60) K. Takayama, N. Qureshi, P. Mascagni, M. A. Nashed, L. Anderson, C. R. H. Raetz, J. Biol. Chem. (submitted).

(61) T. Yasuda, S. Kanagasaki, T. Tsumita, T. Tadakuma, J. Y. Homma, M. Inage, S. Kusumoto, T. Shiba, Eur. J. Biochem. 124 (1982) 405-407.

(62) C. Damais, G. Riveau, M. Parant, J. Gerota, L. Chedid, Int. J.

Immunophrmacol. $\underline{4}$ (1982) 451-462.

(63) D. Migliore-Samour, J. Bouchaudon, F. Floc'h, A. Zerial, L. Ninet, G. H. Werner, P. Jolles, Life Sci. $\underline{26}$ (1980) 883-888.

(64) M. A. Parant, F. M. Audibert, L. A. Chedid, M. R. Level, P. L. Lefrancier, J. P. Choay, E. Lederer, Infect. Immun. $\underline{27}$ (1980) 826-831.

(65) K. Mašek, M. Zaoral, J. Ježek, V. Krchňák, Experientia $\underline{35}$ (1979) 1397-1398.

(66) T. Ogawa, S. Kotani, M. Tsujimoto, S. Kusumoto, T. Shiba, S. Kawata, K. Yokogawa, Infect. Immun. $\underline{35}$ (1982) 612-619.

Further Chemical
and Biological Studies
on Some
Bacterial Lipopolysaccharides

Comparison of the Biological Activities of Lipopolysaccharides Complexed with Outer Membrane Protein

Shiro Kanegasaki, Tatsuji Yasuda, Sonoko Kobayashi, Kenichi Tanamoto
Kazushige Hirosawa
The Institute of Medical Science, The University of Tokyo, Shirokanedai,
Minato-ku, Tokyo 108, Japan

Yasuhiko Kojima
The Kitasato Institute, Shirokane, Minato-ku, Tokyo 108, Japan

Ernst Th. Rietschel
Forschungsinstitut Borstel, D-2061 Borstel, FRG

Hisami Yamada, Shoji Mizushima
Faculty of Agriculture, Nagoya University, Nagoya 464, Japan

Abstract

To minimize the influence of the different physicochemical properties of the various lipopolysaccharides on their activities, we complexed the lipopolysaccharides with an outer membrane protein derived from E. coli and compared biological activities, i.e., induction of interferon from rabbit spleen cells and activation of proclotting enzyme of the horseshoe crab. Differing degrees of activity were exhibited by complexes derived from different lipopolysaccharides, as in the case of the free lipopolysaccharides. The complex derived from lipopolysaccharide of C. violaceum was most active among the lipopolysaccharides tested. These results suggest that substituents not masked after complexing with the outer membrane protein play an important role in the variation of activity.

1. Introduction

Lipopolysaccharide, a constituent of the outer surface of Gram-negative bacteria is known to exert various biological (i.e. endotoxic) effects in mammalians (1). Most of the biological activities of lipopolysaccharide are dependent on the lipid A component of the molecule (1,2). The chemical structure of the lipid A backbone in many Gram-negative bacteria appears to be identical and consists of a β-1,6 linked D-glucosamine disaccharide which carries phosphate residues in position 1 and 4' of the reducing and nonreducing glucosamine residue, respectively, and amide bound D-3-hydroxy fatty acids (2). The hydroxy group at position 3' (or 6') of the nonreducing glucosamine residue is thought to represent the attachment site of the polysaccaharide component. Recent studies revealed that the phosphate groups of the glucosamine disaccharide are (in part) substituted. These phosphate substituents include phosphoryl, phosphorylethanolaminyl, D-glucosaminyl and 4-amino-arabinosyl residues (Table 1). The substituents of the phosphate groups as well as the polysaccharide component are distinct in different bacterial genera and this may affect the biological activities of the respective lipopolysaccharide.

In fact, our preliminary experiments with representative lipopolysaccharides different in the substitution (Table 1) showed that the lipopolysaccharides exhibited different degrees of activity in inhibiting tumor development (assayed by reduction of total packed ascites sarcoma 180 cells (3)), and induction of interferon (see below) as well as induction of fever (4). Lipopolysaccharides of Chromobacterium violaceum were the most active among the lipopolysaccharides tested in all activities (antitumor activity: ED50=0.02 µg/kg and pyrogenicity: minimal effective dose (MED)=0.0003 µg/kg) whereas those derived from Yersinia enterocolitica (pyrogenicity: MED=0.3 µg/kg) and Vibrio cholerae (antitumor activity: ED50=9.8 µg/kg and pyrogenicity: MED=0.3 µg/kg) showed comparatively low activity. Those obtained from Salmonella, Escherichia and Proteus (antitumor activity: ED50=0.2-0.9 µg/kg and pyrogenicity: MED=0.03 µg/kg), all were of similar, moderate activity (Kanegasaki, Tanamoto and Homma, unpublished observations).

It is now widely accepted that isolated lipopolysaccharide in aqueous solution forms various micells and aggregates due to the non-polar interaction of lipid A component as well as intermolecular ionic linkages in the presence of divalent cations or polyamines and that biological activities of lipopolysaccharide are influenced by the physicochemical properties of such micells (2). To compare biological activities of different lipopolysaccharide molecules, it is therefore

Table 1: Substituents of the lipid A backbone in various lipopolysaccharides

	Substituents at			
Lipopolysaccharide derived from	Amino groups of (GlcN)$_2$	Phosphate group of reducing GlcN	Phosphate group of nonreducing GlcN	Hydroxyl group of C'3(C'6)
C. violaceum (S)	D-3-OH-12:0	L-4-AraN	GlcN	KDO[2]
S. minnesota (Re)	D-3-OH-14:0	L-4-AraN[1]	P-Etn[1]	KDO
P. mirabilis (Re)	D-3-OH-14:0	L-4-AraN	—	KDO[2]
E. coli (R-K12)	D-3-OH-14:0	—	P[1]	KDO
Y. enterocolitica (Re)	D-3-OH-14:0	—	—	KDO[2]
V. cholerae (R-95R)	D-3-OH-14:0	—	P-Etn	—[3]

L-4-AraN=4-Amino-4-deoxy-L-arabinose, GlcN=D-Glucosamine
P=Phosphate, P-Etn=Phosphorylethanolamine
1 Partial substitution
2 Presumed to be
3 Other than KDO

desirable to minimize the influence of the different physicochemical properties of each molecule. In previous paper (5,6) we employed liposomal system to compare adjuvant effect and mitogenicity of various synthetic analogues of lipid A. However, the method is not applicable to lipopolysaccharide molecules because of the insolubility of the molecules in organic solvents.

In the present investigation, we complexed lipopolysaccharides from various organisms with outer membrane protein OmpC derived from E. coli, and compared biological activities, i.e., induction of interferon from rabbit spleen cells and activation of proclotting enzyme of the horseshoe crab.

2. Complexing lipopolysaccharides with an outer membrane protein

Outer membrane protein OmpC (O-8 protein) was extracted from peptideglycan sacculi of E. coli YA27 (K12 F⁻ met leu λ⁻) with 2 % sodium dodecylsulfate, 0.5 M NaCl solution and purified by a Sephadex G-200 gel filtration in the presence of sodium dodecylsulfate (7). Purified OmpC contained lipopolysaccharides of less than 5 μg/mg protein. Lipopolysaccharides extracted from various bacteria with phenol/chloroform/ether (8) were mixed with OmpC and dialysed under the conditions described for the reconstitution of membrane vesicle which exhibiting lattice structure (7). The procedure is illustrated in Fig. 1.

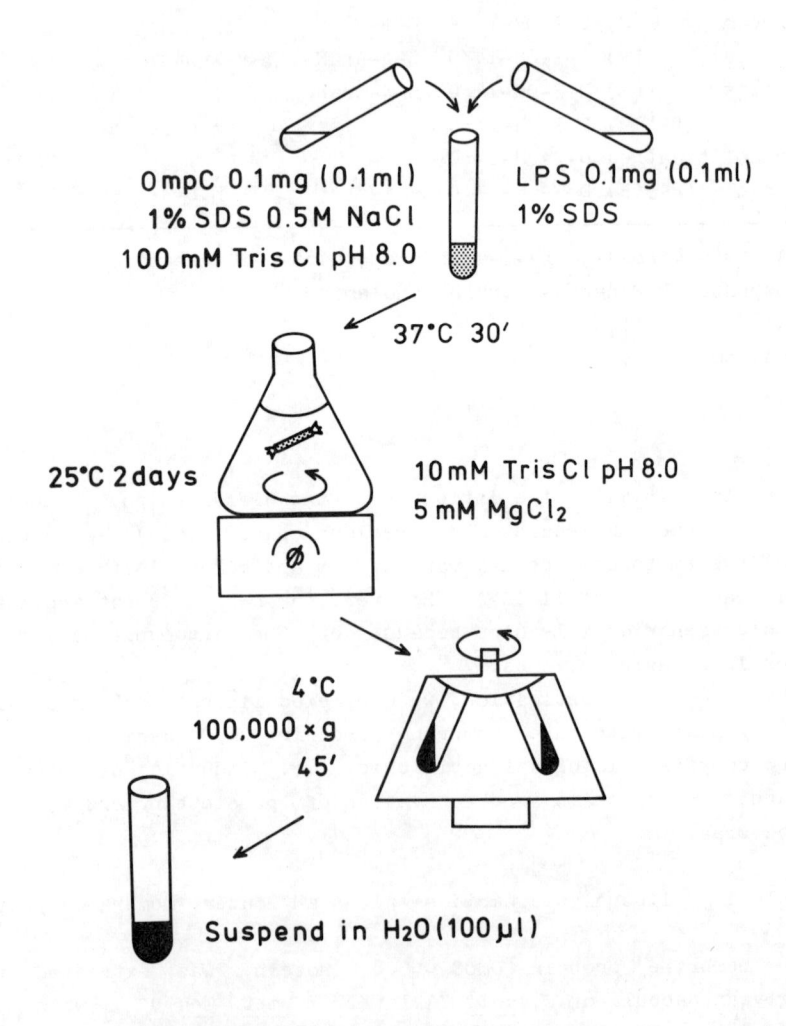

Fig. 1. Method of constitution of lipopolysaccharide-OmpC complex.

In all cases tested, sediment formed which was examined under a Hitachi electron microscope (HU-12A) after negative staining with sodium phosphotungstate (pH 7.2). As shown in Fig 2, OmpC and lipopolysaccharides formed open vesicles in some cases whereas in some other cases they formed sheet-like structures.

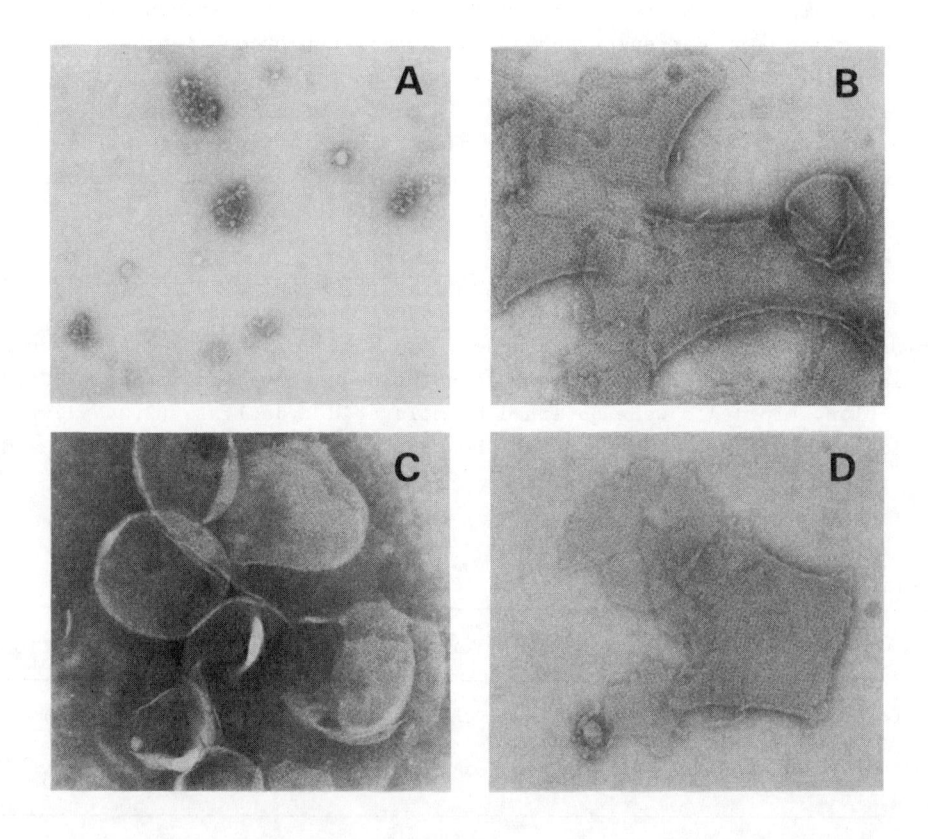

Fig. 2. Ordered lattice structure constituted from OmpC and various lipopolysaccharides
    A OmpC protein alone
    B OmpC+C.violaceum lipopolysaccharide
    C OmpC+S.minnesota lipopolysaccharide
    D OmpC+Y.enterocholitica lipopolysaccharide

In all cases examined, however, the ordered lattice structures resembled those observed in cell envelope treated with sodium dodecylsulfate (9), or reconstituted membrane vesicles (7) were observed, indicating that they had successfully assembled. These complexes were used for the latter experiments.

3. Comparison of interferon inducing activity of various lipopolysaccharides

We compared interferon inducing activity of various lipopolysaccharides in different forms. For interferon production, spleen and lymphnode cells obtained from normal young rabbits were mixed with various doses (10 fold serial dilutions) of lipopolysaccharides or derivatives in a culture medium supplement with 10% of calf serum and incubated at 25° for 24 hours. Interferon was assayed by the plaque reduction method using RK 13 cells of rabbit kidney and vesicular stomatitis virus.

Table 2 shows the interferon inducing activity of the lipopolysaccharides and the lipid A's derived from them. The results are expressed as minimal effective dose (MED). This dose induced interferon which can be detected (more than 50% plaque reduction) after 15 fold dilution. Lipid A's were obtained by mild acid hydrolysis (0.2 N acetic acid in boiling water for 3 hours) following sedimentation and extraction with chloroform.

Table 2: Interferon-inducing activity of lipopolysaccharides from various bacteria[1]

Lipopolysaccharide from		Original form[3] (ng/ml)	Lipid A[4] (x2 pmol phosphate/ml)	Complex with OmpC (ng lipopolysaccharide/ml)
C. violaceum	S	0.1	1	1
S. minnesota	Re	1	1	100
P. mirabilis	Re	1	1	10
E. coli	R(K-12)	1	1-0.1	100
Y. enterocolitica	Re	100	1-0.1	10000
V. cholerae	R(95R)	100	1-0.1	Not done

Minimal effective dose for interferon induction[2]

1 Representative results are shown
2 See text
3 Water suspension (boiled in water bath for 5 min)
4 Triethylamine salt form

In contrast to the differing degrees of interferon inducing activity of lipopolysaccharides derived from various bacteria, no great differences were observed among lipid A preparations, and 0.2-2 pmol phosphate/ml of lipid A's gave significant activity.

The results of the activity of lipopolysaccharide-OmpC complexes, which exhibited lattice structure were also shown in Table 2. Although the activity of the complex was significantly lower than the comparable free lipopolysaccharide (by 1-2 order of magnitude), differing degrees of activity were exhibited by complexes derived from different lipopolysaccharides. As in the case of the free lipopolysaccharides, the complex derived from lipopolysaccharide of C. violaceum was most active and the activity was by several orders of magnitude higher than the complex derived from Y. enterocolitica. Moderate activities were seen by those derived from lipopolysaccharides of Salmonella, Eschrichia and Proteus. These results suggest that substituents released during mild acid treatment and those not masked after constitution of the outer membrane like structure play an important role in the variation of activity.

4. Activation of proclotting enzyme of horseshoe crab by various lipopolysaccharides

In contrast to the results of interferon induction, the degree of activation of proclotting enzyme of the horseshoe crab amebocyte (Tachypleus tridentatus) assayed by ρ-nitroaniline release from a synthetic substrate ρ-N-benzoyl-Leu-Gly-Arg-ρ-nitroaniline ("PYRODICK", Seikagaku Kogyo, Tokyo) (10) was different in each lipid A (Table 3). Lipid A of C. violaceum, as in the case of lipopolysaccharides, was the most active and that of Y. enterocolitica the least active. Those of S. minnesota, P. mirabiris, E. coli and V. cholerae showed intermediate activity.

The results suggest that substituents remaining after mild acid treatment (probably substituents of phosphate group(s)) modified the activity. Lipopolysaccharide-OmpC complexes also showed different degrees of activity (Table 3); the highest and lowest activities were obtained by the complexes of C. violaceum and Y. enterocolitica lipopolysaccharide, respectively.

5. Concluding remarks

In the present study, we compared certain biological activities of selected lipopolysaccharides that were different in the substitutions of lipid A phosphate groups as well as (poly-)saccharide components. No

Table 3: Activation of proclotting enzyme of horeshoe crab by various lipopolysaccharides and lipid A's[1]

Relative incline (absorbance at 405 nm/concentration)

Lipopolysaccharide from		Original form[2]	Lipid A[3]	Complex with OmpC[4]
C. violaceum	(S)	1.00	1.00	1.00
S. minnesota	(Re)	0.63	0.95	0.28
P. mirabilis	(Re)	0.28	0.10	0.39
E. coli	(R-K12)	0.61	0.18	0.10
Y. enterocolitica	(Re)	0.04	0.01	0.02
V. cholerae	(R-95R)	0.27	0.36	Not done

1 Representative results are shown
2 Water suspension (boiled in water bath for 5 min): Control values for CV lipopolysaccharide=0.69 absorbance at 405 nm/0.05ng lipopolysaccharide
3 Triethylamine salt: Control values for CV=0.55 absorbance at 405 nm/10 fmol phosphate lipid A
4 Contral values for CV-OmpC complex=0.54 absorbance at 405 nm/0.1 ng lipopolysaccharide

great difference was observed in interferon-inducing activity by lipid A preparations derived from various lipopolysaccharides. However, they activated proclotting enzyme of horseshoe crab in varying degrees. It may tentatively be concluded that polar head groups (remaining after acid hydrolysis) do not play significant roles in the former activity but do play some role (directly or indirectly) in the latter. Incidentally, no activation of proclotting enzyme was observed by chemically synthesized lipid A analogues carrying no polar substituents on phosphate residues (5,6), whether the analogues were suspended in water as triethylamine or in liposomial form (unpublished observations). It is possible, however, that the presence of polar head groups of lipid A contributes to the solubility of lipid A molecules as well as lipopolysaccharide molecules (1,2) and consequently affects the activation of the enzyme. It is interesting to note that the activities of certain lipopolysaccharides were enhanced greatly by the addition of triethylamine whereas the activities of some other lipopolysaccharides were only slightly influenced (e.g., interferon-inducing activity of lipopolysaccharides from P. milabilis and S. minnesota was enhanced much more than other lipopolysaccharides when triethylamine was present).

To minimize the influence of the different physicochemical properties of the various lipopolysaccharides on their activities, we complexed the lipopolysaccharides with an E. coli outer membrane protein, OmpC to obtain an outer membrane like lattice structure and compared biological activities. Since the fatty acid region of lipopolysaccharide is considered the prime participant in the hexagonal lattice arrangement of the protein (7), lipopolysaccharides may be arranged on the leaflet of the lattice structure in such a way that the hydrophilic (poly-)saccharide and polar portion extends outwards into the aqueous environment. Lipid A substituents, especially the saccharide component, may thus influence biological activities directly, since the constituted structure (the OmpC-lipopolysaccharide complex) exhibited different degrees of biological activity. Unfortunately, information on the precise structure of the saccharide adjacent to lipid A and the physicochemical character of the protein-lipopolysaccharide complex is still limited so the possibility remains that the modifying effect of the above substituents is indirect.

6. References

(1) C. Galanos, O. Luderitz, E.Th. Rietschel, O. Westphal, International Review of Biochemistry, 14 (1977) 239-335.

(2) E. Th. Rietschel, C. Galanos, O. Luderitz, O. Westphal, Immunopharmacology and the regulation of leukocyte function, D. R. Webb (Ed.), Marcel Dekker, New York 1982, pp. 183-229.

(3) K. Tanamoto, C. Abe, J.Y. Homma, Y. Kojima, Eur. J. Biochem. 97 (1979) 623-629.

(4) D.W. Watson, Y.B. Kim, J. Exp. Med. 118 (1963) 425-446.

(5) T. Yasuda, S. Kanegasaki, T. Tsumita, T. Tadakuma, J.Y. Homma, M. Inage, S. Kusumoto, T. Shiba, Eur. J. Biochem. 124 (1982) 405-407

(6) S. Kanagasaki, T. Yasuda, T. Tsumita, T. Tadakuma, J.Y. Homma, S. Kusumoto, T. Shiba, (1983) This book.

(7) H. Yamada, S. Mizushima, Eur. J. Biochem. 103 (1980) 209-218.

(8) C. Galanos, O. Luderitz, O. Westphal, Eur. J. Biochem. 9 (]969) 603-610.

(9) A.C. Steren, B.T. Heggelor, R. Mollor, J. Kistler, J.P. Rosenbusch, J. Cell Biol. 72 (1977) 292-301.

(10) S. Nakamura, T. Morito S. Iwanaga, M. Niwa, K. Takahashi, J. Biochem. 81 (1977) 1567-1569.

Relationship between Biological Activities and Chemical Structure of Pseudomonas aeruginosa Endotoxin

J. Yuzuru Homma

The Kitasato Institute, Shirokane, Minato-ku, Tokyo 108, Japan

Ken-ichi Tanamoto

Department of Bacteriology, Institute of Medical Science, The University of Tokyo, Shirokanedai, Minato-ku, Tokyo 108, Japan

Abstract

Investigation of the relationship between the biological activities of the endotoxin of P. aeruginosa and chemical structure resulted in the dissociation of some of the biological activities of the endotoxin.

Free lipid A was demonstrated to be responsible for the interferon-inducing activity, pyrogenicity and adjuvanticity. For manifesting inhibition of experimental tumor development or activation of proclotting enzyme of horseshoe crab (Tachypleus tridentus), lipid A alone is not sufficient and a saccharide portion such as 3-deoxy-D-manno-octulosonic acid is necessary to link the lipid A portion. That the polysaccharide portion plays a role in the activity of the complete lipopolysaccharide was also supported by the fact that chemical modification on the saccharide reduced the two activities drastically.

Antitumor and interferon-inducing activities, adjuvanticity and activation of proclotting enzyme diminished by complete de-O-acylation by NH_2OH, whereas interferon-inducing activity remained unchanged. Incomplete lipid A containing only amide-linked fatty acid is sufficient to induce interferon in vitro.

None of the activities mentioned here was observed with degraded polysaccharide obtained after extraction of lipid A.

1 Introduction

After the chemical structure of lipid A was thoroughly studied and it was proved that lipid A is an essential compound for various biological activities of the endotoxin (1,2,3), it became important to elucidate whether the endotoxic activities could be dissociated by altering the lipid A and the lipopolysaccharide structures. Pseudomonas aeruginosa endotoxin was chosen for this purpose because several of its biological activities have been found to be extremely strong, rendering them measurable in nanograms or picograms. Investigations on the relationship between the biological activities of the endotoxin and the chemical structure resulted in the dissociation of some of the biological activities of the endotoxin (4,5,6).

2 P. aeruginosa endotoxin: chemical and biological properties

A protein-rich endotoxin (designated as OEP) containing approximately 80% protein was isolated from the autolysate of P. aeruginosa strain N 10. The chemical nature of OEP is shown in Table 1 (7,8).

Table 1. Chemical analyses of protein-lipopolysaccharide complex (OEP) and lipopolysaccharide from the complex (OEP)

Constituent	complex (%) (OEP)	lipopolysaccharide from complex (OEP) (%)
Folin-sensitive material	77	4.0
Total sugars	4.5	20.6
Total fatty acid	8.8	35
Hexosamine	4.5	18.5
3-Deoxy-D-manno-octulosonic acid	1.3	4.0
Phosphorus	1.2	3.0

The protein-lipopolysaccharide complex was isolated from the autolysate of P. aeruginosa N 10 by the method of Homma et al. (1972). Folin-sensitive material was measured by the Folin-Ciocalteu method, total sugars by the anthrone/sulfuric acid reaction, total fatty acids as described by Tanamoto et al. (1979), hexosamine by the Elson-Morgan method, 3-deoxy-D-manno-octulosonic acid by the thiobarbituric acid method and phosphorus by Allen's method.
Data from reference (4).

The protein moiety of the protein-lipopolysaccharide complex is responsible for the activity of the serologically common as well as protective antigen of P. aeruginosa (9,10). This common antigen (OEP) of P. aeruginosa proved to be a common antigen of Vibrio cholerae both serologically and in possessing infection protective properties (11, 12). Antitumor and interferon-inducing activities, nonspecific protective properties and an adjuvant effect of the common antigen have been reported elsewhere (13-16). IgE production to OEP has been reported in mice injected with the antigen incorporated into Al(OH)$_3$ gel (17).

OEP possesses remarkable biological activities as mentioned above. However, after protease digestion, OEP no longer possesses the serologically common properties or common protective antigen of P. aeruginosa (4,8,10). The capacity of OEP to produce IgE antibody is also significantly reduced by protease treatment, and lipopolysaccharide isolated from OEP does not produce IgE antibody by itself, suggesting that the capacity lies in the protein moiety of OEP (17).

The lipopolysaccharide portion isolated from OEP possesses remarkable antitumor and interferon-inducing activities together with adjuvanticity, pyrogenicity and proclotting enzyme-activating activity of horseshoe crab.

3 Regions of the lipopolysaccharide of P. aeruginosa essential for endotoxic activities

A. Materials and Methods

In order to investigate the relationship between the chemical structure of the lipopolysaccharide of OEP and interferon-inducing activities as well as adjuvanticity, pyrogenicity and proclotting enzyme activating activity of horseshoe crab, lipopolysaccharide was isolated using the phenol/water method (1). Highly sensitive assay methods were applied for estimating the biological activities of the lipopolysaccharide and its derivatives. These are discussed briefly below.

As shown in Table 2, when OEP was treated with phenol/water, a large percentage of the protein was removed from the preparation. The remaining material was hydrolyzed and analyzed by chromatography on Dowex 50 H and by amino acid analyzer after complete hydrolysis by acid. By chromatographic analysis with authentic samples four compounds other than amino acids were obtained, i.e. glucosamine (30.3%), galactosamine (20.4%), fucosamine (38.1%) and unidentified substance (11.2%). From the results of total amino acid determination shown in Table 2, a significant amount of alanine and a small amount of lysine were found remaining in the preparation.

Table 2. Comparison of the amino acid composition of the protein-lipopolysaccharide complex and lipopolysaccharide isolated from the complex (OEP)

Amino acid	Amont in	
	protein-lipopolysaccharide complex (OEP)	lipopolysaccharide
	(mol/100 mol)	
Lysine	3.5	4.7
Histidine	2.3	1.0
Arginine	5.3	2.2
Half-cystine	0.10	n.d.
Aspartic acid	6.9	0.7
Threonine	4.6	0.4
Serine	4.2	0.9
Glutamic acid	10.7	0.7
Proline	4.4	0.1
Glycine	8.6	1.2
Alanine	10.7	21.7
Valine	7.4	1.3
Methionine	1.1	n.d.
Isoleucine	5.2	0.5
Leucine	8.3	0.7
Tyrosine	2.9	0.9
Phenylalanine	4.3	0
Tryptophan	0.12	n.d.
Glucosamine	-	22.3
Galactosamine	-	15.0
Fucosamine	-	28.0

Amino acids were estimated after hydrolysis for 24 h at 105°C with 6 M HCl. Methionine was determined after performic acid oxidation. The amounts are expressed as molar percentage of all the amino acids detected. As the positions of the galactosamine and fucosamine peaks corresponded to the positions of tyrosine and phenylalanine, the amounts of galactosamine and fucosamine were corrected from the result of chromatography on Dowex 50 H on the basis of the glucosamine peak, and the amounts of tyrosine and phenylalanine were calculated.

n.d. = not determined

Data from reference (4).

Major neutral sugars in the preparation were rhamnose (12.4%), mannose (2.7%), glucose (66.9%) and three other unidentified sugars. These sugars were the same as those found in the protein-lipopolysaccharide complex (OEP).

The fatty acid composition of lipopolysaccharide obtained from the complex (OEP) is shown in Table 3. It was qualitatively similar to that of the protein-lipopolysaccharide complex (OEP) and lipopolysaccharide obtained from whole cells.

Antitumor activity was determined using 5-week-old female ddN mice weighing 20 g inoculated intraperitoneally with 0.05 ml 3 x 10^7 cells of seven-day-old ascites sarcoma-180. Test samples were injected intraperitoneally once a day for 5 days starting 24 hr after tumor inoculation, and antitumor activity of the various samples was evaluated from the volume ratio of total packed cells from the peritoneum on day 7 (13).

Table 3. Fatty acid composition of lipopolysaccharide from the protein-lipopolysaccharide complex (OEP) before and after deacylation

Fatty acid	Amount in lipopolysaccharide from complex (OEP)			
	Non-treated	NH_2OH-treated	N_2H_4-treated	0.1M NaOH-treated
	(%)			
β-OH $C_{10:0}$	7.3	1.3	0.7	4.2
$C_{12:0}$	38.5	0	0.7	17.7
α-OH $C_{12:0}$	11.5	0	0.2	5.9
β-OH $C_{12:0}$	25.8	29.6	0.2	29.1
$C_{14:0}$	trace	trace	trace	trace
$C_{16:0}$	12.3	0.9	4.8	7.5
$C_{16:1}$	3.4	0	0.4	1.7
$C_{18:1}$	trace	trace	trace	trace
Total	100	31.5	7.0	66.1

The amounts of all compositions are expressed as percentages of the total amount of fatty acid in the lipopolysaccharide; the protein-lipopolysaccharide complex was isolated from the autolysate of P. aeruginosa.
Data from reference (4).

In testing interferon production in vitro, spleen and lymph node obtained from normal young rabbits were mixed with various doses of lipopolysaccharide in a culture medium supplemented with 20% calf serum and incubated at 25°C for 24 hr. Interferon was assayed by the plaque reduction method using RK 13 line cells of rabbit kidney and vesicular stomatitis virus (14).

These are two highly reproducible methods for determining the ED_{50} of the test samples against sarcoma-180 or interferon-inducing activity of the test samples in vitro.

The adjuvant effect was determined by the increase of direct plaque forming cells assay in the spleen of mouse using sheep red blood cells (SRBC) as an antigen. CDF_1 female mice 5-6 weeks old were used (16).

The three activities were compared with pyrogenic activity as well as the activity of activating proclotting enzyme of horseshoe crab (Tachypleus tridentatus), which was assayed by two methods, i.e., p-nitroaniline release from a synthetic peptide, α-N-Benzoyl-Leu-Gly-p-nitroaniline (18) and gelation reaction of hemocyte lysate of Tachypleus tridentatus (19).

The minimum pyrogenic doses of the test samples which caused a 1°F rise in temperature at 3 hr after injection were determined using 3 rabbits for each concentration of the samples (20).

Lipopolysaccharide of Escherichia coli was obtained commercially from Difco Laboratories (Detroit, Mich., U.S.A.). Lipopolysaccharide of Salmonella typhimurium LT 2 and SL 1102 were prepared by the phenol/water/petroleum ether methods (21), respectively. Lipopolysaccharide of Vibrio cholerae (Inaba type) was also extracted by the phenol/water method.

B. Results

1) Inhibition of Tumor Development, Induction of Interferon, Activation of Proclotting Enzyme of Horseshoe Crab, Adjuvanticity and Pyrogenicity

As shown in Table 4, the activities of tumor development inhibition, interferon induction and adjuvanticity of the lipopolysaccharide obtained from the complex (OEP) were found to be much higher than those of the complex (OEP) itself on the basis of sample weight as well as on the basis of lipopolysaccharide content. These results indicate that the protein moiety of the complex (OEP) is not essential for these activities. Pyrogenic activity of the lipopolysaccharide derived from the complex (OEP) by phenol/water extraction was the same as that of the complex (OEP), while the activities of activating proclotting enzyme and tumor development inhibition were found to become significantly stronger than those of the complex (OEP). Interferon induction was also found to become stronger than that of the complex. It is presumed that the protein portion of the complex (OEP) inhibited manifestation of the three activities.

All lipopolysaccharides derived from S. typhimurium LT 2 and SL 1102 (Re mutant) exhibited these activities. The pyrogenicity of the lipopolysaccharide of P. aeruginosa derived from the complex was weak compared with those of the lipopolysaccharides derived from species other than P. aeruginosa, although the other four biological activities were quite strong.

Table 4. Biological activities of LPSs derived from various species of bacteria

Sample	Antitumor activity ED50 µg kg⁻¹ day⁻¹	Interferon-inducing activity µg/ml	Pyrogenicity M.E.D.[a] µg/kg	Release of p-nitroaniline[b]	T.T.lysate[c] gelation µg/ml	Adjuvant index
Protein-LPS complex from autolysate of P. aeruginosa N 10 (OEP)	0.1	0.1	0.3	0.06	10^{-1}	1.47
LPS from above complex (OEP)	0.007	0.01	0.3	1.35	10^{-4}	1.71
E. coli	0.06			1.5	10^{-2}	
S. typhimurium LT 2	<0.1	0.001	0.03	2.4	10^{-3}	
S. typhimurium SL 1102 (Re mutant)	0.04	0.001	0.003	1.6	10^{-3}	
V. cholerae (Inaba type)			0.03	5.7	10^{-4}	

a: Minimum effective dose.
b: Relative O.D. at 405 nm. LPS of E. coli (Difco) was used as a standard.
c: Tachypleus tridentatus.

2) Essential Structural Component of Lipopolysaccharide Responsible for the Activities of Inhibition of Tumor Development, Interferon Induction and Activation of Proclotting Enzyme of Horseshoe Crab, Adjuvanticity and Pyrogenicity

a. Attempts have been made to determine which component of the lipopolysaccharide molecule is responsible for the various activities in question.

Lipid A from lipopolysaccharide derived from the complex (OEP) of P. aeruginosa was obtained with 2% acetic acid at 100°C, 2.5 hr treatment, and solubilized with triethylamine or complexed with bovine serum albumin (22). Polysaccharide was obtained from supernatant hydrolyzed for 2.5 hr after extraction of the lipoidic substances with chloroform.

As shown in Table 5, little or no antitumor activity was found in the lipid A preparations obtained from lipopolysaccharides derived from the complex (OEP) or from Re mutant of S. typhimurium when they were solubilized by triethylamine or complexed with bovine serum albumin. In contrast, all these preparations showed the same interferon-inducing activity as that of the original lipopolysaccharide.

As for adjuvant activity, these lipid A preparations showed almost the same values in adjuvanticity as the control material.

The free lipid A isolated from the complex demonstrated the same potency in pyrogenicity as the original lipopolysaccharide, whereas the activity of activating proclotting enzyme of horseshoe crab (Tachypleus tridentus) estimated by the two methods decreased markedly.

No activity could be demonstrated by these tests in the polysaccharide (protein and lipid-free) derived from lipopolysaccharide by acid hydrolyses.

b. In further experiments, we modified the polysaccharide portion of lipopolysaccharide by either $NaIO_4$, succinic anhydride or phthalic anhydride and tested for the biological activities in order to clarify the participation of the polysaccharide part in the activities. Oxidized lipopolysaccharide was obtained by treatment with 0.1 M $NaIO_4$ at 4°C in the dark for 4 days (pH 2.0). Succinylation and phthalylation were carried out by the method reported by Rietschel et al. (23). The release of ester-linked fatty acid was done by hydroxylaminolysis as described by Snyder and Stephens (24). Lipopolysaccharide derived from S. typhimurium SL 1102 (Re mutant) was used as a reference.

As shown in Table 5, all the treatments significantly reduced the activity of tumor development inhibition. However, interferon-inducing activity was not affected. Both the phthalylated lipopolysaccharide and the lipopolysaccharide oxidized with $NaIO_4$ preserved almost the

Table 5. Biological activities of lipopolysaccharides from P. aeruginosa and S. typhimurium (Re mutant) and their derivatives

Sample	Antitumor activity ED$_{50}$ µg kg^{-1} day^{-1}	Interferon-inducing activity M.E.D. µg/ml	Adjuvant index	Pyrogenicity M.E.D. µg/kg	Release of p-nitro-aniline	T.T. lysate gelation M.E.D. µg/ml
Protein-lipopoly-saccharide from autolysate of P. aeruginosa	0.1	0.1	1.47	0.3	0.06	10^{-1}
Lipopolysaccharide from the complex	0.007	0.01	1.71	0.3	1.35	10^{-4}
lipid A treated with (Et)$_3$N	>10	0.1	1.34	0.03		10^{-2}
BSA	>10	0.01	1.61	0.3	0.02	10^{-1}
oxidized with NaIO$_4$	>10	<0.01	1.50	3.0	0.005	10^{-1}
succinylated	0.9	0.01	1.24	3.0	0.002	10^{-1}
phthalylated	0.6	0.001	1.76	3.0	0.001	10^{-1}
NH$_2$OH	>10	0.01	1.28	3.0	0.01	10^{-1}
N$_2$H$_4$	>10	>1	1.11	>30	0	10^{-1}
0.1M NaOH in 99% EtOH	0.005	0.01	2.02			
polysaccharide	>10	>1	0.81	>30	0	10^{-1}
Lipopolysaccharide from S. typhimurium Re mutant	0.04	0.001		0.003	1.6	10^{-3} - 10^{-4}
lipid A - BSA	2.0	0.001				
phthalylated	0.85	0.001		0.003	0.02	10^{-2}

Data from reference (6).

same adjuvant activity, although the succinylated lipopolysaccharide was found to be less potent than the intact lipopolysaccharide. The above results indicate that the lipid A portion is required for adjuvanticity. Pyrogenicity remained unchanged after these treatments, although it was somewhat reduced in the case of the lipopolysaccharide of P. aeruginosa, while proclotting enzyme activation was drastically reduced by these treatments in the lipopolysaccharide of both P. aeruginosa and S. typhimurium SL 1102 (Re mutant). Thus lipid A is responsible for pyrogenicity, whereas lipid A by itself is insufficient for the manifestation of proclotting enzyme activation.

c. The above results indicate that both polysaccharide and lipid A are required for antitumor activity. To confirm this, the effect of deacylation was studied. When the lipopolysaccharide obtained from the protein-lipopolysaccharide complex (OEP) was treated with hydroxylamine resulting in the removal of ester-linked fatty acids (refer to Table 3), the activity for tumor inhibition was almost completely diminished, whereas the interferon-inducing activity fully remained. In contrast, neither of these activities was observed if all the fatty acids were removed from the lipopolysaccharide by either hydrazinolysis or treatment with 4 M NaOH at 100°C for 5 hr. If the sample was treated by 0.1 M NaOH in ethanol at 37°C for 0.5 hr, however, only about half of the fatty acids was removed and both activities still remained in this preparation (see Tables 3 and 5). These results confirm that both lipid A and the polysaccharide portion, such as 3-deoxy-D-manno-octulosonic acid (see the result of Re lipopolysaccharide, Table 5), are necessary for the enhancement of antitumor activity, whereas lipid A alone is sufficient for inducing interferon.

As for adjuvanticity, only the lipid A portion is required. When the lipopolysaccharide obtained from the complex (OEP) was treated by hydroxylamine or anhydrous hydrazine resulting in the removal of ester-linked fatty acids or all of the fatty acids, the adjuvant effect decreased markedly. If the sample was treated with mild alkaline solution (0.1 M NaOH), however, about half the fatty acids were present and the adjuvant activity was retained. According to these results, the lipid A portion with an adequate amount of ester-linked fatty acids as well as amide-linked fatty acids appears to be essential for adjuvant activity.

The participation of fatty acid in pyrogenicity and in the activation of the proclotting enzyme was also tested. When ester-linked fatty acid was removed, these activities were reduced, and when amide-linked fatty acid was further removed, all activities were completely lost. These findings indicate that fatty acids play an important role in these activities.

4 Summary and Discussion

Because lipid A is difficult to solubilize, the various test for biological activities were conducted using the same sample. The intensity of the biological activities were determined based on the relative activity of the results.

The lipid A portion with an adequate amount of ester-linked and amide-linked fatty acids is shown to be essential for demonstrating adjuvanticity, pyrogenicity and interferon-inducing activity. Incomplete lipid A containing only amide-linked fatty acids is sufficient to induce interferon in vitro.

These results also indicate that both lipid A and the saccharide portion are essential for antitumor activity and activation of proclotting enzyme of horseshoe crab in the case of pseudomonal lipopolysaccharide.

The fact that Re lipopolysaccharide enhances antitumor activity seems to indicate that 3-deoxy-D-manno-octulosonate-lipid A represents the minimal structure active in tumor inhibition.

Mihich et al. (25) reported that free lipid A, complexed with dextran as a carrier, showed regression of transplantable mouse tumors, and recently, Rietschel et al. (3) confirmed that free lipid A did not induce tumor necrosis, but even caused enhacement.

Amano et al. (26) reported that hydroxylaminolyses of the Re lipopolysaccharide led to complete de-O-acylation, and that the products showed complete loss of both antitumor activity and toxicity. Treatment with dilute sodium hydroxide caused partial removal of O-ester-linked fatty acids without loss of these activities. These observations coincide with our results in that at least a part of the fatty acids linked to hydroxyl groups of the disaccharide residue was required for antitumor activity.

However, contrary to our results, the data of Amano et al. indicated that the 3-deoxy-D-manno-octulosonate moiety was not required for antitumor activity. The discrepancy may be caused by different assay methods of tumor inhibition test using different tumor cells.

Our data presented here do not support the existence of a correlation between pyrogenicity and Limulus test on several points. Pyrogenicity of P. aeruginosa lipopolysaccharide was considerably weaker than the other lipopolysaccharides, whereas its activation of the proclotting enzyme was stronger. Lipid A of P. aeruginosa lipopolysaccharide was responsible for the pyrogenicity, but it showed weaker activity of proclotting enzyme activation than the original lipopolysaccharide. Further, when the lipopolysaccharide was chemically modified, the pyrogenicity of the samples remained but the proclotting enzyme activation was reduced or almost lost. As for proclotting enzyme activation, complete lipopolysaccharide itself was the strongest activator.

Accordingly, for manifesting the proclotting enzyme activity as well as antitumor activity, lipid A by itself is not sufficient; the polysaccharide portion is also essential.

The native-protein portion attached to the lipopolysaccharide of P. aeruginosa inhibits both activation of proclotting enzyme activity and antitumor activity of the lipopolysaccharide. Homma et al. reported previously that lipopolysaccharide of the endotoxin masked the activity of the protein portion, e.g., pyocin activity, by complexing the lipopolysaccharide with the protein portion (27,28).

The results of activation of proclotting enzyme using the synthetic substrate paralleled those of the gelation reaction. Accordingly, the method mentioned here is useful for the quantitative determination of lipopolysaccharide.

Our comparative studies of biological activities and structures have been carried out using chemically degraded or modified lipid A preparations as described in the text. However, the exact molecular basis of the structures required for specific biological activities remains to be elucidated, since chemically pure and intact lipid A has never been isolated. Natural lipid A's have always been isolated as a complex mixture of split products and, furthermore, contamination by other cell wall materials could never be excluded (3). The unequivocal identification of primary determinants of endotoxic activities is probably possible only by the biological testing of chemically synthesized substructures and derivatives of lipid A.

5 Reference

(1) O. Westphal, O. Lüderitz, Angew. Chem. 66 (1954) 407-417.

(2) O. Lüderitz, C. Galanos, U. Lehmann, H. Moyer, E. Th. Rietschel, J. Weckesser, Naturwissensch. 65 (1978) 578-585.

(3) E. Th. Rietschel, C. Galanos, O. Lüderitz, O. Westphal, in D. R. Webb (Ed.), Immunopharmacology and the regulation of leukocyte function, Marcel Dekker, New York 1982, pp. 183-229.

(4) K. Tanamoto, C. Abe, J. Y. Homma, Y. Kojima, Eur. J. Biochem. 97 (1979) 623-629.

(5) Y. Cho, K. Tanamoto, Y. Oh, J. Y. Homma, FEBS LETTERS 105 (1979) 120-122.

(6) K. Tanamoto, J. Y. Homma, J. Biochem. 91 (1982) 741-746.

(7) J. Y. Homma, C. Abe, Jpn. J. Exp. Med. 42 (1972) 23-34.

(8) K. Tanamoto, C. Abe, J. Y. Homma, K. Kuretani, A. Hoshi, Y. Kojima, J. Biochem. 83 (1978) 711-718.

(9) J. Y. Homma, C. Abe, K. Okada, K. Tanamoto, Y. Hirano, in Sawai et al. (Ed.), Animal, Plant and Microbial Toxins, Plenum Press, London 1976, pp. 499-508.

(10) C. Abe, K. Tanamoto, J. Y. Homma, Jpn. J. Exp. Med. 47 (1977) 393-402.

(11) Y. Hirao, J. Y. Homma, Infect. Immun. 19 (1978) 373-377.

(12) A. Yamamoto, J. Y. Homma, A. Goda, T. Ishiwara, S. Takeuchi, Jpn. J. Exp. Med. 49 (1979) 383-390.

(13) A. Hoshi, F. Kanzawa, K. Kuretani, J. Y. Homma, C. Abe, GANN 63 (1972) 503-504.

(14) Y. Kojima, J. Y. Homma, C. Abe, Jpn. J. Exp. Med. 41 (1971) 493-496.

(15) G. H. Fukui, J. Y. Homma, C. Abe, Jpn. J. Exp. Med. 41 (1971) 489-492.

(16) M. Sasaki, M. Ito, J. Y. Homma, Jpn. J. Exp. Med. 45 (1975) 335-343.

(17) Y. J. Cho, Y. H. Oh, C. Abe, J. Y. Homma, M. Usui, T. Matsuhashi, Jpn. J. Exp. Med. 48 (1978) 491-496.

(18) S. Iwanaga, T. Morita, T. Harada, S. Nakamura, M. Niwa, K. Takeda, T. Kimura, S. Sakakibara, in P. Brakman (Ed.) Haemostasis, S. Kargel, Basel 1978, pp. 183-188.

(19) S. Nakamura, S. Iwanaga, T. Harada, M. Niwa, J. Biochem. 80 (1976) 1011-1021.

(20) D. W. Watson, Y. B. Kim, J. Exp. Med. 118 (1963) 425-446.

(21) C. Galanos, O. Lüderitz, O. Westphal, Eur. J. Biochem. 9 (1969) 245-249.

(22) C. Galanos, E. Th. Rietschel, O. Lüderitz, O. Westphal, Y. B. Kim, D. W. Watson, Eur. J. Biochem. 31 (1972) 230-233.

(23) E. Th. Rietschel, C. Galanos, A. Tanaka, E. Rushmann, O. Lüderitz O. Westphal, Eur. J. Biochem. 22 (1971) 218-224.

(24) F. Snyder, N. Stephens, Biochim. Biophys. Acta 34 (1959) 244-245.

(25) E. Mihich, O. Westphal, O. Lüderitz, E. Neter, Proc. Soc. Exp. Biol. Med. 107 (1961) 816-819.

(26) K. Amano, E. Ribi, L. Cantrell, J. Biochem. 93 (1983) 1391-1399.

(27) J. Y. Homma, N. Suzuki, J. Bacteriol. 87 (1964) 630-640.

(28) J. Y. Homma, Z. Allg. Mikrobiol. 8 (1968) 227-248.

Biological Activities of Klebsiella 03 Lipopolysaccharide

Izumi Nakashima, Michio Ohta, Nobuo Kido, Yasuaki Fujii, Takashi Yokochi, Fumihiko Nagase, Takaaki Hasegawa, Masashi Mori, Ken-ichi Isobe, Kenji Mizoguchi, Mitsuru Saito, Nobuo Kato
Departments of Immunology and Bacteriology, Nagoya University School of Medicine, 65 Tsurumai-cho, Showa-ku, Nagoya, Aichi 466, Japan

Nobuhiko Kasai
Department of Microbiology, Showa University, School of Pharmaceutical Sciences, 1-5-8, Hatanodai, Shinagawa-ku, Tokyo 142, Japan

Abstract

　Results of current studies on a peculiar adjuvant activity of Klebsiella 03 lipopolysaccharide (KO3-LPS) extracted from the culture supernatant of Klebsiella strain Kasuya (03-K1) or its decapsulated mutant strain LEN-1 are reviewed. KO3-LPS, which was a complex of a mannan as a O-specific polysaccharide and lipid A, exhibited a strong adjuvant activity for triggering the immune response to otherwise non- or weakly immunogenic antigens such as deaggregated serum proteins and syngeneic tissue extracts, when injected into mice subcutaneously together with the antigens. LPS prepared from Escherichia coli O55 showed this activity only weakly, although this LPS shared many other activities with KO3-LPS to the same extents. The peculiar adjuvant activity of KO3-LPS required synergistic actions of the mannan and the lipid A of Klebsiella. One of the direct cellular target of KO3-LPS for the adjuvant action seemed to be the macrophage, which released much interleukin 1 after stimulation by KO3-LPS in vitro. KO3-LPS induced the neutrophil infiltration and the retardness of the lymph flow at the local site of injection of LPS, which would modify the immunogenic property of the antigen. A hypothetical model of interaction between the KO3-LPS and the macrophage cell surface was presented.

1 Introduction

Since 1971 we have reported that a polysaccharide-rich fraction termed CPS-K which is extracted from the culture supernatant of capsulated Klebsiella strain Kasuya (O3:K1) by the procedure of Batshon (1) originally for preparation of capsular polysaccharide exhibits a strong adjuvant activity on antibody responses to otherwise non- or weakly immunogenic antigens such as deaggregated serum proteins (2-6) and syngeneic tissue extracts (3,7). The lipopolysaccharide (LPS) extracted by the Westphal's method from Escherichia coli O55 or O111 bacterial cells shows such an activity only weakly (3,5,6).

Earlier studies provided an evidence that the active substance in CPS-K which is responsible for the strong adjuvant action is not the acidic type-specific capsular polysaccharide but can be the neutral polysaccharide (8). The neutral polysaccharide fraction of CPS-K prepared by Eriksen's method (9) released a small amount of lipids when hydrolyzed in 1N HCl (8). Later studies have however revealed that a noncapsulated mutant of the strain Kasuya (strain LEN-1) produced as much active substance as the original strain (10), and the active substance is serologically identical to the Klebsiella O3 antigen (11). The purified active substance from the culture supernatant of Klebsiella strain LEN-1 released a definite amount of lipid A under more suitable conditions (12). We have concluded from these results that the active substance in the original CPS-K for the strong adjuvant activity is Klebsiella O3 LPS (KO3-LPS) (12).

Our studies of the biological activities of KO3-LPS as compared with those of other LPS have indicated that LPSs of different bacterial sources show dissociated activities. Our data suggest that special chemical structures in the O-specific polysaccharide and lipid A moieties of KO3-LPS are responsible for the strong adjuvant activity to trigger the immune response, whereas the remaining activities of KO3-LPS depend fully on the structure of the lipid A which is common to various LPSs (12). Evidence has been presented that the strong adjuvant activity for triggering the immune response requires the mannan which should be organized on the appropriate lipid A. Here we briefly review our data about biological activities of KO3-LPS with special attention to its peculiar adjuvant activity.

```
  3      1,3      1,2      1,2      1,2     1
── Man─── Man─── Man─── Man─── Man───  ─ ─ ─  (Core)─ ─ ─ ─  Lipid A
         α        α        α        α       α
```

    Mannan consisting of
 a pentasaccharide repeating unit
          (O3 antigen)

Fig. 1. The chemical structure of KO3-LPS.

2 Chemical structures of KO3-LPS

KO3-LPS was extracted from the culture supernatant of Klebsiella strain Kasuya or its decapsulated mutant LEN-1 according to the procedure described elsewhere (1,2,8,9). Briefly, the LPS was precipitated by adding ethanol into the culture supernatant, dissolved and dialyzed agaist water and deprotenized by shaking with several portions of chroloform and butanol mixtures or by treating with hot phenol. For some experiments KO3-LPS was prepared from bacterial cells by Westphal's hot phenol technique. LPS was also extracted from other O groups of Klebsiella, Escherichia and Salmonella by either of the 2 methods described above, or was purchased from Difco Laboratories, Detroit, Michigan. Preliminary studies have shown that the method of preparation of LPS does not affect much the activity of LPS. The defined chemical structure of KO3-LPS is shown in Fig. 1. A highly purified KO3-LPS prepared from Klebsiella strain LEN-1 liberated a lipid A in 20 % when hydrolyzed in 1 % acetic acid (12). The lipid A of KO3-LPS was serologically cross-reactive to that of E. coli LPS (12). The chemical structure of the O-specific polysaccharide of KO3-LPS has been defined as a mannan consisting of a pentasaccharide repeating unit (13), which corresponds to the result of earlier investigators about LPS extracted from bacterial cells (14). The chemical structure of the lipid A moiety of KO3-LPS has been characterized (15). The backbone structure of the Klebsiella lipid A is similar to that of Salmonella minnesota R595 lipid A, but the contents of 4-amino-L-arabinose, ethanolamine and fatty acid components in the 2 lipid A are different.

3 Biological activities of KO3-LPS

Current studies to compare activities of KO3-LPS and LPS from E. coli O55 (EO55), E. coli O111 (EO111) or Salmonella enteritidis (SENT) have revealed the following. KO3-LPS exhibits an activity

Table 1. A list of class 1 activities of KO3-LPS (activities that are not stronger than those of EO55, EO111 or SENT LPS)

	Class 1 activities of KO3-LPS	Reference
In rabbits	Pyrogenicity	17
	Skin-preparatory property for dermal Shwartzman phenomenon	17
In mice	Lethal toxicity	17
	Activities for	
	Infection-promotion	21
	Interferon and cytotoxin production in BCG-infected mice	22
	Mitogenic activity for B lymphocyte in vitro	18
	Adjuvant actions on	
	Antibody response to aggregated protein antigens injected intravenously	19
	Antibody response to particulate antigens injected intraperitoneally	20

that is definitely stronger than that of other LPS for triggering the immune response (3,5,6,16) whereas KO3-LPS shares many conventional activities with other LPS. This has indicated that KO3-LPS displays 2 activities which should be mediated by different portions of the structure of LPS. We tentatively designated the latter activity as the class 1 and the former as the class 2. Examples of the 2 classes of activities of KO3-LPS are presented in Tables 1 and 2.

Many conventional activities of LPS are included in the class 1 (Table 1). The pyrogenicity and skin-preparatory property for the dermal Shwartzman phenomenon in rabbits of KO3-LPS are rather weaker than those of Salmonella LPS compared by minimal effective doses (17). In mice KO3-LPS and other LPS show similar levels of lethal toxicity (17), mitogenic activity for B lymphocytes in vitro (18) and conventional adjuvant activities for antibody response (19). The adjuvant activities of this category develop for antibody responses to particulate antigens such as sheep red blood cells (20) and aggregated forms of protein antigens (19) that are injected intraperitoneally or intravenously. KO3-LPS as well as other LPS also exhibit activities to promote bacterial infections in mice (21) and

Table 2. A list of class 2 activities of KO3-LPS (activities that are definitely stronger than those of EO55, EO111 or SENT LPS)

Class 2 activities of KO3-LPS	Reference
Adjuvant actions	
On antibody response to deaggregated protein antigens injected subcutaneously	3-6
For induction of delayed type hypersensitivity to deaggregated protein antigens injected subcutaneously	16,23
On antibody response to particulate antigens injected subcutaneously	20
Activities for	
Promoting antigen retention in the draining lymph node (retardation of lymph flow) (early time)	19
Induction of neutrophil infiltration in the draining lymph node (early time)	26
Induction of macrophage-like cells and IgG secreting plasma cells in the draining lymph node (late time)	26-27
Stimulation of macrophages in vitro for production of interleukin 1 activity	24

to induce the production of interferon and cytotoxin in mice infected with Mycobacterium bovis BCG (22).

Table 2 collects class 2 activities of KO3-LPS that are definitely stronger than those of other LPS. They include strong adjuvant activities on helper (6,19) and delayed type hypersensitivity (16,23) T cell responses to otherwise nonimmunogenic deaggregated protein antigens which are injected subcutaneously (this route is crucial for development of the adjuvant action (19)). An additional type of the class 2 adjuvant activity is the augmentation of antibody response to sheep red blood cells injected subcutaneously, which affects the late stage of the immune response and probably involves a mechanism other than that for T cell activation (a second stage of the class 2 adjuvant activity) (20). Other activities included in this Table possibly explain some of the mechanism of the class 2 adjuvant activities. The in vitro activity to stimulate macrophages for

Table 3. A list of successful induction of experimental autoimmune diseases by use of the strong class 2 adjuvant activity of KO3-LPS

Experimental autoimmune diseases	Reference
Ophthalmitis	28,37
Thyroiditis	7,28,29
Pancreatitis	30
Hepatitis	31
Orchitis	32
Hemolytic anemia	T. Kato, submitted
Carditis	N. Kato, unpublished

production of the interleukin 1 (24), and the in vivo activities to retard the lymph flow (25) and to induce a severe infiltration of neutrophils in the draining lymph node (26) at an early time after injection of LPS could be included in the mechanism of the initial T cell triggering. The activity to induce a marked proliferation of macrophage-like cells and IgG producing plasma cells in the lymph node at a later time (26,27) might relate to the second stage of the class 2 adjuvant activity.

4 The biological significance of the class 2 adjuvant activity of KO3-LPS

It should be stressed that the class 2 adjuvant activity probably plays a significant role in switching the direction of immune response, by changing otherwise nonimmunogenic antigens to be highly immunogenic. The use of KO3-LPS in SMA mice as a high responder to the class 2 activity has thus made us establish definite models of experimental autoimmune diseases which are severer than those ever induced by use of Freund's complete adjuvant (3,28,29). Table 3 illustrates examples of the successful induction of severe autoimmune diseases by immunization with the mixture of KO3-LPS and syngeneic tissue extracts (28-32). In addition, hybridomas secreting monoclonal autoantibodies to tissue-specific antigens have been established in this experimental system (32).

Table 4. Relationship between the class 2 adjuvant activity and the chemical structure of LPS

Lipid A	O-specific polysaccharide	Class 2 adjuvant activity
Klebsiella	KO1	±
	KO3 (mannan)	++++
	KO5 (mannan$^{(a)}$)	++++
E. coli	EO8 (mannan$^{(b)}$)	+++
	EO9 (mannan) = KO3	+++
	EO26	±
	EO55	±
	EO111	±
	EO127	±
	EO128	++
Salmonella	SENT	±
	Minnesota (no O-polysaccharide)	±

(a) Mannan consisting of a pentasaccharide ($-\text{M}\frac{3}{\alpha}\text{M}\frac{1,2}{\alpha}\text{M}\frac{1,3}{\alpha}\text{M}\frac{1,2}{\alpha}\text{M}\frac{1}{\alpha}-$) repeating unit.

(b) Mannan consisting of a trisaccharide ($-\text{M}\frac{3\ 1,2}{\alpha}\text{M}\frac{1,2}{\alpha}\text{M}\frac{1}{\alpha}-$) repeating unit.

5 Identification of the structures of KO3-LPS that are responsible for the class 2 adjuvant activity

We compared the levels of the class 2 adjuvant activity of various preparations of LPS from Klebsiella, Escherichia and Salmonella. The summary of the results is shown in Table 4. LPS from Klebsiella O3 or O5 or from E. coli O8 or O9, whose O-specific polysaccharides consist of mannans (33-35), showed a much stronger class 2 adjuvant activity than other LPS, although some LPS that had not been reported to carry mannans showed an intermediate level of the activity (13,16). This suggests that the mannan structure is closely related to the class 2 activity. In addition, LPS extracted from Klebsiella O3 or O5 showed a slightly stronger class 2 activity than that from E. coli O8 or O9 (13,16). This suggests that the class 2 activity also depends on the structure of the lipid A which is peculiar to Klebsiella. It can be that the physical property rather than the chemical structure of LPS primarily determines the level of

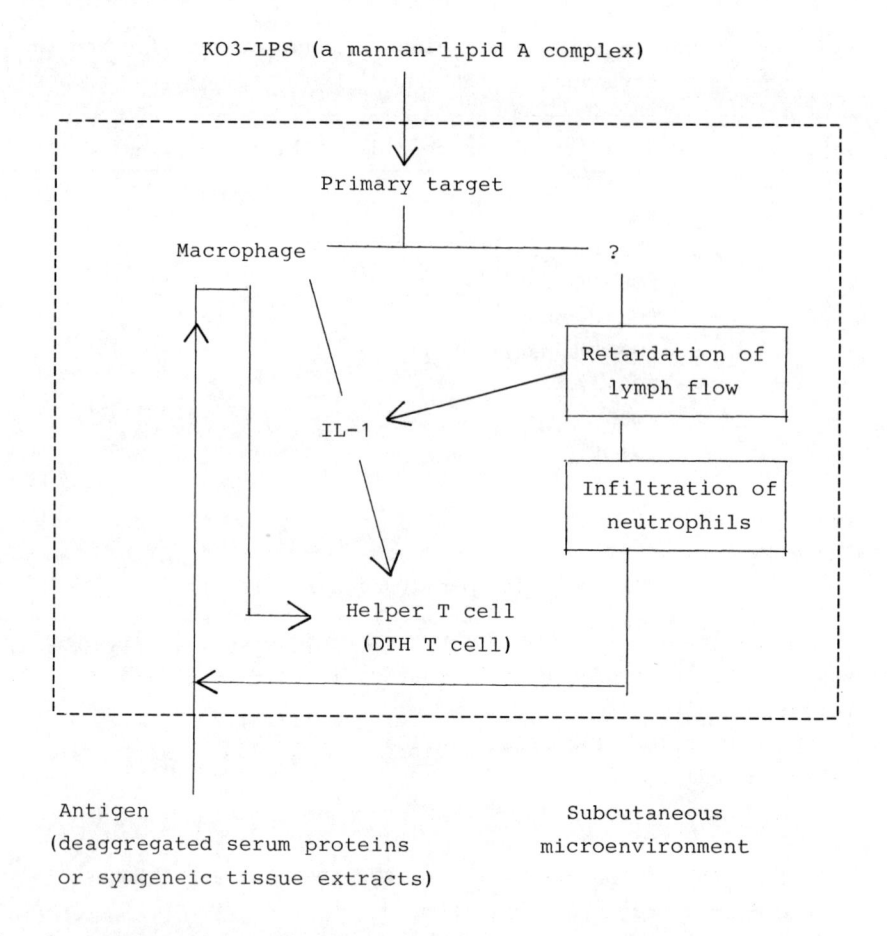

```
                    KO3-LPS (a mannan-lipid A complex)
                                    │
   ┌────────────────────────────────┼───────────────────────────────┐
   │                                ▼                               │
   │                         Primary target                         │
   │                                │                               │
   │         Macrophage ────────────┴──────────── ?                 │
   │              ▲  \                            │                 │
   │              │   \                  ┌────────┴────────┐        │
   │              │    \                 │  Retardation of │        │
   │              │     \                │   lymph flow    │        │
   │              │    IL-1 ◄────────────┤                 │        │
   │              │      \               └────────┬────────┘        │
   │              │       \                       │                 │
   │              │        \              ┌───────┴─────────┐       │
   │              │         \             │ Infiltration of │       │
   │              │          \            │   neutrophils   │       │
   │              │           \           └───────┬─────────┘       │
   │              │            ▼                  │                 │
   │              └──►  Helper T cell             │                 │
   │                     (DTH T cell)             │                 │
   │                                              │                 │
   └──────────────────┬───────────────────────────┘                 │
                      │
         Antigen                           Subcutaneous
   (deaggregated serum proteins          microenvironment
    or syngeneic tissue extracts)
```

Fig. 2. A tentative mechanism of the first stage of the class 2 adjuvant activity at the cellular level

the class 2 activity. Recent studies have ruled out this possibility; conjugation of different cations to the low or high active LPS after electrodialysis (36), which changed much the solubility of the LPS, did not affect seriously the class 2 adjuvant activity (unpublished). It appears therefore that the class 2 adjuvant activity depends on the special chemical structures of the polysaccharide and lipid A moieties of LPS. Further studies have revealed that neither the polysaccharide fraction alone nor the lipid A fraction alone which has been solubilized by conjugating bovine serum albumin does show a strong class 2 activity, suggesting that the O-specific polysaccharide (mannan) and lipid A moieties of LPS act synergistically to create the class 2 activity (12).

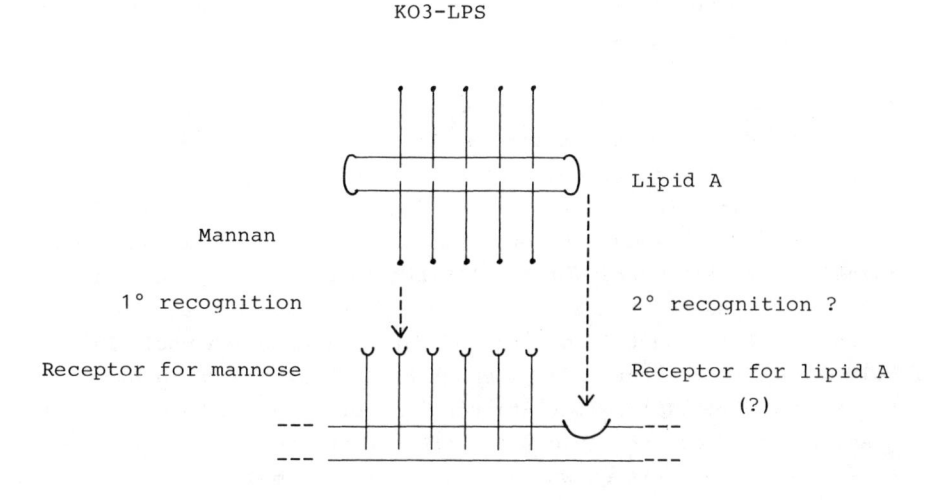

Fig. 3. A hypothetical model for interaction between KO3-LPS and the cell membrane of the macrophage

6 The mechanism of the class 2 adjuvant activity of KO3-LPS at the cellular level

What is the mechanism at the cellular level of the class 2 adjuvant activity of KO3-LPS ? The recent finding that the class 2 adjuvant activity correlates at least in part to the action of LPS to stimulate macrophages in vitro for producing interleukin 1 (24) could explain the main part of the mechanism of the class 2 adjuvant activity. Another action of KO3-LPS to retard the lymph flow (25) would retain the interleukin 1 produced at the local site to be effective for T cell activation in vivo. The KO3-LPS-induced neutrophil infiltration (26) and retardation of lymph flow retaining antigens at the local site (25) would modify the properties of the antigen to be more immunogenic. A tentative mechanism at the cellular level of the first stage of the class 2 adjuvant activity is shown in Fig. 3. It is not known in this model whether or not the macrophage is the only primary target of LPS for development of the class 2 activity. In addition, the first stage of the class 2 adjuvant activity should be followed by the second stage where LPS would further stimulate B cells for proliferation and differentiation (20,38,39) which had first been activated antigen-specifically at the initial stage.

7 The possible mechanism of the class 2 adjuvant activity of KO3-LPS at the molecular level

Much less is known about the mechanism of the class 2 adjuvant activity of KO3-LPS at the molecular level. Recent studies have demonstrated that the treatment of KO3-LPS with Con A which specifically binds to the mannose residue inactivates the class 2 but not the class 1 adjuvant activity of the LPS (unpublished). Our conclusion that the mannan of KO3-LPS should play an important role in the class 2 adjuvant activity remind us the report that the macrophage has a receptor for the mannose residue which could be involved in the structure of a lymphokine such as the macrophage activating factor (40). A tentative model for interaction between the mannan organized on the lipid A and the macrophage carrying the receptor for mannose is shown in Fig. 2. In this model the recognition of the mannan by the receptor for mannose might control the recognition of the lipid A by its undetermined receptor for generation of a signal to activate the cell. A promising aspect of our view is that the interaction between LPS and the macrophage would be a variant of membrane-membrane interaction among different mammalian cells which should be regulated by the recognition of the polysaccharide on one membrane by its receptor on another.

8 References

(1) B.A. Batshon, H. Baer, M. F. Shaffer, J. Immunol. 90 (1963) 121-126.

(2) I. Nakashima, T. Kobayashi, N. Kato, J. Immunol. 107 (1971) 1112-1121.

(3) I. Nakashima, J. Immunol. 108 (1972) 1009-1017.

(4) I. Nakashima, N. Kato, Japan. J. Microbiol. 19 (1975) 277-285.

(5) I. Nakashima, F. Nagase, T. Yokochi, M. Ohta, N. Kato, Cell. Immunol. 46 (1979) 69-76.

(6) I. Nakashima, F. Nagase, A. Matsuura, N. Kato, Cell. Immunol. 49 (1980) 360-371.

(7) I. Nakashima, N. Kato, Japan. J. Microbiol. 19 (1975) 13-18.

(8) I. Nakashima, N. Kato, Japan. J. Microbiol. 17 (1973) 461-471.

(9) J. Eriksen, Acta Pathol. Microbiol. 17 (1973) 461-471.

(10) M. Ohta, M. Mori, T. Hasegawa, F. Nagase, I. Nakashima, N. Naito, N.Kato, Microbiol. Immunol. 25 (1981) 939-948.

(11) N. Kato, M. Ohta, T. Hasegawa, M. Mori, K. Yamaki, K. Mizuta, I. Nakashima, Microbiol. Immunol. 25 (1981) 1317-1325.

(12) N. Kato, I. Nakashima, Japan. J. Bacteriol. 38 (1983) 70-71 (Abstract).

(13) N. Kato, M. Ohta, T. Hasegawa, M. Mori, I. Nakashima, F. Nagase, T. Yokochi, Y. Fujii, Immunomodulation by Microbial Products and Related Synthetic Compounds, Exerpta Medica, Amsterdam 1981, pp. 253-256.

(14) M. Curvall, B. Lindberg, J. Lönngren, W. Nimmich, Acta Chem. Scand. 27 (1973) 2645-2649.

(15) N. Kasai, K. Egawa, J. Mashita, M. Sasaki, M. Yoshida, N. Kato, I. Nakashima, Japan. J. Bacteriol. 38 (1983) 391 (Abstract).

(16) M. Ohta, I. Nakashima, N. Kato, Immunobiol. 163 (1982) 460-469.

(17) N. Kato, I. Nakashima, M. Ohta, Japan. J. Microbiol. 20 (1976) 173-181.

(18) I. Nakashima, F. Nagase, N. Kato, Z. Immun.-Forsch. 153 (1977) 204-216.

(19) I. Nakashima, F. Nagase, A. Matsuura, T. Yokochi, N. Kato, Cell. Immunol. 52 (1980) 429-437.

(20) T. Yokochi, I. Nakashima, F. Nagase, N. Kato, M. Ohta, Y. Fujii, K. Mizoguchi, K. Isobe, M. Saito, Microbiol. Immunol. 26 (1982) 843-852.

(21) N. Kato, O. Kato, I. Nakashima, Japan. J. Microbiol. 20 (1976) 163-172.

(22) N. Kato, I. Nakashima, M. Ohta, S. Naito, T. Kojima, Microbiol. Immunol. 23 (1979) 383-394.

(23) M. Ohta, I. Nakashima, N. Kato, Cell. Immunol. 66 (1982) 111-120.

(24) N. Kido, I. Nakashima, N. Kato, Japan. J. Bacteriol. 38 (1983) 225 (Abstract).

(25) I. Nakashima, A. Matsuura, F. Nagase, T. Yokochi, N. Kato, Cell. Immunol. 57 (1981) 477-485.

(26) Yokochi, T., I. Nakashima, N. Kato, J. Asai, S. Iijima, Microbiol. Immunol. 24 (1980) 933-944.

(27) T. Yokochi, I. Nakashima, N. Kato, Microbiol. Immunol. 24 (1980) 141-154.

(28) I. Nakashima, T. Yokochi, N. Kato, J. Asai, Microbiol. Immunol. 21 (1977) 279-288.

(29) T. Yokochi, I. Nakashima, N. Kato, J. Asai, Microbiol. Immunol. 22 (1978) 619-630.

(30) K. Yamaki, M. Ohta, I. Nakashima, A. Noda, J. Asai, N. Kato, Microbiol. Immunol. 24 (1980) 945-956.

(31) J. Kuriki, H. Murakami, S. Kakumu, N. Sakamoto, T. Yokochi, I. Nakashima, N. Kato, Gastroenterology 84 (1983) 596-603.

(32) Y. Fujii, I. Nakashima, N. Kato, Clin. Immunol. 14 (1982) 206-210.

(33) B. Lindberg, J. Lönngren, W. Nimmich, Acta Chem. Scand. 26 (1972) 2231-2236.

(34) P. Prehm, B. Jann, K. Jann, Eur. J. Biochem. 67 (1976) 53-56.

(35) K. Reske, K. Jann, Eur. J. Biochem. 31 (1972) 320-328.

(36) C. Galanos, O. Lüderitz, Eur. J. Biochem. 54 (1975) 603-610.

(37) I. Nakashima, F. Nagase, T. Yokochi, M. Ohta, Y. Fujii, N. Kato, Immunomodulation by Microbial Products and Related Synthetic Compounds, Excerpta Medica, Amsterdam 1981, pp. 257-260.

(38) I. Nakashima, N. Kato, Immunology 27 (1974) 179-193.

(39) I. Nakashima, T. Kojima, N. Kato, Immunology 30 (1976) 229-240.

(40) Y. Fukazawa, K. Kagaya, H. Miura, Microbiol. Immunol. 25 (1981) 1163-1172.

Lipopolysaccharides of the Family Vibrionaceae

Kazuhito Hisatsune, Seiichi Kondo, Takehiro Iguchi, Fumihiro Yamamoto, Makoto Inaguma, Shigeru Kokubo, and Shigeri Arai

Department of Microbiology, School of Pharmaceutical Sciences, Josai University, Sakado, Saitama 350-02, Japan

Abstract

A chemotaxonomical study was carried out on the sugar composition of lipopolysaccharides (LPS) isolated from representative strains of members of the family Vibrionaceae including all of the constituting genera, i.e., Vibrio, Aeromonas, Photobacterium, Plesiomonas and Lucibacterium. More than 100 strains were examined. It was found that, with the exception of Vibrio parahaemolyticus 06 and Plesiomonas shigelloides, 2-keto-3-deoxyoctonate (KDO), known generally as a component sugar in the core region of the usual gram-negative bacterial LPS, was virtually absent from LPS of the Vibrionaceae strains examined. Instead, some periodate-thiobarbituric acid test-positive KDO-like substances were found in LPS from V. parahaemolyticus 07 and 012, three strains of V. alginolyticus, Vibrio ("Beneckea") nereida (ATCC 25917) and some strains of Pl. shigelloides including the type strain (ATCC 14029).

1 Introduction

According to the eighth edition of Bergey's Manual of Determinative Bacteriology (1), the family Vibrionaceae consists of five genera: Vibrio, Aeromonas, Photobacterium, Plesiomonas and Lucibacterium. The

---

Abbreviations: LPS: Lipopolysaccharide; KDO: 2-keto-3-deoxyoctonate.

present paper reports a comparative study on the sugar composition of
O-antigenic lipopolysaccharides (LPS) isolated from representative
strains of the members of the family Vibrionaceae including all of the
constituting genera carried out from the chemotaxonomical point of view.
More than 100 strains were examined, many of which were type or neotype
strains or recommended as working strains in the eighth edition of
Bergey's Manual of Determinative Bacteriology. 2-Keto-3-deoxyoctonate
(KDO) is generally known as a component sugar of the usual gram-negative
bacterial LPS; its absence in LPS from the genus Vibrio was first
pointed out by Jackson and Redmond (2) and then by Jann et al. (3) with
Vibrio cholerae 569B (Inaba). In the present study, we demonstrated for
the first time that, with the exception of Vibrio parahaemolyticus O6
and genus Plesiomonas, KDO is virtually absent from LPS of all of the
Vibrionaceae strains examined. Instead, some periodate-thiobarbituric
acid-positive substances were found in LPS isolated from V. parahaemolyticus O7, O12, three strains of V. alginolyticus, Vibrio ("Beneckea")
nereida ATCC 25917 and some strains of Plesiomonas shigelloides
including the type strain ATCC 14029. The KDO-like substances yielded
a color with maximal absorption at 549 nm in the periodate-thiobarbituric acid test, and the spectra were identical to that of KDO.
However, they were not identical to KDO in behaviors, at least in
high-voltage paper-electrophoresis and thin-layer chromatography.

2 Materials and Methods

The sources and the growth conditions of microbes were described in
our preceding paper (4). LPS were isolated from the acetone-dried cells
by the phenol-water technique of Westphal et al. (5) and purified by
washing three times by ultracentrifugation then treatment with ribo-
nuclease, and finally by washing three more times by ultracentrifugation. Sugar analyses were carried out by the gas-liquid chromatographic and colorimetric techniques described in our previous paper (6).
Fructose was determined after hydrolysis in 0.2 M acetic acid at 100°C
for 8 hr by gas-liquid chromatography as O-acetylated-O-methyloxime
derivatives prepared according to Mawhinney et al. (7). KDO and KDO-like
substances were estimated by the method of Weissbach and Hurwitz, i.e.,
the periodate-thiobarbituric acid test (8). KDO was also identified by
gas-liquid chromatography as trimethylsilyl (TMS) derivatives (9) as
well as alditol acetate as described previously (4). The carbazol/H_2SO_4
test (10) was used for colorimetric assay for uronic acid. Authentic
D-glucuronic acid was used as a standard.

3 Absence of KDO

The sugar composition of LPS isolated from members of the family Vibrionaceae is presented in Table 1. Glucose, L-glycero-D-mannoheptose and glucosamine were found as common component sugars. It was particularly noted that mannose was not found in Vibrionaceae LPS with the exception of only one strain, Aeromonas hydrophila subsp. anaerogenes ATCC 15467. This is, however, presently not regarded as a member of the genus Aeromonas because of its (G+C) content, RNA homology and other taxonomical characteristics that are not consistent with those for this genus (unpublished data of T. Shimada and R. Sakazaki, National Institute of Health, Tokyo, Japan). Furthermore, rhamnose was rarely present; only eight strains contained this sugar in their LPS.

The most striking feature revealed for the sugar composition of Vibrionaceae LPS was that, with the exception of V. parahaemolyticus 06 and genus Plesiomonas, KDO, a regular sugar component in the core region of the usual gram-negative bacterial LPS, is virtually absent from LPS of all of the Vibrionaceae strains examined. In our preceding paper (4) we reported that KDO is lacking in LPS of Plesiomonas as in those of the other members of Vibrionaceae except V. parahaemolyticus 06, on the basis of the behavior in thin-layer chromatography of periodate-thiobarbituric acid test-positive substances released from their LPS by mild acid hydrolysis (1% acetic acid, 100°C, 2 hr). Rf values of these substances were slightly but definitely different from those of standard KDO. However, gas-liquid chromatographic analysis carefully performed on both TMS-derivatives and alditol acetates of the acidic components contained in the same hydrolysates revealed that significant amounts of KDO are definitely present in LPS isolated from at least five strains of Pl. shigelloides including the type strain ATCC 14029. Since the sugar composition of LPS of Pl. shigelloides has not hitherto been reported, analytical data obtained not only for KDO but also for neutral and amino sugar components of LPS of Pl. shigelloides are presented in Table 2. Although KDO was detected in the mild acid hydrolysates of Pl. shigelloides LPS by gas-liquid chromatography, precise quantitation of its content by this technique was difficult so that only preliminary data for it were presented. Thus, we came to the conclusion that, with the exception of V. parahaemolyticus 06 and Pl. shigelloides, KDO is lacking in LPS of all members of the family Vibrionaceae so far tested. It is known that, in the periodate-thiobarbiturate reaction of KDO, a positive reaction will be observed in cases in which the OH groups on C-4 and C-5 are free and in which the OH group on C-5 is substituted, but those on C-4, C-6 and C-7 are

Table 1. Sugar composition of lipopolysaccharides isolated from Family Vibrionaceae.

Strain	Glc	Gal	Man	Fru	Rha	Ara	Fuc	L-D Hep	D-D Hep	KDO	KDO-like	Uronic acid	GlcN	GalN	ManN	FucN	QuiN	PerN	3-N-3,6-dideoxy Glc	3-N-3,6-dideoxy Gal	4-N-4,6-dideoxy Glc	Unknown amino sugar Y
Vibrio																						
1. V. cholerae																						
1a. biotype classical																						
Inaba 35A3	+	–	–	+	–	–	–	+	–	–	–	–	+	–	–	–	+	+	–	–	–	–
569B	+	–	–	+	–	–	–	+	–	–	–	–	+	–	–	–	+	+	–	–	–	–
P1418-UV601	+	–	–	+	–	–	–	+	–	–	–	–	+	–	–	–	+	+	–	–	–	–
Ogawa NIH 41	+	–	–	+	–	–	–	+	–	–	–	–	+	–	–	–	+	+	–	–	–	–
NIH 90	+	–	–	+	–	–	–	+	–	–	–	–	+	–	–	–	+	+	–	–	–	–
P 1418	+	–	–	+	–	–	–	+	–	–	–	–	+	–	–	–	+	+	–	–	–	–
1b. biotype eltor PE1	+	–	–	+	–	–	–	+	–	–	–	–	+	–	–	–	+	+	–	–	–	–
SLH 22	+	–	–	+	–	–	–	+	–	–	–	–	+	–	–	–	+	+	–	–	–	–
Ubon 13	+	+	–	+	–	–	–	+	+	–	–	–	+	–	–	–	+	+	–	–	–	+
ATCC 14033	+	+	–	+	–	–	–	+	+	–	–	–	+	–	–	–	–	+	–	–	–	+
1c. NAG 03 4715	+	–	–	+	–	–	–	+	–	–	–	–	+	–	–	–	–	–	–	–	–	–
722-75	+	–	–	+	–	–	–	+	–	–	–	–	+	–	–	–	–	–	–	–	–	–
05	+	–	–	+	–	–	–	+	–	–	–	–	+	–	–	–	–	–	–	–	–	–
07	+	–	–	+	–	–	–	+	–	–	–	–	+	–	–	–	–	–	–	–	–	–
08	+	–	–	+	–	–	–	+	–	–	–	–	+	–	–	–	–	–	–	–	–	–
1d. R-mutant 35A3-R	+	+	–	+	–	–	–	+	–	–	–	–	+	–	–	–	–	–	–	–	–	–
NIH 41-R	+	–	–	+	–	–	–	+	–	–	–	–	+	–	–	–	–	–	–	–	–	–
CA385	+	–	–	+	–	–	–	+	–	–	–	–	+	–	–	–	–	–	–	–	–	–
4715-R	+	+	–	+	–	–	–	+	–	–	–	–	+	–	–	–	–	–	–	–	–	–
05-R	+	–	–	+	–	–	–	+	–	–	–	–	+	–	–	–	–	–	–	–	–	–
07-R	+	–	–	+	–	–	–	+	–	–	–	–	+	–	–	–	–	–	–	–	–	–
08-R	+	–	–	+	–	–	–	+	–	–	–	–	+	–	–	–	–	–	–	–	–	–

Vibrionaceae Lipopolysaccharides

1e. biotype proteus
    V. proteus NCTC 8563
    V. metschnikovii NCTC 8443
1f. biotype albensis
    V. albensis NCMB 41
2. V. parahaemolyticus
    O1
    O2
    O3
    O4
    O5
    O6
    O7
    O8
    O9
    O10
    O11
    O12
3. V. alginolyticus
    505-78
    905-78
    1013-79
    53-79
    1027-79
4. V. fisheri NCMB 1143
5. V. costicola NCMB 701
6. Beneckea ('Vibrio')
    V. nereida ATCC 25917
    V. campbelli ATCC 25920
    V. natrigens ATCC 14018
7. F group vibrio
    (V. fluvialis) 308-77
                   210-73
                   305-77
                   306-77
8. Lactose positive vibrio
    (V. vulnificus) ATCC 27562
                    C 4123

191

Chemistry and Biology of Some LPS

	Glc	Gal	Man	Fru	Rha	Ara	Fuc	L-D Hep	D-D Hep	KDO	KDO-like	Uronic acid	GlcN	GalN	ManN	FucN	QuiN	PerN	3-N-3,6-dideoxy Glc	3-N-3,6-dideoxy Gal	4-N-4,6-dideoxy Glc	Unknown amino sugar Y
9. V. anguillarum NCMB 6	+	–	–	–	+	–	–	+	–	–	–	–	+	–	+	–	–	–	–	–	–	–
NCMB 828	+	–	–	–	+	–	–	+	–	–	–	–	+	–	+	–	–	–	–	–	–	–
Aeromonas																						
1. A. hydrophila																						
1a. subsp. hydrophila ATCC 7966	+	–	–	–	–	–	–	+	–	–	–	–	+	–	–	–	–	–	–	–	–	–
1b. subsp. anaerogenes ATCC 15467	+	–	+	–	–	–	–	+	–	–	–	–	+	+	–	–	–	–	–	–	–	–
1c. subsp. proteolytica ATCC 15338	+	+	–	–	–	–	–	+	–	–	–	–	+	+	–	–	–	–	–	–	–	–
2. A. punctata																						
2a. subsp. punctata NCMB 74	+	–	–	–	+	–	–	+	–	–	–	–	+	–	–	–	–	–	–	–	–	–
2b. subsp. caviae ATCC 15468	+	–	–	–	+	–	–	+	–	–	–	–	+	–	–	–	–	–	–	–	–	–
3. A. salmonicida																						
3a. subsp. salmonicida NCMB 1102	+	+	–	–	+	–	–	+	–	–	–	–	+	+	+	–	–	–	–	–	–	–
3b. subsp. achromogenes NCMB 1110	+	+	–	–	–	–	–	+	–	–	–	–	+	+	–	–	–	–	–	–	–	–
3c. subsp. masoucida ATCC 27013	+	+	–	–	+	–	–	+	–	–	–	–	+	+	+	–	–	–	–	–	–	–
A. sobria CIP 202	+	–	–	–	–	–	–	+	+	–	–	–	+	–	–	+	–	–	+	–	–	–
CIP 210	+	–	–	–	–	–	–	+	+	–	–	–	+	–	–	–	–	–	–	–	–	–
CIP 208	+	+	–	–	–	–	–	+	–	–	–	–	+	–	–	+	–	–	+	–	–	–
CIP 217	+	+	–	–	+	–	–	+	–	–	–	–	+	–	–	–	–	–	+	–	–	–

Lucibacterium
  1. V. harveyi ATCC 14126
Photobacterium
  1. P. phosphoreum NCMB 844
  2. P. mandapamensis NCMB 391
Plesiomonas
  1. Pl. shigelloides
     ATCC 14029
     O11
     O17
     O22
     O23

Abbreviations: Glc: glucose; Gal: galactose; Man: mannose; Fru: fructose; Rha: rhamnose; Ara: arabinose; Fuc: fucose; L-D Hep: L-glycero-D-mannoheptose; D-D Hep: D-glycero-D-mannoheptose; KDO: 2-keto-3-deoxy-octonate; GlcN: glucosamine; GalN: galactosamine; ManN: mannosamine; FucN: fucosamine; QuiN: quinovosamine; PerN: perosamine; 3-N-3,6-dideoxy glc: 3-amino-3,6-dideoxy glucose; 3-N-3,6-dideoxy gal: 3-amino-3,6-dideoxy galactose; 4-N-4,6-dideoxy glc: 4-amino-4,6-dideoxy glucose; KDO-like: KDO-like substance. ?: not confirmed by thin-layer chromatography and high-voltage paper electrophoresis.

Table 2. Sugar composition of lipopolysaccharides isolated from Plesiomonas shigelloides (%, w/w).

Strain	Glc	Gal	L-D Hep	KDO[a]	KDO-like	KDO + KDO-like[b]	Uronic acid	GlcN	GalN	Unknown amino sugar
Pl. shigelloides										
ATCC 14029	1.7	3.3	4.7	1.2	+	5.6	4.6	7.7	-	-
011	3.0	3.9	5.6	1.8	?	5.8	3.9	6.3	6.2	-
017	2.7	3.2	4.8	1.4	?	5.9	3.8	6.5	-	-
022	3.3	3.1	5.8	1.5	?	4.2	4.5	4.9	6.1	-
023	2.9	3.0	6.1	1.3	+	5.9	3.7	10.0	6.5	(Ps1 / Ps2)

Abbreviations: See the footnote to Table 1. a) Estimated by gas-liquid chromatography. b) Estimated as KDO by colorimetric method.

free (11). However, C-4 and/or C-5 O-substituted KDO will give a positive reaction in the semicarbazide reaction for α-keto acid. As far as we have determined so far, no positive reaction has been obtained in this semicarbazide test with mild-acid hydrolysates of LPS except those from V. parahaemolyticus 06 and genus Plesiomonas.

4 Chemotaxonomy of Vibrio cholerae based on the sugar composition of LPS

A chemotaxonomically interesting feature was also disclosed in the sugar composition of Vibrio cholerae LPS. Fructose was present exclusively in LPS of V. cholerae (both 01 and non-01 groups and both classical and eltor biotypes) with the exception of one strain, Photobacterium phosphoreum NCMB 844. Furthermore, a pair of rarely occurring amino sugars, perosamine and quinovosamine, was found only in LPS from 01 group V. cholerae regardless of either the biotype (classical or eltor) or the serotype (Inaba or Ogawa), whereas this pair is not present in non-01 group V. cholerae (the so-called NAG vibrios). This feature was confirmed with LPS from more than 30 strains of 01 group V. cholerae isolated from patients. In contrast, neither perosamine nor quinovosamine was present in LPS isolated from R mutants of 01 V. cholerae regardless of the biotype (classical or eltor) or the serotype (Inaba or Ogawa), whereas they contain fructose. Based on the quantitative analytical data on the sugar composition of LPS presented

here, the following characteristics were found for V. cholerae.
Within the 01 group of V. cholerae, the classical biotype is
distinguished from biotype eltor by the fructose content of LPS; it is
more than 3% in the case of the classical biotype, but less than 1.5%
in the case of eltor. Thus, we demonstrated that 01 classical V.
cholerae, 01 eltor V. cholerae and non-01 group (the so-called NAG
vibrios) are distinguished from each other on the basis of the pattern
of the sugar composition of their LPS. It should be kept in mind,
however, that the pair perosamine and quinovosamine is not present in
LPS isolated from R mutant strains of 01 group V. cholerae (12).

5 Occurrence of Uronic Acid

The occurrence of uronic acid as a sugar constituent of LPS in
Vibrionaceae was demonstrated for the first time. Of the five genera
constituting the family Vibrionaceae, Plesiomonas, Photobacterium and
Lucibacterium contained uronic acid in LPS of all of their constituting
members examined, while it was totally lacking in the Aeromonas LPS
tested. Only the members of genus Vibrio were found to be divided into
uronic acid-containing and -lacking groups; V. parahaemolyticus, V.
alginolyticus, V. fisheri, V. costicola, Vibrio ("Beneckea") and V.
fluvialis belonged to the uronic acid-containing group, while all four
biotypes of V. cholerae regardless of their serotypes, V. vulnificus
and V. anguillarum belonged to the uronic acid-lacking group. The
uronic acid residues were found to be present in LPS in two forms, one
in 1% acetic acid-labile linkage and the other in acetic acid-stable
linkage. The uronic acid residue released from V. parahaemolyticus 01,
04, 07, 010 and 012 LPS by heating in 5% acetic acid at 100°C for 2 hr
was identified as galacturonic acid; in particular, that from 012 LPS
was characterized as D-galacturonic acid.

6 KDO-like Substances

As mentioned above, the KDO-like substances were found in LPS of
several members of Vibrionaceae. These components were strongly
positive in the periodate-thiobarbituric acid test, yielding a color
with maximal absorption at 549 nm, that is characteristic for alkali-
labile β-formyl pyruvate chromogen (8). The spectra were identical to
that of standard KDO as illustrated in Fig. 1. However, on the basis of
the data presented in Table 3, these KDO-like substances are not
identical to KDO. These KDO-like substances have already been isolated

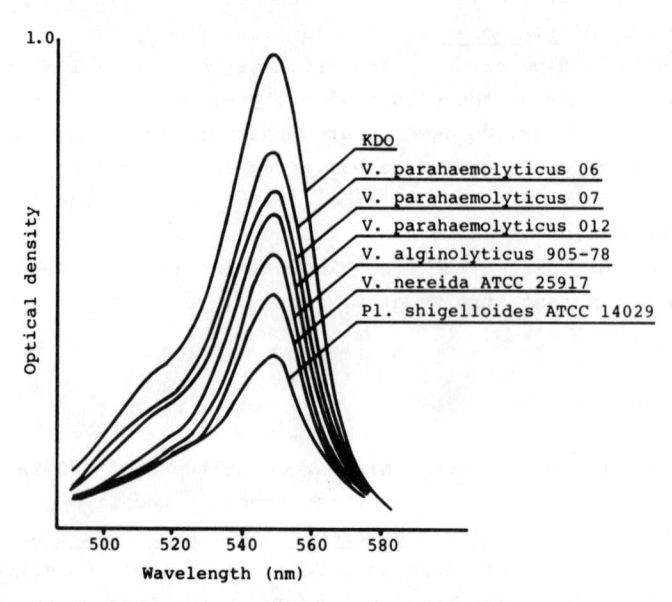

Fig. 1. Absorption spectra of color produced by KDO and KDO-like substances present in LPS isolated from Vibrionaceae strains in periodate-thiobarbituric acid test.

Table 3. Behavior of KDO and KDO-like substances in high-voltage paper electrophoresis and thin-layer chromatography.

	High-voltage[a] paper electrophoresis	Thin-layer chromatography[b] solvent system A	B
KDO	1.00	1.00	1.00
KDO present in V. para-haemolyticus 06 LPS	1.00	1.00	1.00
KDO-like substances present in LPS from			
V. parahaemolyticus 07	1.21	1.31	nt[c]
012	1.21	1.58	1.49
V. alginolyticus 905-78	1.21	1.17	nt
V. nereida ATCC 25917	0.81	0.68	nt
Pl. shigelloides ATCC 14029	0.98	0.81	0.93
011	0.98	0.70	0.95
017	0.97	0.75	0.93
022	0.97	0.84	0.69
023	0.98	0.79	0.63

a) Electrophoresis was performed with the solvent system; pyridine/acetic acid/water, 10:4:86, v/v/v, at 45 V/cm for 40 min. b) Thin-layer chromatography was performed on Avicel plate (Funakoshi Chemical Co., Tokyo) with solvent system A: pyridine/ethyl acetate/acetic acid/water, 36:36:7:21, v/v/v/v, B: butanol/acetic acid/water, 6:2:2, v/v/v.
c) nt: not tested.

from these various sources and their chemical characterization is going on in our laboratory.

7 Elution Profile of Degraded Polysaccharide in Gel Filtration

In general, the polysaccharide portion ("degraded polysaccharide") of LPS is released by heating in dilute acetic acid (1% acetic acid, 100°C, 1.5 - 2 hr). Typically, the elution profile of degraded polysaccharide of the usual gram-negative bacterial LPS in gel filtration on a Sephadex G-50 column shows three distinct peaks as shown in Fig. 2A. The first peak (Frc. I), eluted in void volume, represents a high-molecular-weight component corresponding to "whole polysaccharide" (i.e., polymeric O-specific side chain with attached core stubs). The second peak (Frc. III) represents an intermediate fraction (core polysaccharide) and the third peak (Frc. IV), eluted almost in bed volume, represents a low-molecular-weight fragment (chiefly KDO and orthophosphate). Fig. 2B presents the elution profile for degraded polysaccharide of 01 V. cholerae LPS. In this case, three peaks were also eluted; the second and third peaks in the same volumes as those of the corresponding peaks in the case of Salmonella typhimurium and Escherichia coli degraded polysaccharides, whereas the first peak was not eluted in the void volume but in the imbibed volume considerably later than the former. The third peak (Frc. IV) contained, in addition to orthophosphate, free fructose instead of KDO. Of particular interest was that the degraded polysaccharides of LPS isolated from V. parahaemolyticus 01, 04, 06, 07, 010 and 012, V. alginolyticus, V. nereida, the so-called lactose positive vibrio V. vulnificus, and Aeromonas punctata yielded only two peaks (Fig. 2C), corresponding to the second (Frc. III) and third (Frc. IV) peaks, respectively, observed for S. typhimurium, E. coli and 01 V. cholerae. No peak was observed that represents a high-molecular-weight component (i.e., polymeric O-specific side chain with attached core). If the results of this degradation study can be interpreted in the same way as those of similar experiments of Müller-Seitz et al. (13) or Fensom and Meadow (14), the first peak (Frc. III) corresponds to core oligosaccharide and the second peak (Frc. IV) to low-melocular-weight fragments (i.e., monosaccharide and orthophosphate). Moreover, this elution pattern was found to be identical to that of the degraded polysaccharide isolated from LPS of R-mutants of 01 V. cholerae as seen in Fig. 2C. In general, LPS isolated from R mutants of gram-negative bacteria lacks an O-specific side chain composed of repeating units of oligosaccharide. Since these 11 vibrios and A. punctata are not rough strains, but are

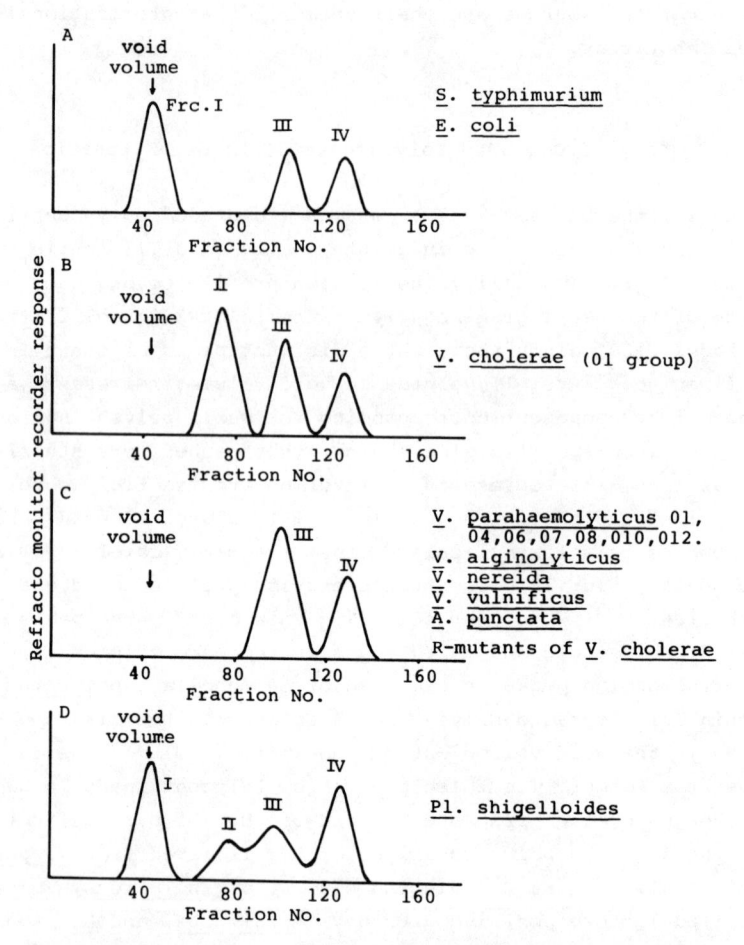

Fig. 2. Sephadex G-50 gel-filtration profiles of "degraded polysaccharide" derived from Vibrionaceae LPS.

definitely known to possess serologically O-specificity, the results of this degradation study are compatible with the interpretation that the O-specificities of these vibrios are associated with the structure and composition of the core region of their LPS molecules and/or their O-specific side chains are very short. This may be also considered to be one of the characteristics of Vibrionaceae LPS. Our results are consistent with those observed by Shaw's group with A. hydrophila (15).

The elution profile shown in Fig. 2D was observed for Pl. shigelloides degraded polysaccharide; it was quite different from those for the degraded polysaccharide of 01 V. cholerae, these vibrios and Aeromonas.

8 V. parahaemolyticus O6 LPS

V. parahaemolyticus O6 was found to be the only strain that contains KDO in its LPS among the Vibrionaceae except for genus Plesiomonas. It remains to be determined whether or not the KDO residue present in this strain is located between the heptose residues of the core polysaccharide and lipid A, connecting these two portions as in the usual gram-negative bacterial LPS, or in a different portion; for instance, in the O-antigenic polysaccharide region as an immunodominant sugar as in the case of the KDO residue found in Shigella sonnei Phase II LPS.

Fig. 3 presents the partial structure of the hexose region of Sh. sonnei Phase II LPS bearing KDO as a possible immunodeterminant, proposed by Kontrohr (16). One piece of evidence supporting this structural concept is in the passive hemolysis-inhibition test, where Salmonella minnesota R595 glycolipid exerted considerable inhibition in the [sonnei Phase II LPS - sensitized SRBC / anti-sonnei Phase II antiserum] passive hemolysis system. As shown in Fig. 3, this glycolipid contains KDO as the sole terminal residue in the "polysaccharide moiety". Passive hemolysis inhibition analysis was carried out on these two LPS and V. parahaemolyticus O6 LPS.

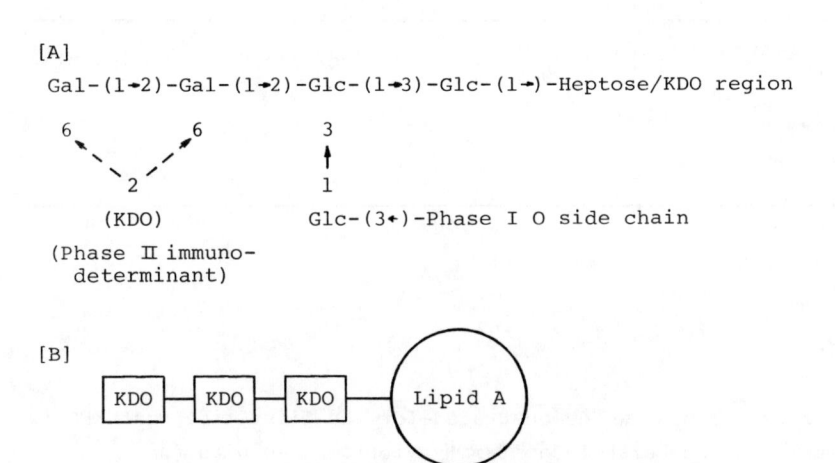

[A]
Gal-(1→2)-Gal-(1→2)-Glc-(1→3)-Glc-(1→)-Heptose/KDO region

```
  6       6           3
   \     /            ↑
    \   /             1
     \ /
      2
    (KDO)        Glc-(3←)-Phase I O side chain
 (Phase II immuno-
   determinant)
```

[B]
KDO — KDO — KDO — (Lipid A)

Fig. 3 Structure of the hexose region of Sh. sonnei Phase II lipopolysaccharide with 2-keto-3-deoxyoctonate (KDO) as possible immunodeterminant (A) and S minnesota R595 (B).

Sheep blood cells were sensitized with LPS and homologous passive hemolysis system consisting of LPS-sensitized SRBC, guinea pig complement and homologous antiserum was inhibited by homologous and heterologous LPS (17). As seen in Table 4, there was naturally almost complete inhibition in the homologous system, and, furthermore, a significant extent of inhibition was definitely observed in the heterologous system; 50% inhibition dose values (27 and 57 µg/ml) of S. minnesota R595 glycolipid for both heterologous systems were compatible with those (98 µg/ml) observed in Kontrohr's experiment for the Phase II / anti-Phase II system (16). In analogy with Kontrohr's study on Sh. sonnei Phase II LPS, this serological cross-reactivity between these three LPS may be an indication of the possibility that KDO is located in V. parahaemolyticus 06 LPS molecule as an immunodeterminant sugar on the non-reducing end of the polysaccharide portion, regardless of whether it is also present between the outer core and lipid A.

Table 4. 50% inhibition doses of LPS isolated from V. parahaemolyticus 06, Sh. sonnei Phase II and S. minnesota R595 for passive hemolysis system.

Inhibitor	Passive hemolysis system (antigen / anti serum)		
	06 LPS/ anti 06	Phase II LPS/ anti Phase II	R595 LPS/ anti R595
V. parahaemolyticus 06 LPS	0.40 (µg/ml)	82 (µg/ml)	-[a] (µg/ml)
Sh. sonnei Phase II LPS	130	0.07	-
S. minnesota R595 LPS	27	57	2.9

a) > 1000 µg/ml.

9 Conclusion

In the present study, we demonstrated for the first time that KDO is absent from LPS of members of the family Vibrionaceae with the exception of V. parahaemolyticus 06 and Plesiomonas. This result indicates that the structure of the core region, in particular the inner core region, of LPS of this family might be fundamentally different from that of the usual gram-negative bacteria. Our results are also compatible with the interpretation that the virtual absence of KDO in LPS can be one of the taxonomical characteristics of Vibrionaceae, in addition to those such as (G+C) content, DNA (or RNA) homology, numerical analysis data and other biological and biochemical properties.

10 References

(1) R. E. Buchanan, N. E. Gibbons (Ed.), Bergey's Manual of Determinative Bacteriology, 8th ed., The Williams and Wilkins Co., Baltimore 1965.

(2) G. D. F. Jackson, J. W. Redmond, FEBS Lett. 13 (1971) 117-120.

(3) B. Jann, K. Jann, G. O. Beyaert, Eur. J. Biochem. 37 (1973) 581-584.

(4) K. Hisatsune, S. Kondo, T. Iguchi, M. Machida, S. Asou, M. Inaguma, F. Yamamoto, Microbiol. Immunol. 26 (1982) 649-664.

(5) O. Westphal, O. Lüderitz, R. Bister, Z. Naturforsch. 7b (1952) 148-155.

(6) K. Hisatsune, A. Kiuye, S. Kondo, Microbiol. Immunol. 25 (1981) 127-136.

(7) T. Mawhinney, M. Feather, R. R. Martinetz, G. J. Barbero, Carbohydr. Res. 75 (1979) C21-C23.

(8) A. Weissbach, J. Hurwitz, J. Biol. Chem. 234 (1959) 705-709.

(9) D. T. Williams, M. B. Perry, Can. J. Biochem. 47 (1969) 691-695.

(10) T. Bitter, H. M. Muir, Anal. Biochem. 4 (1962) 330-334.

(11) F. M. Unger, Adv. Carbohydr. Biochem. 38 (1981) 323-388.

(12) K. Hisatsune, S. Kondo, Biochem. J. 185 (1981) 77-81.

(13) E. Müller-Seitz, B. Jann, K. Jann, FEBS Lett. 1 (1968) 311-314.

(14) A. Fensom, P. M. Meadow, FEBS Lett. 9 (1970) 81-84.

(15) D. H. Shaw, H. J. Hodder, Can. J. Microbiol. 24 (1978) 864-868.

(16) T. Kontrohr, B. Kocsis, Eur. J. Biochem. 88 (1978) 267-273.

(17) K. Hisatsune, S. Kondo, K. Kobayashi, Japn. J. Med. Sci. Biol. 31 (1978) 181-184.

Action of Endotoxin on Tumors and Related Aspects

Antitumor Effects of Endotoxin: Possible Mechanisms of Action

Herbert F. Oettgen, Lloyd J. Old, Michael K. Hoffmann, Malcolm A. S. Moore

Memorial Sloan-Kettering Cancer Center
1275 York Avenue, New York, New York 10021, USA

Abstract

Bacterial preparations containing endotoxin were first reported 100 years ago to cause regression of human cancer. Therapeutic activity has been confirmed in recent controlled trials in patients with leukemias and lymphomas. Research originating from these observations has resulted in the identification of three endogenous mediators of possible relevance to the antitumor effect of endotoxin - tumor necrosis factor, interleukin 1 and granulocyte macrophage differentiation factor. Considering the advances in recombinant DNA technology, large scale production of these mediators for clinical study is a realistic prospect.

1 Introduction

Bacterial preparations containing endotoxin were first used in the treatment of cancer by William Coley (1). Following his observation of tumor regression in human patients after acute bacterial infections, he treated cancer patients with bacterial vaccines. Those who have scrutinized Coley's records have little doubt that the bacterial preparations that came to be known as Coley's toxin were in some instances highly effective. This form of therapy later fell into disuse, no doubt because of high hopes raised by the introduction of radiation therapy, and later chemotherapy, which were more predictable and comprehensible, and whose mechanisms seemed more easily amenable to scientific inquiry. Interest was kept alive in the laboratory, especially by Shear who isolated a polysaccharide now known to be endotoxin from S. marcescens, one of the components of Coley's toxin, and showed that it caused necrosis of tumors in mice (2).

The hemorrhagic necrosis induced in certain experimental tumors by endotoxin derived from the cell wall of Gram negative bacteria is one of the most striking phenomena of tumor biology. Within a few hours of intravenous endotoxin injection the tumor mass begins to undergo a progressive darkening in color indicative of tumor cell death and hemorrhage, leading in many instances to complete tumor regression. This reaction has long been viewed as the experimental counterpart of clinical observations of tumor regression in humans following acute bacterial infection or administration of mixed bacterial vaccines. Although extensively investigated over the past 40 years, the way endotoxin causes tumor destruction is not known.

2 Therapeutic effects of bacterial preparations containing endotoxin in modern clinical trials

Taking a new look at the possible usefulness of endotoxin in the treatment of human cancer some 10 years ago, we attempted first to determine if endotoxin was effective in controlled therapeutic trials. In the first of these trials, patients with acute myeloblastic leukemia were randomized into two groups. One group received a vaccine prepared from Pseudomonas aeruginosa containing large amounts of endotoxin (3), the other group did not. Both groups were treated according to the same chemotherapy protocol, the L-6 protocol. In this protocol, remissions were induced and consolidated with arabinosylcytosine and thioguanine, and then maintained with vincristine, Methotrexate, BCNU, thioguanine, Cytoxan, hydroxyurea and Daunorubicin (4). The results are shown in Figure 1. While the frequency and severity of Pseudomonas infections was not affected by the vaccine, the duration of remission from leukemia was much longer in the vaccinated group. Six of 13 patients who achieved remission in the vaccine group are still in complete remission. Of the 17 patients achieving remission in the group that was not treated with the vaccine, only 1 is still in remission.

The second therapeutic trial involved patients with advanced non-Hodgkin lymphoma including nodular poorly differentiated lymphoma, nodular mixed lymphoma and nodular histiocytic lymphoma. The patients were randomized to receive or not receive a mixed bacterial vaccine (MBV) containing heat-killed Streptococcus pyogenes and Serratia marcescens, the components of Coley's toxin. Both groups were placed on the NHL-4 chemotherapy protocol which consisted of two regimens, regimen 1 - thio-tepa, vincristine, chloambucil and prednisone, followed by regimen 2 - Cytoxan, adriamycin, melphalan and prednisone. Radiation therapy was administered to initial areas of bulky disease or to nodal and extranodal sites responding partially to chemotherapy (5).

Figure 2 shows the duration of response in the two treatment groups. In the MBV group, 17 of 22 patients showed a response, with a median response duration of more than 4 years. In the group which did not receive the vaccine, only 10 of 23 patients showed a response, with a median response duration of approximately 2 years.

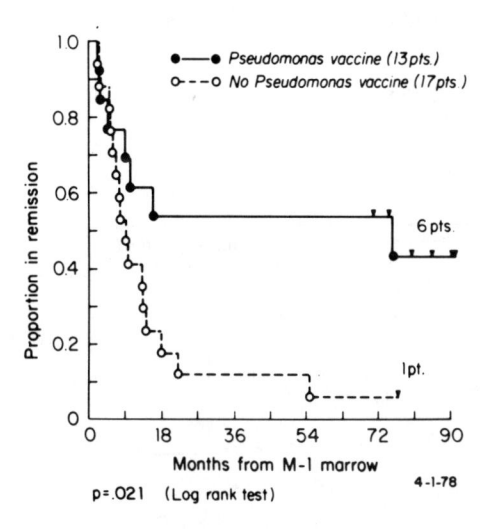

Fig. 1. Duration of remission in patients with acute myeloblastic leukemia treated according to the L-6 protocol.

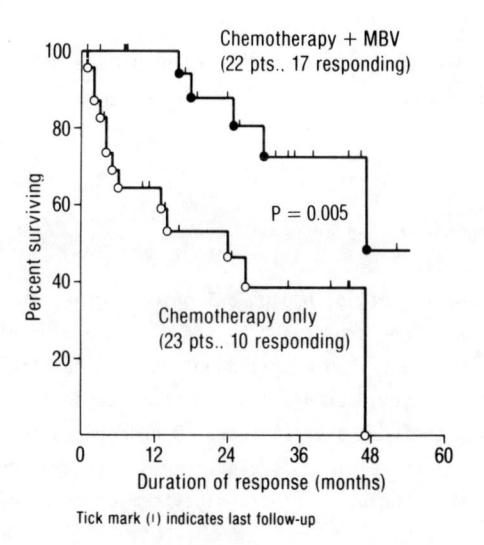

Fig. 2. Treatment of non-Hodgkin lymphomas according to the NHL-4 protocol; duration of response.

Figure 3 shows the results in terms of survival. In the vaccine group, all patients are still alive. In the group which received no vaccine, 7 of 25 patients have died. Thus we have evidence, developed in controlled clinical trials, that bacterial preparations containing endotoxin have therapeutic activity in patients with acute myeloblastic leukemia or non-Hodgkin lymphoma. In the light of these clinical observations, it seemed important to learn more about the mechanisms by which endotoxin affects tumor cells.

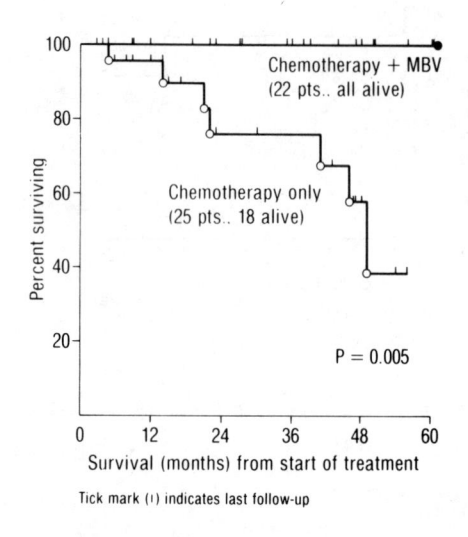

Fig. 3. Treatment of non-Hodgkin lymphomas according to the NHL-4 protocol; duration of survival.

3 Endotoxin-induced endogenous mediators

Considering the great variety of biological effects caused by endotoxin, it seemed highly unlikely that they are caused by the direct action of the endotoxin molecule. It seemed much more probable that endotoxin induces the release of endogenous mediators, and that diverse endogenous mediators cause the diversity of biological effects. Over the past several years, 3 endogenous mediators have been identified whose action may be relevant to endotoxin's antitumor activity. They are tumor necrosis factor, interleukin 1, and granulocyte-macrophage differentiation factor.

4 Endotoxin-induced release of tumor necrosis factor

The way endotoxin causes tumor destruction is not known. A favorite mechanism following earlier work has been that endotoxin-induced systemic hypotension leads to collapse of the tumor vasculature with resulting tumor cell anoxia and death (6). In contrast to this view, recent work has led to the conclusion that endotoxin causes the release of an endogenous factor, of macrophage origin, that is itself directly responsible for tumor cell killing. This factor has been called tumor necrosis factor of TNF (7).

The criterion adopted as a standard for assaying TNF in the serum is the degree of necrosis of a well-established (7-day) subcutaneous transplant of BALB/c sarcoma Meth A in CB6 mice. The response is graded by visual observation. In the maximum (+++) response, the major part of the tumor graft is destroyed, leaving only a narrow peripheral rim of viable tumor tissue. Tumors which show a +++ response often regress completely. An important point is that mice receiving TNF positive serum show no marked signs of toxicity, in contrast to mice receiving endotoxin itself.

Conditions for the production of TNF have been well established. They involve one or another means of macrophage activation, followed by administration of endotoxin. As shown in Table 1, tumor necrosis factor was demonstrable in the serum of BCG infected mice given endotoxin, but not in the serum of mice given either BCG alone or endotoxin only (7). An inoculum of 2×10^7 viable units of BCG was chosen because it gives maximal stimulation of the reticuloendothelial system and sensitization to endotoxin lethality, the peak occurring 2 to 3 weeks after injection. The optimal time for collecting serum was 2 hours after injection of endotoxin. Although at this time the mice were in acute shock, their circulating blood volume was still sufficient for a good yield. Clotting of blood containing TNF was minimal or absent. Corynebacterium parvum and zymosan were as effective as BCG as priming agents for the release of TNF endotoxin. They have in common their capacity to produce marked hyperplasia of the reticuloendothelial system. The question arises of course if residual endotoxin could be responsible for the tumor necrotizing activity of tumor necrosis serum. Two assays were used to detect residual endotoxin in tumor necrosis serum, the standard pyrogenicity assay in rabbits and the limulus assay. The amount of residual endotoxin was less than .1 to 1% of the amount necessary to produce comparable hemorrhagic necrosis of sarcoma Meth A.

Table 1. Necrosis of sarcoma Meth A* produced in vivo by serum from BCG-infected CD-1 Swiss mice treated with endotoxin

Serum† from mice treated with:		TNF assay: Necrotic response			
BCG§	Endo-toxin¶	+++	++	+	−
			Number of mice		
−	−				9
+	−			2	7
−	+				9
+	+	171	109		

* 7-day subcutaneous transplants of BALB/c sarcoma Meth A in (BALB/c x C57BL/6)F$_1$ mice; initial inoculum 2 x 10^5 cells; approximate diameter of tumor mass at time of assay, 7-8 mm.
† Pooled sera from female CD-1 Swiss donors; 0.5 ml iv per tumor-bearing recipient.
§ Viable organisms (2 x 10^7) iv per mouse 14 days before exsanguination.
¶ Twenty-five micrograms iv per mouse 2 hr before exsanguination.

 The activity of TNF was not restricted to sarcoma Meth A. A high degree of sensitivity was also observed with other transplanted tumors, sarcomas S180 and BP8, leukemias EL4, ASL1, RADA1, RL♂1 and EARAD1, and mastocytoma P815. The reticulum-cell sarcoma RCS5, which disseminates widely, was resistant. Among primary spontaneous neoplasms, AKR leukemias showed intermediate sensitivity (indicated by reduction of the size of spleen and lymph nodes) and mammary tumors of C3H origin were only slightly responsive (7).

 TNF destroying the mouse sarcoma Meth A could also be produced in rats and rabbits. As in the mouse, both BCG and endotoxin were required to induce appreciable amounts of TNF. Rabbits are particularly sensitive to endotoxin, so the dose was adjusted accordingly. The serum of BCG-infected rabbits that died less than an hour after endotoxin showed little or no TNF activity (7).

 In contrast to endotoxin itself, TNF was found to be cytotoxic or cytostatic for tumor cells in vitro. This was tested with L cells and cells of the sarcoma Meth A. The L cells proved most sensitive, Meth A sarcoma cells somewhat less sensitive and normal mouse fibroblasts were virtually insensitive. The criterion employed was the count of viable cells after 48 hour exposure. L cells die within the 48 hour test period. The toxicity is delayed; no effect of TNF is demonstrable in the first 16 hours of exposure. Measurable toxicity for L cells was demonstrable with dilutions of TNF positive serum as high as 1 to 10,000. Toxicity was not abolished by heating the TNF serum to 56° for 30 minutes. Sera from normal mice, or mice treated with either BCG or endotoxin alone, tested under the same conditions as TNF sera, showed no toxicity (Fig. 4). Endotoxin itself in concentrations as high as 500 micrograms/liter was not toxic for L cells. Rabbit and rat tumor necrosis factor sera showed the same pattern of toxicity as mouse tumor necrosis factor (7).

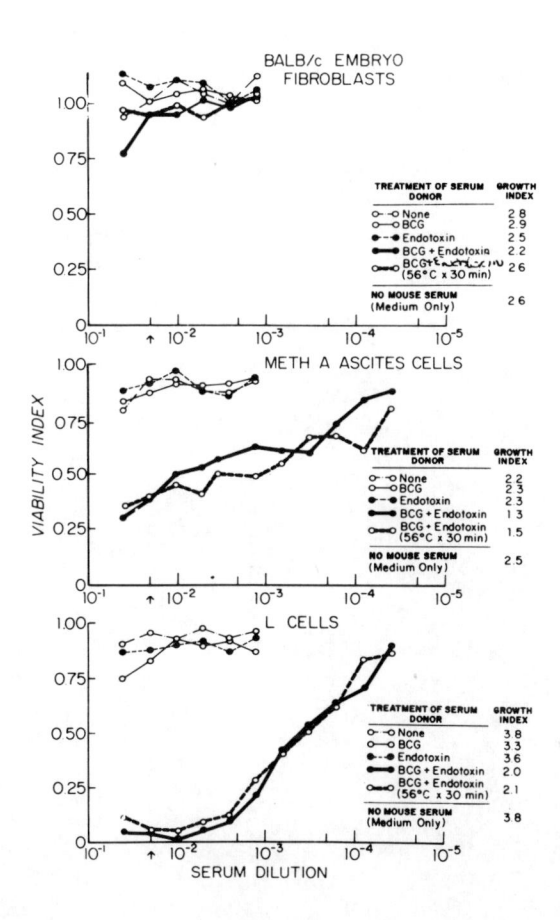

Fig. 4. Inhibition of growth of cultured cell lines by TNF positive serum. L cells (NCTC Clone 929) and BALB/c embryo fibroblasts (MEF) were grown as monolayers, and Meth A cells in suspension. Viability index = number of viable cells present in culture with test serum, divided by number in culture with medium alone. Insets: growth index = total number of cells (viable and dead) after 48 hr in culture in 1/50 mouse serum (↑) divided by number of cells plated.

An important feature of TNF is that its action is not restricted to mouse tumors but extends to human tumor cells. A large number of human tumor cell lines have now been tested. Table 2 shows initial results of these studies. Eight of 10 breast cancer cell lines, 13 of 18 melanoma cell lines, 5 of 9 renal cancer cell lines, 3 of 6 leukemia cell lines and 12 of 19 cell lines derived from other tumors showed cytotoxic or cytostatic effects when exposed to mouse TNF in vitro. By contrast, cell lines derived from normal tissues did not show growth inhibition when cultured in TNF (L. J. Old, unpublished observation).

Table 2. Cytotoxic or cytostatic effects of mouse TNF on human cancer cells in vitro

Cell lines	Number tested	Cytotoxic effect	Cytostatic effect	No effect
Tumor cells				
Astrocytoma	2	1		1
Burkitt's lymphoma	1			1
Bladder cancer	3	1		2
Breast cancer	10	6	2	2
Cervix cancer	2	1	1	
Colon cancer	3	1	1	1
Leukemia	6	1	2	3
Hepatoma	1			1
Lung cancer	2		1	1
Melanoma	18	2	11	5
Osteogenic sarcoma	2	1	1	
Ovarian cancer	1	1		
Renal cancer	9	4	1	4
Teratoma	1	1		
Uterine cancer	1		1	
Normal cells				
Fibroblasts	4			4
Bladder	1			1
Kidney	3			3
Skin	2			2
Monkey kidney	1			1

Recently, human cell lines of hematopoietic origin have been tested for production of TNF. B-cell lines transformed by Epstein-Barr virus release a factor (referred to as hTNF) that is cytotoxic for mouse L cells sensitive to mouse TNF but not for L cells resistant to mouse TNF. Partially purified hTNF induces cross-resistance to mouse TNF in vitro, and causes hemorrhagic necrosis of the Meth A mouse sarcoma in the standard in vivo mouse TNF assay. Tests with a panel of 23 human cancer cell lines showed that hTNF is cytotoxic for 7 cell lines, cytostatic for 5, and has no effect on 11 (8).

5 Endotoxin-induced release of interleukin 1

Tumor necrosis serum has potent immunological effects. Immunological studies have focused mainly on antigen-dependent antibody production, as tested in the Mishell-Dutton culture system (9). In this system, mouse spleen cells are incubated with red blood cell antigen for several days and then mixed with agar containing the same red blood cell antigen and exposed to complement. Spleen cells which secrete antibody cause lysis of the red blood cells in their immediate vicinity. The number of lytic plaques represent the number of antibody secreting cells, plaque forming cells or PFC. An advantage of the system is that it permits removal of selected cell populations - T cells, B cells, macrophages and their subsets - and assessment of their role in the production of antibody.

Initial experiments showed that tumor necrosis serum augments the production of antibody to sheep red blood cells in vitro. Table 3 shows the number of plaque forming cells after 3 or 4 days of culture. Tumor necrosis serum produced an increase in the number of plaque forming cells, serum from untreated mice did not (10).

Table 3. Effects of tumor necrosis serum on the production of antibody to SRBC in vitro

Addition to culture	Anti-SRBC PFC per culture	
	Day 3	Day 4
None	65	3,900
TNS 2%	1,850	13,000
Normal mouse serum	28	4,000
LPS serum *	108	6,500
BCG serum **	78	4,000

*   Serum from mice given LPS only.
** Serum from mice given BCG only.

Further analysis showed that tumor necrosis serum increased antibody production by replacing a function of T lymphocytes. Figure 5 shows an example of experiments that led to this conclusion. The spleen cells used in this experiment were obtained from congenitally athymic nu/nu mice. As these mice lack helper T cells, their spleen cells do not produce antibody to red blood cells in vitro. Addition of tumor necrosis serum to the culture medium permitted a response of the same magnitude as that of spleen cells from littermate nu/+ mice which do not lack T cells. The effect was restricted to the antigen to which the spleen cells had been sensitized - in this case burro red blood cells (BRBC) - and did not extend to an unrelated antigen - in this experiment sheep red blood cells (SRBC). Addition of tumor necrosis serum was effective in this system even if delayed as long as two days. Normal mouse serum, serum from mice treated with BCG only or serum from mice treated with endotoxin only, had no effect (10).

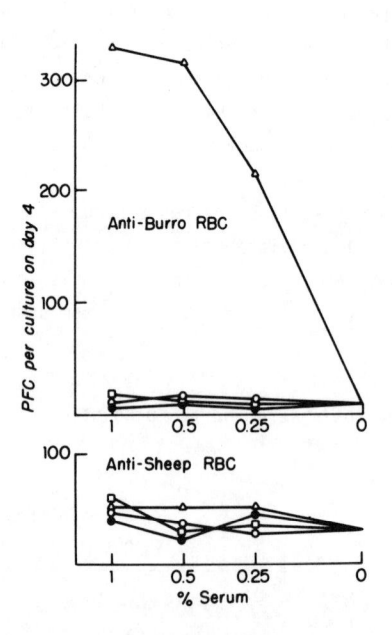

Fig. 5. Restoration of antibody production by spleen cells from nu/nu mice with TNS. Cultures (in MEM + 5% FCS) were immunized with BRBC. Anti-BRBC (a) and anti-SRBC PFC (b) were assayed on day 4. Δ, TNS; ●, normal mouse serum; O, serum from mice treated with BCG only; □, serum from mice treated with LPS only.

An important question in this work has been whether or not the observed effect was due to residual endotoxin. While this was unlikely efforts were made to prove the point. Figure 6 shows an example. The experiment has two essential features. First, two mouse strains were used, the endotoxin-unresponsive mouse strain C3H/HeJ and its congenic LPS-responsive counterpart C3HeB/FeJ. Second, the antigen, trinitrophenyl-conjugated mouse erythrocytes, is not itself immunogenic in the mouse but becomes immunogenic when given with endotoxin. The effects of tumor necrosis serum on the production of antibody against trinitrophenyl-conjugated mouse erythrocytes were compared with those of endotoxin. In the absence of tumor necrosis serum or endotoxin no antibody response was seen on day 4. Addition of endotoxin caused a response in cultures of endotoxin-responsive C3HeB/FeJ spleen cells but not in cultures of endotoxin-unresponsive C3H/HeJ spleen cells. Addition of tumor necrosis serum, on the other hand, facilitated antibody production by cultures of spleen cells from both the endotoxin-responsive and the endotoxin-unresponsive substrain (11).

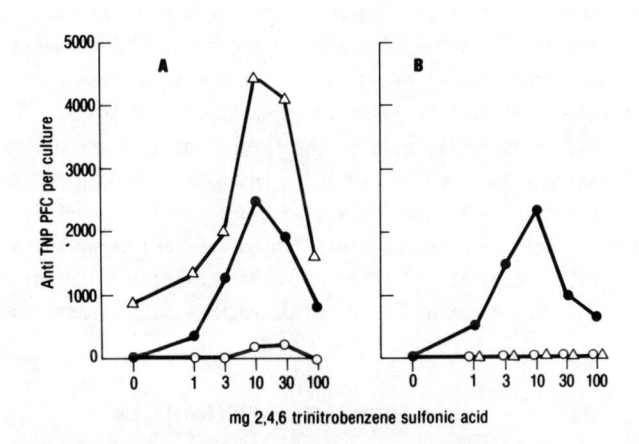

Fig. 6. Facilitation of anti-TNP response by LPS or by TNS. 10^7 spleen cells from C3HeB/FeJ mice (A) or C3H/HeJ mice (B) were immunized with 10^6 MRBV conjugated with TNP at different densities. Amount of 2,4,6-trinitrobenzene sulfonic acid per 10 ml buffer containing 10^8 MRBC is indicated on the abscissa. Additions to cultures: O, none; ●, TNS 1%; Δ, LPS 10 µg/ml.

Further studies have shown that tumor necrosis serum has striking effects on a variety of immunological reactions in vitro, some of which mimic endotoxin activity, and some of which are distinct. Similar to endotoxin, tumor necrosis serum induces maturation of B cells and substitutes for helper T cells. Unlike endotoxin, tumor necrosis serum inhibits the proliferative response of T and B cells to mitogens and reverses the suppression of antibody production in vitro induced by Concanavalin A or by specific antibody. Tumor necrosis serum also differs from endotoxin in that it is not a mitogen for B cells, does not induce maturation of T cells and does not induce polyclonal antibody production (12).

While the antitumor effects of TNF in vivo and in vitro and the effects on immunological functions were not dissociated in impure preparations, further purification has shown that the immunologically active factor in tumor necrosis serum is distinct from TNF. It was first described as B cell differentiating factor (13) and is now designated interleukin 1 (IL-1). IL-1 plays a central role in the regulation of the immune response. It is released by macrophages in response to a T cell signal. This signal can be replaced by endotoxin. IL-1 induces maturation of T and B cell precursors, and stimulates T cells to produce interleukin 2, also known as T cell growth factor. Thus, IL-1 acts at the beginning of a sequence which leads to effector cell function - cytotoxic activity of B cells and antibody production by B cells. If we return to the phenomenon of tumor necrosis and regression in the mouse, we can speculate that TNF and IL-1, both induced by endotoxin, may be mediators of two characteristic phases of the tumor response. TNF causes acute necrosis of tumor tissue, the initial event which is known to be independent of an immune response; it can be elicited in severely immunosuppressed hosts. IL-1, on the other hand, may play an essential role in the second phase of the process, complete regression of residual tumor cells, which is known to depend on an intact immune system. Now that these molecules have been characterized, some key questions can be answered such as their relation to other lymphokines and serum factors, the basis for the tumor cell selectivity of tumor necrosis factor, and its possible involvement in the cytotoxic effect of activated macrophages and other classes of killer cells.

6 Endotoxin-induced granulocyte-macrophage differentiation factor

A third endogenous mediator released in response to endotoxin induces leukemia cells to differentiate into mature granulocytes or macrophages. When it was first reported that serum from mice injected with endotoxin induced granulocyte and macrophage (GM) differentiation of the mouse myeloid leukemia cell line M1 (14) it was unclear whether the differentiation-inducing factor was colony stimulating factor (CSF,MGI-1). It was shown subsequently that a pure preparation of GM-CSF has some capacity to induce differentiation of murine WEHI-3 myelomonocytic leukemia cells (15). More recently biochemical analysis has shown that the differentiation factor could be separated from GM-CSF (MGI-1) - it was termed MGI-2 (16), or could be separated from the bulk of serum CSF but coeluted with a minor species of CSF that stimulated only granulocyte colony formation (17). Differentiation of leukemia cell lines after exposure to GM-CSF containing preparations was associated with a suppression of their capacity to produce progressively growing leukemia cell populations after transplantation to syngeneic recipients (18, 19).

Induction of the GM differentiation factor (GM-DF) by endotoxin in the mouse is shown in Figure 7. The target cells are WEHI-3B murine myelomonocytic leukemia cells, cultured in semi-solid agar. Shown are the number of colonies as indicator of proliferative activity, and the number of differentiated colonies as indicator of differentiation. The left panel shows the effects of serum from mice that had been injected with 5 µg endotoxin. All leukemic colonies converted from a tight, undifferentiated pattern to diffuse differentiated colonies composed of mature granulocytes or macrophages when exposed to 5 to 10% of post-endotoxin serum. Differentiation was still noted at a 1:64 dilution of the serum. Significantly, colony inhibition was not observed even at high serum concentrations. The panel in the middle shows the effects of tumor necrosis serum - unfractionated serum from mice treated with Corynebacterium parvum and endotoxin. Leukemic colony growth was markedly inhibited when the serum was used at a 10% concentration, but this presumed tumor necrosis factor-dependent inhibition rapidly titrated out. Significant levels of differentiation-inducing activity, comparable to levels in post-endotoxin serum on the left, were also noted. To define further the independent actions of tumor necrosis factor and differentiation factor, tumor necrosis serum was subjected to fractionation by DEAE and Sephadex G-100 chromatography. The right panel shows that the peak rich in tumor necrosis factor activity was strongly inhibitory for leukemia cell proliferation but contained no significant differentiation inducing activity (20).

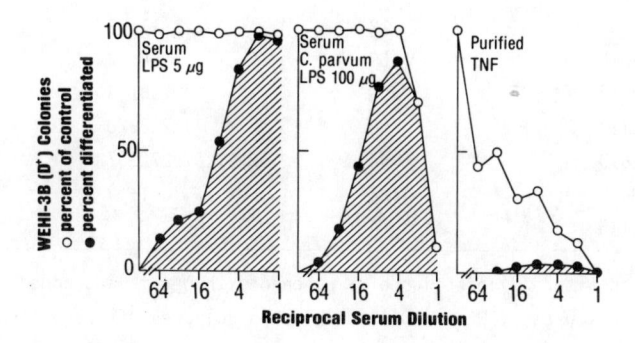

Fig. 7. Differentiating and growth inhibitory activities in mouse serum after injection of endotoxin. Left panel: serum from unprimed mice 3 hours after injection of 5 µg endotoxin. Middle panel: serum from C. parvum primed mice 3 hours after injection of 100 µg endotoxin (tumor necrosis serum). Right panel: purified TNF.

We have extended the study of GM-DF induction to cancer patients treated with a highly purified endotoxin preparation, Novo-Pyrexal, produced by the Hermal Company in Germany from S. abortus equi according to a procedure developed by Dr. Galanos in Professor Westphal's group in Freiburg, Germany. Serum samples were obtained from patients at various intervals after intravenous administration of Novo-Pyrexal 1 µg/m^2. The serum was tested for its capacity to induce differentiation of WEHI-3B cells, and to induce myeloid colony formation by the patient's own bone marrow obtained before and 24 hours after endotoxin administration. Table 4 shows an example of results obtained in one of these patients. Maximal differentiation inducing activity of the serum was seen 2 to 6 hours after endotoxin treatment. Serum obtained prior to endotoxin injection and 24 hours after endotoxin injection had no significant differentiation inducing activity. In tests on the patient's own bone marrow, colony stimulating activity was observed between 4 and 8 hours after endotoxin injection. The bone marrow obtained before endotoxin administration showed greater responsiveness than the bone marrow obtained 24 hours after endotoxin administration, reflecting most likely partial refractoriness after in vivo stimulation by endotoxin (21).

Table 4. CSF and differentiation-inducing activity in human post-endotoxin sera *

Hours post-endotoxin	WEHI-3(D+) Colonies/300 cells	WEHI-3(D+) Colonies % diff.	Autologous marrow colonies/10^5 Pre-endotox.	24 h post-endotox.
0	32±2	2	0	4± 1
0.5	47±4	22	11±17	4± 3
2	30±6	50	4± 2	1± 1
4	25±6	59	198± 4	46± 4
6	21±3	50	202±20	41±12
8	26±3	44	272±37	27± 6
12	34±9	26	1± 2	0
24	41±5	4	0	1± 2

* Sera collected after IV injection of 1 µg endotoxin per m^2 and added at 10% v/v to 1 ml agar cultures of WEHI-3(D+) cells or to cultures of the patients bone marrow obtained immediately prior to endotoxin treatment or 24 h post-endotoxin.

Induction of the differentiation factor by Novo-Pyrexal was found to be readily reproducible. Figure 8 shows the results obtained in 5 consecutive patients with advanced cancer. Serum samples were obtained at various intervals during 24 hours after intravenous injection of 1 µg/m^2 of Novo-Pyrexal. The serum was assayed on

WEHI-3B leukemia cells. Pretreatment sera had no significant effect on cloning efficiency or colony differentiation. A serum activity which induced up to 85% of leukemia colonies to differentiate to mature granulocytes or macrophages showed its peak between 2 and 4 hours after Novo-Pyrexal administration, and disappeared by 12 to 24 hours after injection. An acute decrease in the white blood cell count was observed in every instance as early as 30 minutes after Novo-Pyrexal administration, followed by return to normal and subsequent rebound above normal at 24 hours. It should be noted that serum collected 30 minutes after the Novo-Pyrexal administration was inhibitory for leukemia cell colony formation, but was not active in inducing leukemia cell differentiation. Whether or not this inhibitory activity is related to TNF remains to be seen (21).

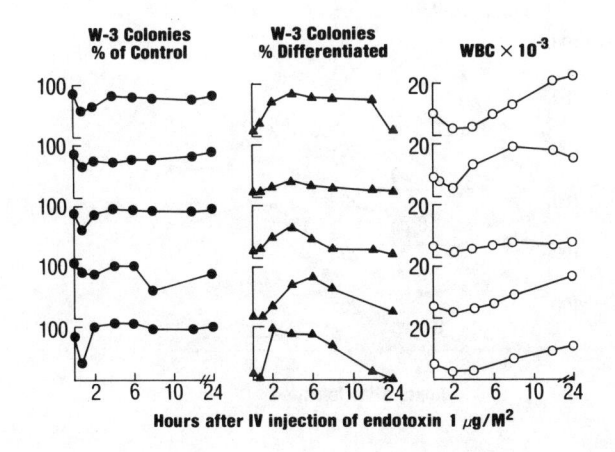

Fig. 8. Induction of a leukemia differentiation inducing factor (GM-DF) in the sera of 5 patients with advanced cancer following IV injection of 1 μg of endotoxin (Novo Pyrexal) per m^2. Serum DF assayed on a target population of WEHI-3B(D+) leukemic cells. Note peak induction of differentiation factor 2-4 h post endotoxin. Serum colony inhibitory activity (left panel) was maximal 30 min following endotoxin (note, 1-10 μg of endotoxin added directly to cultures of WEHI-3B did not inhibit cloning nor induce differentiation).

The ability to induce serum activities such as GM-DF reproducibly by repeated endotoxin administration may be affected by the development of endotoxin tolerance. In an attempt to overcome tolerance in patients receiving repeated doses of Novo-Pyrexal, dose escalation was studied. The results of one such study in a patient with advanced cancer are shown in Figure 9. Sera obtained before and at various intervals after the administration of Novo-Pyrexal were assayed for ability to induce differentiation of WEHI-3B cells. The kinetics of induction were similar to those seen after single injection. Leukemia differentiation inducing activity rose from undetectable levels to a peak 2-6 hours after injection of endotoxin, and the escalation of the endotoxin dose at twice weekly intervals to a maximum of 30 µg/m^2 produced consistent re-induction of this response. The patients showed a regular biphasic leukocyte response with an initial leukepenia followed by leukocytosis (21).

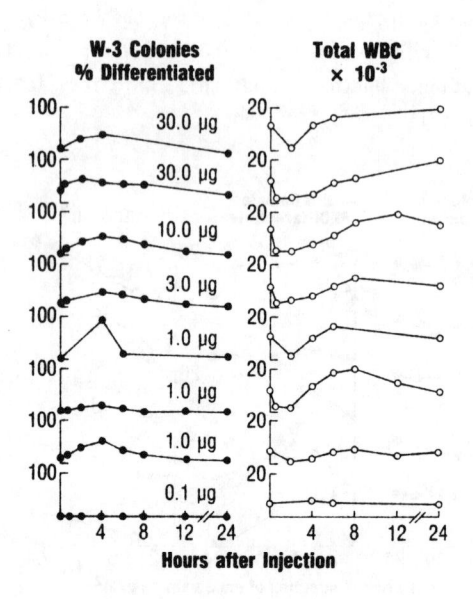

Fig. 9. Induction of a leukemia differentiation inducing factor (GM-DF) in the serum of a patient with advanced malignancy. Biweekly intravenous injections of increasing doses (expressed as dose per m^2) of highly purified endotoxin (Novo Pyrexal) to the patient induced a regular biphasic leukocyte response with initial leukopenia followed by a leukocytosis. Serum (10% v/v) was added to agar cultures containing 300 WEHI-3B(D+) leukemic cells and the % of colonies becoming diffuse (differentiated) by 7 days was scored. Note the induction of GM-DF activity 2-4 h after endotoxin treatment.

7 Conclusions

In attempts to define the mechanisms of endotoxin's action on tumors, three endogenous mediators of possible relevance have been identified. TNF, a glycoprotein of 40,000-60,000 M_r, is cytotoxic or cytostatic for a range of cancer cells and causes hemorrhagic necrosis of transplanted tumors. Il-1, a protein of 15,000 M_r, plays a central role in the regulation of antibody production and T cell cytotoxicity. GM-DF, a protein of 28,000 M_r causes differentiation of leukemia cells to granulocytes or macrophages and reduces the leukemogenicity of transplanted leukemia cells. These mediators should now be tested in the clinic to determine if they have anticancer effects not accompanied by the toxic effects that limit the use of endotoxin. Considering the advances that have been made in recombinant DNA technology, large scale production for clinical study is a realistic prospect.

8 References

(1) H. C. Nauts, W. E. Swift, B. L. Coley, Cancer Res. 6 (1946) 205-216.

(2) M. J. Shear, J. Natl. Cancer Inst. 4 (1944) 461-476.

(3) S. Hanessian, W. Regan, D. Watson, T. Haskell, Nature New Biol. 229 (1971) 209-210.

(4) B. D. Clarkson, M. D. Dowling, T. S. Gee, I. B. Cunningham, J. H. Burchenal, Cancer 36 (1975) 775-795.

(5) S. Kempin, C. Cirrincione, D. S. Straus, T. S. Gee, Z. Arlin, B. Koziner, C. Pinsky, L. Nisce, J. Myers, B. J. Lee, III, B. D. Carkson, L. J. Old, H. F. Oettgen, Proc. Amer. Soc. Clin. Oncol. 22 (1981) 514.

(6) G. H. Algire, F. Y. Legallais, B. F. Anderson, J. Natl. Cancer Inst. 12 (1952) 1279-1295.

(7) E. A. Carswell, L. J. Old, R. L. Kassel, S. Green, N. Fiore, B. Williamson, Proc. Natl. Acad. Sci. USA 72 (1975) 3666-3670.

(8) B. D. Williamson, E. A. Carswell, B. Y. Rubin, J. S. Prendergast, L. J. Old, Proc. Natl. Acad. Sci. USA 80 (1983) 5397-5401.

(9) R. I. Mishell, R. W. Dutton, J. Exp. Med. 126 (1967) 423-442.

(10) M. K. Hoffmann, S. Green, L. J. Old, H. F. Oettgen, Nature 263 (1976) 416-417.

(11) M. K. Hoffmann, C. Galanos, S. Koenig, H. F. Oettgen, J. Exp. Med. 146 (1977) 1640-1647.

(12) M. K. Hoffmann, H. F. Oettgen, L. J. Old, R. S. Mittler, U. Hammerling, J. Reticuloendothel. Soc. 23 (1978) 307-319.

(13) M. K. Hoffmann, H. F. Oettgen, L. J. Old, A. F. Chin, U. Hammerling, Proc. Natl. Acad. Sci. USA 74 (1977) 1200-1203.

(14) E. Fibach, L. Sachs, J. Cell Physiol. 89 (1976) 259-266.

(15) D. Metcalf, J. Cell Physiol. (Suppl.) 1 (1982) 175-183.

(16) J. Lotem, J. H. Lipton, L. Sachs, Int. J. Cancer 25 (1980) 763-771.

(17) A. W. Burgess, D. Metcalf, Blood 56 (1980) 947-958.

(18) Y. Honma, T. Kasukabe, J. Okabe, M. Hozumi, Cancer Res. 39 (1979) 3167-3171.

(19) D. Metcalf in J. H. Burchenal, H. F. Oettgen (eds.), Cancer - Achievements, Challenges and Prospects for the 1980s, Grune and Stratton, New York 1981, pp. 465-475.

(20) M. A. S. Moore, J. Cell Physiol. (Suppl.) 1 (1982) 53-64.

(21) M. A. S. Moore, J. Gabrilove, A. P. Sheridan, Blood Cells 9 (1983) 125-137.

Induction and Properties of the Tumor Necrosis Factor

Akihiro Yamamoto[#1], Barbara Williamson, Elizabeth C. Richard, Nancy Fiore and Lloyd J. Old

Memorial Sloan-Kettering Cancer Center, new York, N.Y., 10021

Abstract

Mouse tumor necrosis factor (TNF) was examined for its effects on human cell lines derived from tumors and normal tissues.

TNF showed effects on about two-thirds of 62 cell lines (15 different kinds of tumor). A cytocidal effect was observed in 19 out of 62 cell lines; breast tumor lines in particular sensitive to TNF. TNF had a cytostatic effect on 21 out of the remaining 42 cell lines. The typical antiproliferative effect was seen when the melanoma cell lines were incubated with TNF.

In contrast, all the cell lines derived from normal human tissues, regardless of their origin, were completely resistant to both the cytocydal and cytostatic effects of TNF. These effects of TNF were confirmed by a plating efficiency test.

1 Introduction

The injection of bacterial lipopolysaccharide (LPS) in mice causes the release of biological mediators into the serum.

Tumor Necrosis Factor (TNF) is a substance which causes necrosis of some transplantable tumors (1) and cytotoxic or growth inhibitory in

---

[#1] Present address: Chugai Pharmaceutical Co., Ltd., Takada 3-41-8, Toshima-ku, Tokyo, 171, Japan
Requests for reprints should be addressed to Dr. A. Yamamoto.

vitro to a number of cell lines (2). TNF is released into the blood after endotoxin injection in animals (mice, rats and rabbits) which have been pretreated with agents such as Bacillus Calmette Guerin (BCG) or Corynebacterium parvum which induce macrophage hyperplasia.

Helson et al. reported that TNF was strongly inhibitory for an established line of human melanoma, whereas three other human cell lines (two neuroblastomas and one normal fibroblast) were not affected.

In this paper we report on a study made of an extensive series of established human cell lines derived from tumors and normal tissues to determine their susceptibility to TNF.

2 Materials and Methods

Preparation of TNF: CD-1 female Swiss mice weighing 35-40g were injected intraperitoneally with 1mg of formalin-killed Corynebacterium parvum. After 11 days, 10 µg of endotoxin from Escherichia coli was injected intravenously and the mouse blood was collected 1.5 h later. Serum was separated by centrifugation and ultracentrifuged at 113,000 x G at 5°C for 20 h. The top one-third of the serum (lipid rich layer) containing TNF-inactive protein was discarded. TNF containing serum (TNS) was purified by using DEAE A-50 Sephadex ion exchange chromatography, Sephadex G-100 and Sephadex G-200 column chromatography. These partially purified TNF preparations do not contain interferon and are not pyrogenic for rabbit. Pooled normal mouse serum was treated in the same manner as TNS and used as the normal control.

In vivo TNF assay: (BALB/c x C57BL/6) F_1 mice with intradermal 7-day transplants of sarcoma Meth-A averaging 7-8 mm in diameter, are injected with TNF. Twenty-four hours later, the degree of necrosis of the tumor is assessed visually:

grade -   = no change
grade +   = slight necrosis
grade ++ = moderate necrosis (central necrosis extending over approximately 50% of the tumor surface)
grade +++ = extensive necrosis (massive necrosis leaving at most only a small rim of viable tumor tissue)

In vitro TNF assay: Established human cell lines derived from human tumors and normal tissues were obtained from Dr. J. Fogh of Memorial Sloan-Kettering Cancer Center. L-929 was provided by NCTC and maintained in our laboratory.

The cells were grown as monolayers in culture medium consisting of Eagle's minimum essential medium, nonessential amino acid, 10% heat inactivated fetal calf serum, penicillin (100 U/ml) and streptomycin

(100 μg/ml). Equal volumes of cell suspension (1 x 10^5 cells/ml) and serially diluted TNF were incubated in wells of Linbro plates in 5 % CO_2 in air at 37° C. The cytostatic effect (growth inhibition) and the cytotoxic effect were calculated as follows:

$$\text{per cent growth inhibition} = \left(1 - \frac{\text{total cell no. of treated group}}{\text{total cell no. of control group}}\right) \times 100$$

$$\text{per cent cytotoxicity} = \left(\frac{\text{no. of dead cells}}{\text{no. of total cells}}\right) \times 100$$

3 Results

Production of TNF

As Table 1 shows, in the preparation of mice for production of TNF two steps are critical. The first involves priming mice with agents, such as BCG, that cause macrophage hyperplasia; the second involves eliciting TNF release by small amounts of endotoxin. TNF was not detected in the serum of BCG-infected mice or mice given endotoxin alone.

Both rats and rabbits produced TNF; as in the mouse, both BCG and endotoxin were required to induce appreciable amounts of TNF.

Table 1 Necrosis of Meth-A sarcoma produced in vivo by TNF serum

Serum from mice treated with:		TNF assay : Necrotic response [3]			
		+++	++	+	−
BCG [1]	Endotoxin [2]	Number of mice			
+	+	17	11		
+	−			2	7
−	+				9
−	−				9

1) Viable organisms (2×10^7) ip per mouse 14 days before exsanguination.
2) Twenty-five mg iv per mouse 2 hr before exsanguination.
3) For scoring of the necrotic response, see Materials and Methods.

As shown in Table 2, C. granulosum, C. parvum and Zymosan are effective as BCG as priming agents for TNF release by endotoxin. These agents have in common the capacity to produce marked hyperplasia of the reticuloendothelial system. Although nude mice respond well to C. parvum in terms of macrophage hyperplasia in the spleen and liver, TNF cannot be elicited by LPS in these mice. This suggests the participation of T cells in the process of TNF release.

Brucella abortus, which also contains endotoxin, did not elicit TNF in BCG-primed mice. This correlates with the observation that B. abortus endotoxin is not very toxic in other systems. Old tuberculin was also ineffective in inducing TNF.

Table 2. TNF activity in the serum of mice treated with various priming and eliciting agents

Treatment of serum donors [1]		Necrotic response [2]			
Priming agent	Eliciting agent	+++	++	+	-
		Number of mice			
BCG (2×10^7 viable cells)	Endotoxin (25 µg)	171	109		
"	Mixed bacterial vaccine (5 µl)	10	6	3	
"	Old tuberculin (50 mg)				9
"	B. abortus (1×10^8 viable cells)				11
"	BCG (1×10^8 viable cells)			2	9
C. granulosum (700 µg)	Endotoxin (25 µg)	8			
C. parvum (1 mg)	"	24	7		
Zymosan (2 mg)	"	4			

1) CD-1 Swiss mice recieved the eliciting agent 14 days after the priming agent (both iv) and were exsanguinated 2 hr later.

2) For scoring of the necrotic response, see Materials and Methods.

The strain of mouse is an important factor in TNF production. TNF can be produced in BCG-infected mice by endotoxin in endotoxin-sensitive mice (C3H/HeN) but not in endotoxin-insensitive mice (C3H/HeJ).

In the Limulus assay, endotoxin levels estimated at 1 µg/ml were found in the serum of endotoxin-treated mice, whether pretreated with BCG or not. These amounts of residual endotoxin in TNF-positive sera are less than 0.1 - 1 % of the amount necessary to produce comparable hemorrhagic necrosis in Meth-A.

Effect of TNF on human cell lines

Next experiment was performed to determine the effect of TNF on BT-20, a cell line derived from human breast cancer. After 7 days' incubation in 5 % CO_2 in air, the cells in the supernatant and monolayer cells detached by trypsinization were observed and counted under a phase microscope. As shown in Fig. 1, in the absence of TNF, the number of BT-20 cells increased progressively, whereas the multiplication of BT-20 cells was markedly inhibited by the addition of TNF. TNF-treated cells were flattened and shrunken, and at the highest level of TNF activity, most of them became detached from the cell layer. The supernatant contained many dead cells and cell debris. These observations indicated that cytostasis and cytolysis of BT-20 occurred in the presence of TNF. Viability of BT-20 cells was also investigated using trypan blue to stain dead cells. The experimental values obtained by this staining method agreed well with those obtained by using the phase microscope to distinguish live and dead cells.

The concentration of TNF required to produce 50 % cytotoxicity (CD_{50}) of BT-20 was obtained by a dilution of 1/625. This dilution has a protein content of 2.6 µg/ml. In contrast, a 50 fold dilution (protein 32 µg/ml) of normal mouse serum preparation had no cytotoxic or cytostatic effects against BT-20 cells and the cells maintained a normal appearance.

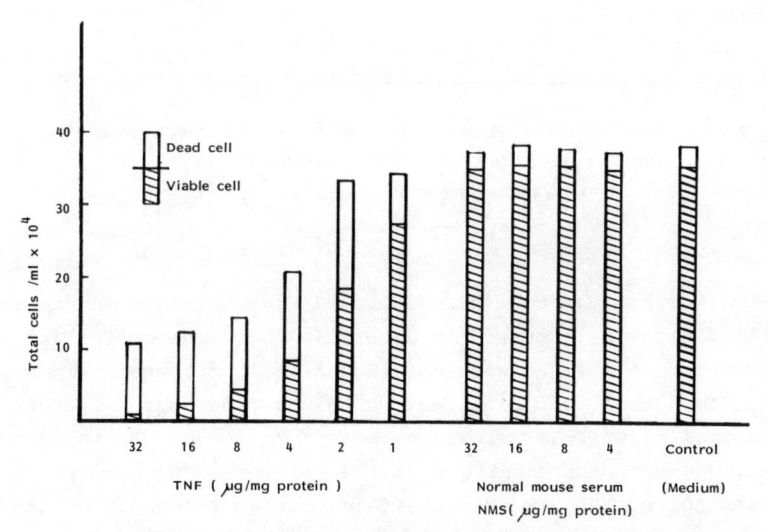

Fig. 1. Effect of TNF on BT-20 (breast tumor)

BT-20 cells (10×10^4 cells/0.5 ml) were incubated with 0.5 ml of various dose level of TNF or normal mouse serum.

The susceptibility of BT-20 cells to TNF was compared with that of L-929 cells, a transformed cell line derived from the C3H strain of mouse. As shown in Table 3, using L-929 cells, the cytotoxic activity of TNF as compared to that of TNS indicated an approximately 265 fold purification of the active factor. Using BT-20 cells an approximately 275 fold purification was indicated. At each purification step, the increase of specific activity of TNF against BT-20 was proportional to that of TNF against L-929 cells. Normal mouse serum treated in the same manner as TNS did not show any effect on BT-20 and L-929 cells at any step in purification procedure.

Table 3. Comparison of specific activity of TNF at different stage of purification

Purification stage	L-929 cells [1] Unit of TNF [2]	Specific activity [3]	BT-20 cells [1] Unit of TNF [2]	Specific activity [3]
TNF Serum	5.3 μg	189	713 μg	1.4
DEAE A-50 Sephadex	182 ng	5,495	22.4 μg	44.1
Sephadex G-100	74 ng	13,514	8.7 μg	114.9
Sephadex G-200	20 ng	50,000	2.6 μg	384.6

1) L-929 and BT-20 cells were cultured with TNF for 2 and 7 days respectively.
2) Amount of protein required to produce 50 % cell cytotoxicity
3) Expressed as TNF units per mg material

Sixty-two human tumor cell lines derived from patients and twelve normal cell lines were prepared to determine their susceptibility to TNF. Those tumor cell lines were obtained from 15 different kinds of tumors and the normal cell lines were obtained from four different kinds of normal tissue. TNF showed an effect on 40 out of the 62 tumor cell lines. However, the magnitude of the cytotoxic and the cytostatic effects obtained by TNF treatment differed markedly among the tumor cell lines tested in this study. BT-20 and SK-BR-3 derived from breast tumors and SK-MEL-29 from a melanoma were very susceptible to both the cytocidal and the cytostatic effects of TNF. As Fig. 2a shows, BT-20 cells were inhibited in their cell growth and 85 - 95 % of these died in the presence of TNF.

As shown in Fig. 2b, the cytocidal effect of TNF was observed in ME-180 (cervical tumor), although not as much growth inhibition was observed in this cell line.

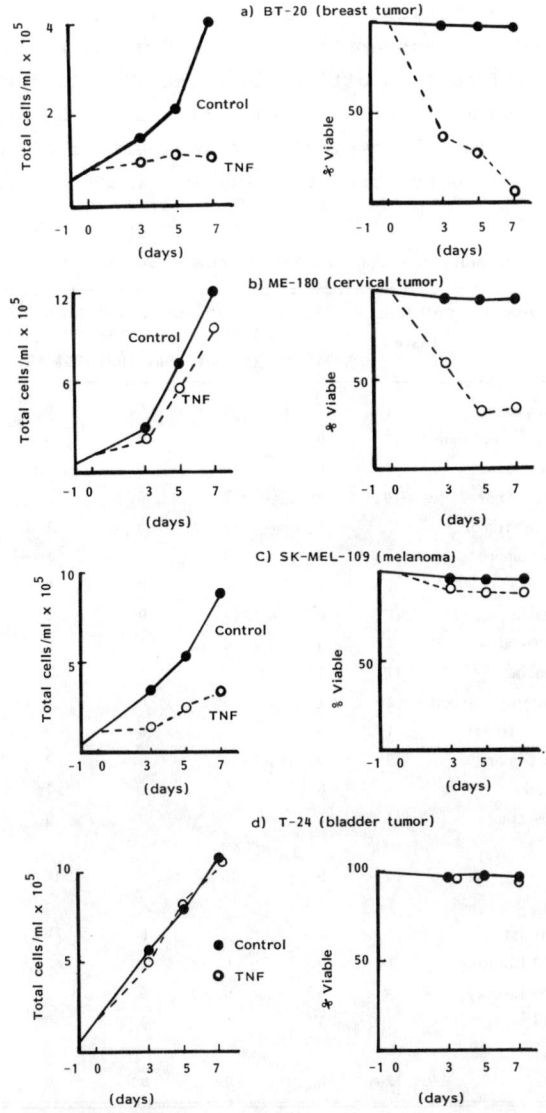

Fig. 2. Effect of TNF on human tumor cell lines

Each human cell line (10 x 10^4 cells/0.5 ml) was incubated with 32 μg protein/0.5 ml of TNF (O) or normal mouse serum (●). Total number of cells and % viable were determined after 5 days incubation.

As Fig. 2c shows, SK-MEL-109 from a melanoma, which was sensitive to the antiproliferative effect of TNF, was relatively insensitive to the cytocidal effect of TNF.

Twenty-two tumor cell lines, that is, about one-third of all the tumor cell lines examined in this study, were resistant to the cytostatic and the cytotoxic effects of TNF (Fig. 2d).

Results from these studies are further tabulated in Table 4. There were some differences in susceptibility to TNF among the same kind of tumor cell lines derived from different patients. In general, however the effect of TNF was most markedly demonstrated in the breast tumor and melanoma cell lines. The typical cytocidal effect of TNF was observed in six out of the ten breast tumor cell lines. A typical

Table 4. Susceptibility of human cell lines to TNF (Summary)

	Cell type	Cell lines tested	Cytotoxic 50%-100%	Cytotoxic 25%-50%	Cytostatic 50%-100%	Cytostatic 25%-50%	Resistant
Tumor cell lines	Astrocytoma	2	1	0	0	0	1
	Burkitt lymphoma	1	0	0	0	0	1
	Bladder tumor	3	0	0	0	1	2
	Breast tumor	10	4	2	0	2	2
	Cervical tumor	2	1	0	0	1	0
	Colon tumor	3	0	1	0	1	1
	Hepatoma	1	0	0	0	0	1
	Leukemia	6	0	1	0	2	3
	Lung tumor	2	0	0	1	0	1
	Melanoma	18	1	1	4	7	5
	Osteogenic sarcoma	2	1	0	0	1	0
	Ovarian tumor	1	0	1	0	0	0
	Renal tumor	9	0	4	0	0	5
	Teratoma	1	0	0	0	1	0
	Uterus tumor	1	1	0	0	0	0
	Total	62	9	10	5	16	22
Normal cell lines	Fibroblast	5	0	0	0	0	4
	Normal bladder	1	0	0	0	0	1
	Normal kidney	4	0	0	0	0	3
	Normal skin	2	0	0	0	0	2
	Total	12	0	0	0	0	12

1) A cell lines is defined as being susceptible if TNF has a cytotoxic or cytostatic effect on 25 % or more of the cells. When both cytotoxic and cytostatic effects were observed in the same line, only the stronger effect is noted in the Table.

antiproliferative effect was obtained when melanoma cell lines were incubated with TNF. The growth of 11 out of 18 melanoma cell lines was strongly inhibited by the addition of TNF.

Twelve normal human cell lines cultured with TNF to determine their susceptibility were completely resistant to both the cytotoxic and cytostatic effects of TNF regardless of their origin.

The cytotoxic effect of TNF was also investigated by a plating efficiency test. After BT-20 cells were exposed to 1/100 dilution (16 μg protein) of TNF for 3 days, about 35 % of the cells were killed (Fig.3a). The viable cells harvested from this culture did not recover their normal growth pattern in fresh medium without TNF: they were flattened and shrunken after one day, and most of them became detached from the cell layer after two days. More than 80 % of TNF-treated cells died during the 2 days' incubation in the TNF free medium, whereas normal mouse serum-treated cells doubled their cell number within 48 hours (Fig.3b). A similar result was obtained with ME-180 cells which were susceptible to the cytotoxic effect but not to the cytostatic effect of TNF.

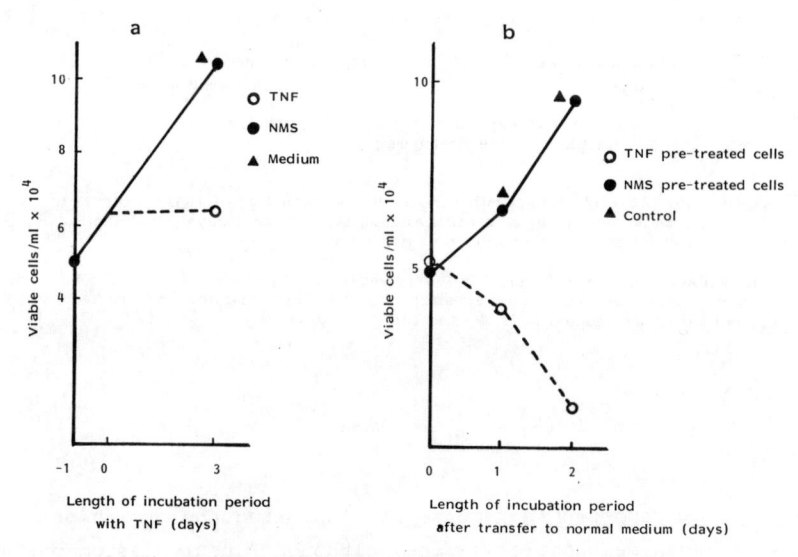

Fig. 3. Effect of TNF exposure on the plating efficiency of BT-20 (breast tumor)

(a) BT-20 cells (10 x 10^4 cells/1 ml) were incubated with 1 ml of 16 μg protein of TNF (○), normal mouse serum (●) or control medium (▲). After 3 days' incubation viable cells were harvested by trypsinization and counted.

(b) The viable cells (5 x 10^4 cells/1 ml) were replated into fresh medium without TNF or normal mouse serum and counted after 24 and 48 h incubation. TNF treated cells (○), normal serum treated cells (●), control medium (▲).

The growth of SK-MEL-109 cells was 40 % inhibited after 3 days' incubation with the same dose of TNF (Fig.4a). These TNF-treated cells increased in number to the same extent as the control cells after the TNF was removed (Fig.4b). Even after a 4-week-incubation period with TNF they demonstrated the cell growth rate as the control group, when transferred to TNF free medium.

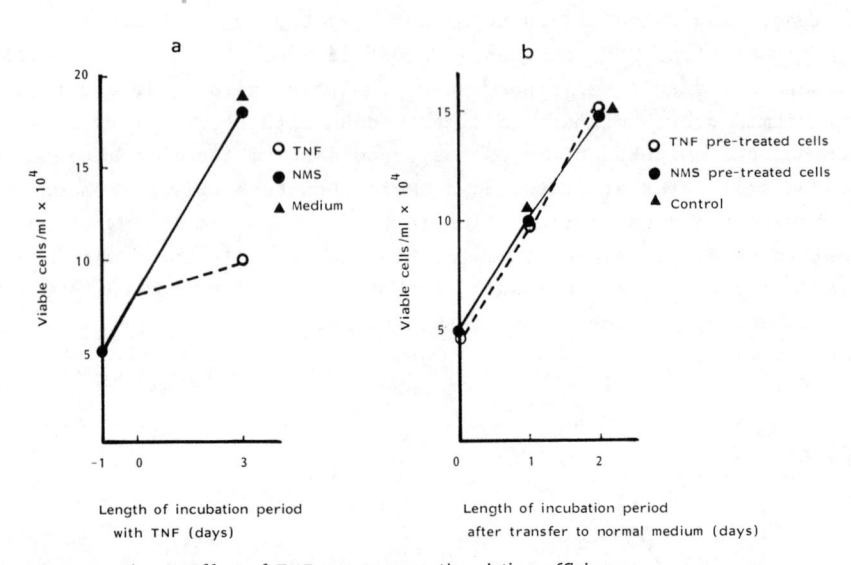

Fig. 4. Effect of TNF exposure on the plating efficiency of SK-MEL-109 (melanoma)

(a) SK-MEL-109 (10 x 10^4 cells/1 ml) were incubated with 1 ml of 16 µg protein of TNF (O), normal mouse serum (●) or control medium (▲). After 3 days' incubation viable cells were harvested by trypsinization and counted.

(b) The viable cells (5 x 10^4 cells/1 ml) were replated into fresh medium without TNF or normal mouse serum and counted after 24 and 48 h incubation. TNF treated cells (O), normal serum treated cells (●) and control medium (▲).

4. Discussion

Helson et al. reported that mouse TNF inhibited the growth of an established human melanoma cell line, although 2 neuroblastomas were not affected (3). It is very important, therefore, to determine whether mouse TNF has an effect on other human cell lines. In this paper we demonstrate that mouse TNF has an effect (cytotoxic and/or cytostatic) on 40 out of 62 human cell lines derived from various tumors. In addition, as shown in Table 3, at each purification step, the increase of specific activity of TNF against a human tumor cell line (BT-20) was completely proportional to that of TNF against a mouse transformed cell line (L-929). These results demonstrate that the action of mouse TNF is not restricted to mouse tumors.

In many studies in vitro, the effect of TNF on L-cells was estimated by its cytotoxic ability (6-8). We found, however, in preliminary experiments that the mode of susceptibility of human cell lines to TNF differed between various cell lines. Some cell lines were inhibited in their cell growth and many dead cells were observed in the supernatant fluid, whereas in other cell lines only inhibition of cell growth was observed. We, therefore, designed an experiment to estimate both the cytotoxic and cytostatic effects of TNF on human cell lines. The cytotoxic effect of TNF was observed in about one-third of the tumor cell lines examined in this study. When these cell lines were exposed to TNF, they did not recover their cell growth ability after being replated in fresh medium without TNF.

On the other hand, TNF-treated SK-MEL-109 cells showed the same growth rate as the control cells when removed to TNF free medium after a 4-week-exposure to TNF. These results suggest that the susceptibility of tumor cell lines to TNF depends neither on the strength of the TNF nor on length of exposure time to TNF. There were also considerable differences in susceptibility to TNF among the same kind of tumor cell lines derived from different patients. These results agree with those obtained in an in vivo study by Carswell et al.(1).

The cell growth form does not seem to correlate with their susceptibility to TNF, for TNF showed an effect on leukemia cell lines grown in suspension as well as on other cell lines grown as a monolayer. The variable effect of TNF on tumor cells, the reason for which is unknown at this time, may be accounted for by the fact that tumor growth is most likely a manifestation of a disparate number of diseases.

The most important feature of TNF is that it does not appear to have any effect on normal cells. Our results demonstrate that 12 normal human cell lines from 4 different kinds of normal tissue were completely resistant to both the cytocidal and cytostatic effects of TNF. In this respect TNF is totally different from lymphotoxin which has cytostatic and cytotoxic effects on L-cells, and also has lytic activity for both normal and neoplastic cells (9-11).

The TNF effect observed in tissue cultures was confirmed in animal experiment. TNF induced a hemorrhagic necrosis of ME-180 subcutaneously transplanted into nude mice (to be published elsewhere). These results suggest that tumor cells which have high susceptibility to the in vitro cytocidal effect of TNF could be most favorable candidates for TNF therapy.

5. References

(1) E. A. Carswell, L. J. Old, R. L. Kassel, S. Green, N. Fiore and B. Williamson, Proc. Natl. Aca. Sci. USA 72 (1975) 3666-3670.

(2) S. Green, U. A. Dobrajansky, E. A. Carswell, R. L. Kassel, L. J. Old, N. Fiore and M. K. Schwarts, Proc. Natl. Acad. Sci. USA 73 (1976) 381-385.

(3) L. Helson, S. Green, E. A. Carswell and L. J. Old, Nature (London) 258 (1975) 731-732.

(4) M. Ito and R. F. Buffett, JNCI 66 (1981) 819-825.

(5) M. R. Ruff and G. E. Gifford, Infec. Immun. 31 (1981) 381-385.

(6) F. C. Jr. Kull and P. Cuatrecasas, J. Immunol. 124 (1981) 1264-1279.

(7) N. Matthews, H. C. Ryley and M. L. Neale, Br. J. Cancer 42 (1980) 416-422.

(8) M. R. Ruff and G. E. Gifford, J. Immunol. 125 (1980) 1670-1677.

(9) G. Kunitomi, W. Rosenau, G. C. Burke and M. L. Goldberg, Am. J. Pathol. 80 (1975) 249-260.

(10) W. Rosenau, M. L. Goldberg and G. C. Burke, J Immunol. 111 (1973) 1128-1135.

(11) J. Sawada and T. Osawa, Jpn. J. Exp. Med. 47 (1977) 87-92.

Effects of Endotoxin Administration on Tumor and Host: An Experimental Observation on Tumor-Bearing Rabbits

Naoto Aoki, Wataru Mori

Department of Pathology, School of Medicine, University of Tokyo, 7-3-1 Hongo, Bunkyo-ku, Tokyo, Japan (113)

Abstract

Based on the clinical observation of patients with cancer, we presumed that not only tumor tissue but also tumor-bearing host might be in a preparative state to the Shwartzman reaction. Thus we performed an experiment on cancer-bearing rabbits which were injected single dose of endotoxin, to see whether the Shwartzman reaction occurred in a generalized form or in a localized form. As a result, in addition to the hemorrhagic necrosis of the tumor tissue, multiple fibrin thrombi in main organs and a high incidence of liver cell necroses were observed. Therefore it is clearly indicated that the generalized Shwartzman reaction was provoked with single injection of endotoxin in some cancer-bearing rabbits. Participation of the Shwartzman mechanism in the pathogenesis of the fulminant hepatitis in patients with cancer, was also suggested.

1 Introduction

The relationship between tumor and Shwartzman reaction has

Abbreviations: GSR: Generalized Shwartzman reaction; LSR: Local Shwartzman reaction; DIC: Disseminated intravascular coagulation; PTAH: Phosphotungstic acid hematoxylin; TNF: Tumor necrosis factor; RES: Reticuloendothelial system.

attracted much attention since Shwartzman (1) found the phenomenon that the rabbit previously received intradermal injection of bacterial culture filtrate developed hemorrhagic necrosis of the injected skin after second intravenous injection of the filtrate. In the meantime, Sanarelli (2) noticed the phenomenon that the animals previously received intravenous injection of Vibrio cholerae, died from shock with severe hemorrhage 24 hrs. after coli filtrate injection. Later, Apitz (3) reviewed and re-examined these phenomena and draw a conclusion that, although these phenomena were different in a topological distribution, the basic mechanism must be very similar. Therefore he designated the phenomenon that Sanarelli found as a generalized Shwartzman reaction (GSR) and the Shwartzman's original skin reaction as a local Shwartzman reaction (LSR).

Then after, the detail mechanism and the influence of drugs and chemicals to the phenomenon were studied. The active principle of the bacterial culture filtrate was attributed to the endotoxin of the bacteria. The rabbits that are injected thorotrast or trypan blue dye develop GSR with only one shot of intravenous injection of endotoxin, namely these substances bear preparative potential (4). Heparin and sodium warfarin are known to have inhibitory effect to the Shwartzman reaction (5,6). Corticosteroids, depending on the timing of administration, have both the inhibitory and preparative effects (7,8). As Apitz (3) exquisitely pointed out, the general rule that a true Shwartzman reaction have to be produced by two injections given at a short interval has two exceptions, i.e., tumor tossie to LSR and pregnancy to GSR. In other words, one local injection of culture filtrate is able to provoke hemorrhagic necrosis of a tumor tissue without any previous treatment and one intravenous injection of culture filtrate precipitate GSR in a pregnant rabbit.

In this article, we would like to present another exception to the above mentioned rule, it is the host bearing malignant neoplasms.

It has long been noticed that the patients with malignancy often develop blood coagulation abnormalities such as venous thrombosis or disseminated intravascular coagulation (DIC) (9,10). And even without overt manifestations, low grade coagulation abnormalities which sometimes corresponded to a state called chronic DIC were pointed out (10-12). As to the relationship between DIC and GSR, they have been known to share common mechanism and to be quite similar phenomena (13). In fact, GSR has been used as an animal model of human DIC, though some differences are pointed out. Therefore, we presumed that not only the tumor tissue but also the tumor-bearing hosts themselves might be in a preparative state to the Shwartzman reaction. In order to examine our presumption and to investigate the influence of the endotoxin to the interactions between tumor and host, we performed an experiment using carcinoma-bearing rabbits.

2 Materials and methods

1) Carcinoma-bearing rabbits

Twenty Vx-2 carcinoma-bearing cottontail rabbits (14) were used as cancer bearing animals. They were used for the experiment 3 to 4 weeks after inoculation of carcinoma to their thigh muscle, when the tumors had grown up to 5 cm in diameter. Fifteen rabbits were used as an experimental group which was injected the endotoxin. Five rabbits were used as a control group which was injected saline instead of endotoxin.

2) Endotoxin

In order to provoke the Shwartzman reaction, E.coli endotoxin (E. coli lipopolysacharide, 0111:B4, Difco Laboratory) dissolved in normal saline was used. According to the endotoxin dosage, rabbits in the experimental group were subdivided into 3 groups. The dosage of the endotoxin which was dissolved in 1 ml of normal saline to group A was 0.1 mg, to group B was 0.5 mg, to group C was 1.0 mg. Control group was given 1 ml of normal saline injection.

3) Methods of observation

The rabbits of the experimental group were sacrificed with pentobarbital overdose 36 to 48 hrs. after a provocative endotoxin injection. After macroscopic examination, main organs and tumor tissue were fixed immediately with formalin immersion. These specimens were processed for a light microscopic examination. Sections were stained with hematoxylin-eosin, phosphotungstic acid hematoxylin (PTAH) and with the Mallory's Azan staining. These histologic changes were compared with those of normal rabbits that were injected endotoxin only one time in our previous experiment.

3 Results

1) The tumor tissue

Vx-2 carcinoma had grown up to 6 cm in diameter in the thigh muscle. The hemorrhagic necrosis of the tumor was found 7 out of 15 rabbits of the experimental group (Fig. 1). Microscopically, these necroses are intense hemorrhagic necroses with polymorphonuclear leucocyte infiltration involving almost whole mass of the tumor. Fibrin microthrombi were found in the small blood vessels of the surrounding tissue of the tumor. These microthrombi were stained blue-purple with PTAH staining (Fig. 2).

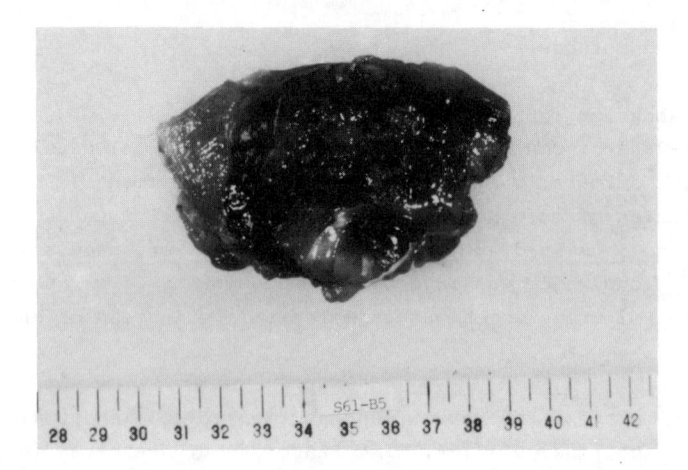

Fig. 1. Macroscopical appearance of hemorrhagic necrosis (dark area in the center) of the tumor.

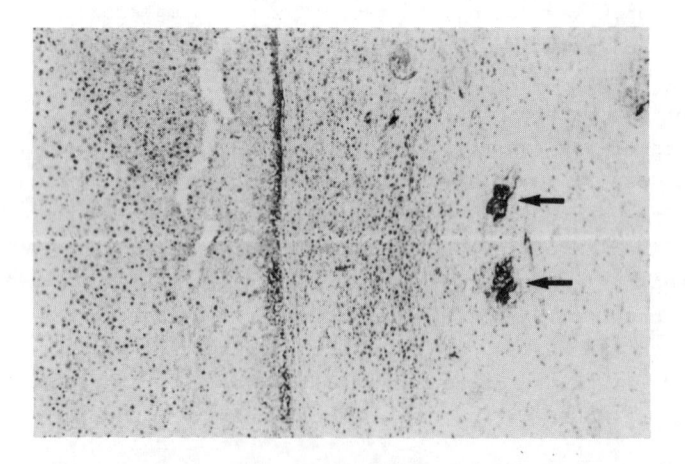

Fig. 2. Fibrin thrombi (arrows) in the small blood vessels in the vicinity of tumor. PTAH staining, x 400.

In the control group, tumor tissues were associated more or less with degeneration or necrosis. These were not hemorrhagic ones but rather lytic ones. No hemorrhagic necrosis was found in the control group. These findings are summarized in Table 1.

Table 1. Hemorrhagic necrosis of tumor (a).

Group	Ent. dose	Rabbit	Hemorrhagic	necrosis	Total
A	0.1mg/1ml	A 1 A 2 A 3 A 4 A 5	− − − + +	2/5	
B	0.5mg/1ml	B 1 B 2 B 3 B 4 B 5	− − − + +	2/5	7/15
C	1.0mg/1ml	C 1 C 2 C 3 C 4 C 5	− + + − +	3/5	
T	Saline, 1ml	T1−T5	−	0/5	0/5

(a) Abbreviations: Ent.: Endotoxin; +: positive; −: negative; Group A, B, C: experimental group; Group T: control group. Figures indicate number of rabbits.

2) The kidneys

Bilateral cortical necrosis, the hallmark of GSR, was found in one rabbit which was injected 1 mg of endotoxin (Fig. 3). Microscopically, this change was associated with extensive coagulation necrosis of the renal cortex with numerous fibrin thrombi in the glomerular tufts, in the arterioles and in the interlobar arteries (Fig. 4). In other two rabbits of the experimental group, numerous fibrin thrombi were found in the glomeruli though they were not associated with cortical necrosis. No microthrombi were found in the kidneys of the control group.

3) The liver

Various degrees of necrosis of the liver were found in 13 out of 15 rabbits of the experimental group. On macroscopic examination, foci of necrosis were recognized as yellowish white flecks of the liver (Fig. 5). Of the liver that showed slight degree of necrosis, the foci were limited within a part of one lobe. Of the liver with marked degree of necrosis, they were distributed all over the lobes, which were often confluent to each other. Microscopically, these were foci of coagulation necrosis of the liver tissue with some lytic necrosis (Fig. 6). They consist of several liver cells up to several lobules of the liver. When a lobule were involved central area often became necrosis and periportal area was preserved. Marked neutrophilic leucocyte infiltration was noticed along the margin of the necrotic area. Microthrombi were found in the sinusoid, portal vein and hepatic vein of the necrotic area and its neighboring tissue.

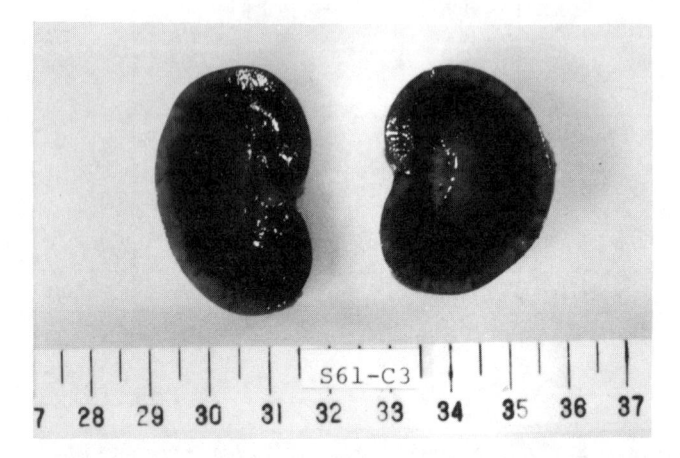

Fig. 3. Bilateral cortical necrosis of the kidneys. The necrotic part is the whitish area in the outer portion of the kidneys.

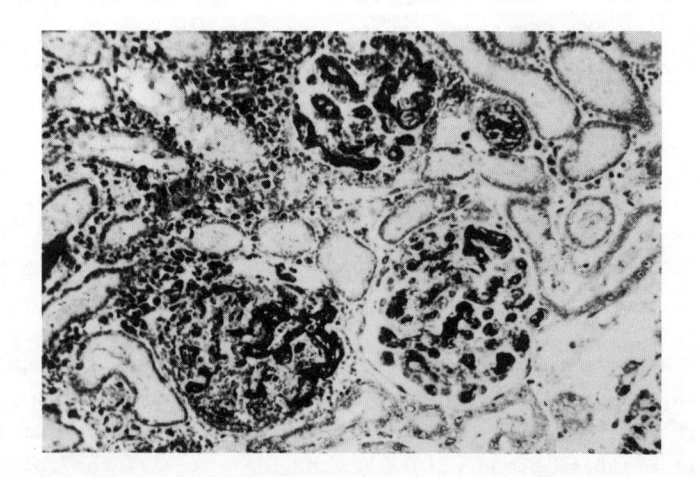

Fig. 4. Microscopical appearance of the kidney in Fig. 3. showing numerous fibrin thrombi (dark portion). PTAH staining, x 400.

Effects of Endotoxin on Tumor and Host

Fig. 5. Macroscopical appearance of necrosis of the liver. They appear as confluent yellowish white flecks (arrows).

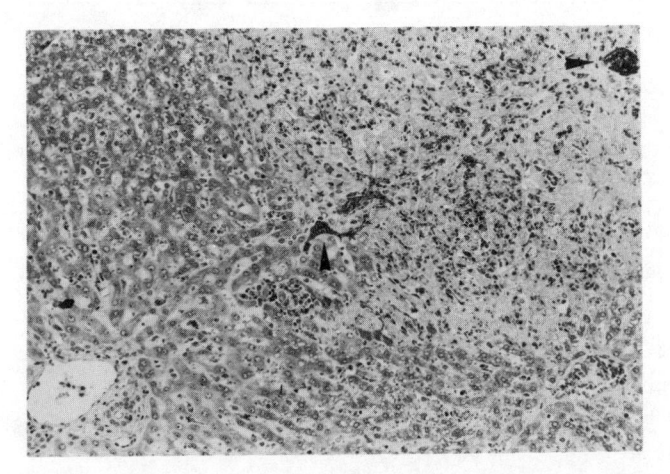

Fig. 6. Microscopical appearance of the liver cell necrosis (right upper area) with fibrin thrombi (arrow heads). PTAH staining, x 100.

The degree of the liver cell necrosis is summarized in Table 2. It was graded according to the following criteria.

Table 2. Liver cell necrosis (b).

Group	Ent. dose	Rabbit	Degree of necrosis	Frequency of necrosis	Total (Frequency)
A	0.1mg/1ml	A 1	M	5/5	
		A 2	++		
		A 3	++		
		A 4	++		
		A 5	+++		
B	0.5mg/1ml	B 1	+	5/5	13/15
		B 2	+		
		B 3	++		
		B 4	+++		
		B 5	+		
C	1.0mg/1ml	C 1	—	3/5	
		C 2	+		
		C 3	+++		
		C 4	—		
		C 5	+		
T	Saline, 1ml	T1—T5	—	0/5	0/5

(b) Abbreviations used are the same as those in Table 1. Criteria for the degree of necrosis are shown in the text.

- M: Microscopical necrosis.
- +: Flecked necrosis, mild and restricted in a limited area of the liver.
- ++: Flecked necrosis, moderate and involving all the lobes of the liver.
- +++: Submassive necrosis, diffuse and often confluent to each other.
- ++++: Massive necrosis.

No liver cell necrosis was found in the control group.

4) The lungs

Microthrombi of the capillaries were found in 6 out of 15 rabbits of the experimental group. Four of these 6 rabbits were associated with metastases of the tumor to the lungs and microthrombi were conspicuous around the metastatic foci. One rabbit that was given 1 mg of endotoxin was associated with marked pulmonary hemorrhage in addition to the tumor metastasis and microthrombi (Fig. 7).

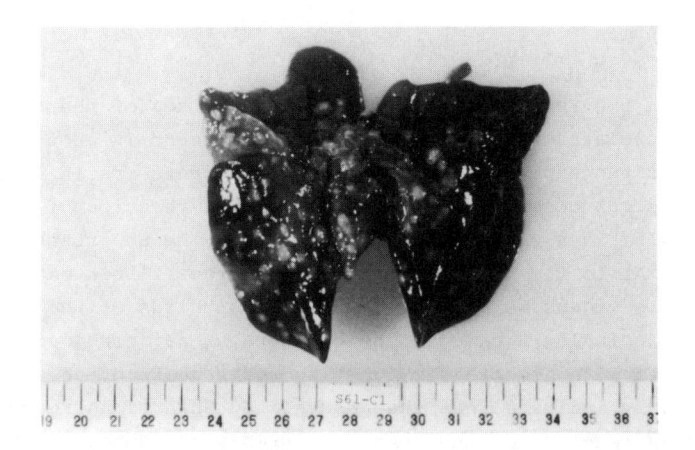

Fig. 7. Macroscopical appearance of the lungs with marked hemorrhage (dark area) and multiple metastatic tumor (whitish nodules).

5) Findings of other organs

Of the spleen, microthrombi were found in 10 out of 15 rabbits of the experimental group. These were located in the capillary around the lymph follicles and the sinusoid. Neither necroses nor microthrombi were found in the pituitary glands or adrenal glands. The findings about fibrin thrombi are summarized in Table 3.

Table 3. Frequency of fibrin thrombi of the organs (c).

	Group A	Group B	Group C	Total (A+B+C)	Group T (control)
Ent.dose	0.1mg/1ml	0.5mg/1ml	1.0mg/1ml		0mg/1ml
Kidney	0/5	1/5	2/5	3/5	0/5
Liver	4/5	5/5	3/5	12/15	0/5
Spleen	4/5	4/5	2/5	10/15	0/5
Lung	2/5	2/5	3/5	7/5	0/5

(c) Abbreviations used are the same as those in Table 1.

4 Discussion

As shown in the results, the hemorrhagic necroses of the tumor which were associated with fibrin thrombi were found in about a half of the rabbits of the experimental group. These results are in accordance with the classical observation of the Shwartzman reaction of the tumor-bearing animals. Thus we succeeded in provoking the Shwartzman reaction of the tumor. As to the general reaction, numerous fibrin microthrombi of the multiple organs were found in about two thirds of the experimental group including a case of the bilateral renal cortical necrosis, which coincided with hemorrhagic tumor necrosis quite often. These findings clearly indicate that the generalized Shwartzman reaction and the Shwartzman reaction of the tumor occurred simultaneously in a rabbit with single intravenous injection of endotoxin. This would bolster our presumption that not only the tumor tissue but also the tumor-bearing hosts are in a preparative state to the Shwartzman reaction.

The association between the malignant neoplasms and the Shwartzman reaction was pointed out exquisitely by Shwartzman himself (15). He described the hemorrhagic necrosis of the malignant neoplasms of the animals when bacterial culture filtrate was given directly into the tumor or intraperioneally or intravenously. Meanwhile, application of bacterial extracts to the treatment of malignant neoplasms had been attempted and numerous literature concerning clinical trials and basic researches on treatment for cancer with the use of bacterial extracts has been reported (16-19). On the other hand, the venous thrombosis, DIC and bleeding tendency have been well known as a clinical manifestation of patients with cancer. According to the hematological examination of cancer patients, they often have low grade coagulation abnormalities without overt manifestation of DIC (20). The abnormalities of coagulation factors such as fibrinogen, plasmin, fibrinogen degradation products, of tumor-bearing animals were also confirmed (21-23). Although the connection between the DIC and GSR was not recognized at the beginning, a close relationship and similalities between them has been repeatedly pointed out since the 1950s. And it is described in a standard textbook that DIC is the human counterparts of GSR in animal (24). In fact, Hjort (25) who compared blood coagulation abnormalities among the different kinds of diseases with GSR had an insight into the similarity between GSR and coagulation abnormalities in patients with cancer, exemplifying the intravascular coagulation, venous thrombosis and bilateral renal cortical necrosis. These observations indirectly suggest that cancer-bearing hosts became preparative to GSR in some instances. Thus our experiment using cancer-bearing rabbits in combination with endotoxin provided direct and strong evidence to our presumption that not only the tumor tissue but also the tumor-bearing hosts might be in a preparative state to GSR.

As to the mechanism contributing to the hemorrhagic necrosis induced by the bacterial extracts, the Shwartzman reaction has long been thought to be one of the major factors. But recent studies on the endotoxin-induced tumor necroses, revealed tumor necrosis factor (TNF) which is directly toxic to the tumor cells and secreted by endotoxin-stimulated macrophages (26). It would be natural to assume that not only the Shwartzman mechanism but also TNF is responsible for the endotoxin-induced tumor necrosis. The relationship between TNF and the Shwartzman mechanism has not clarified yet, it would be reasonable to assume both the two mechanisms operate upon the endotoxin-induced hemorrhagic necrosis of the tumor in vivo.

Concerning about the mechanism that bring the tumor-bearing host to the hypercoagulative state or the Shwartzman reaction preparative state, numerous mechanisms are suggested. For instance, in the case of mucin producing carcinoma it is known that mucus itself activates the coagulation (27). In general, tissue thromboplastins released from destructed tissue by tumor invasion activate the coagulation cascades and tumor cell inflow into the blood stream also stimulates the coagulation factors (28). Based on these hypercoagulability or low grade intravascular coagulation, triggering factors such as endotoxin will precipitate wide-spread coagulation and following necrosis of the tissue and organs. These tissue destruction will enhance the further release of tissue thromboplastin.

One of the interesting observations in our experiment is that the incidence of fibrin thrombi of the liver and spleen was high relative to that of the kidneys. The reticuloendothelial system (RES) which plays an important role as scavengers of coagulation products and other cell fragments occupies a great mass in the liver and spleen (29). The phagocytic activity might be altered in the tumor-bearing state (30,31) and by the consumption of cold insoluble globulin which effects the phagocytic activity of the macrophage (32). Another point we were interested in was the high incidence of liver cell necrosis (87 % in experimental group). We previously succeeded producing massive liver cell necrosis with one injection of endotoxin through the bile duct in tumor-bearing rabbits (33). The necroses produced in this experiment were histologically very similar to the previous one. Therefore we presume that these changes are the result of the Shwartzman reaction in the liver. The hepatic changes in GSR were described by Apitz and others. Apitz (34) found the liver cell necrosis in 2 out of 20 rabbits in classical GSR with two intravenous bacterial filtrate injection. Sakuma (35) described the ultrastructural change of the liver in GSR, which occurred quite often without visible change at the light microscopic level. Although the liver is a target organ in GSR, the described incidence of the liver cell necrosis is relatively low. We infer that the high incidence of the liver cell necrosis in our experiment might be ascribed to the difference of the preparative state, i.e., the tumor-bearing state. Different preparative state, i.e., one intravenous injection of endotoxin, pregnancy, injection of RES blocking agent or tumor-bearing state, could produce the difference of target organ when GSR is provoked. Taking the clinical observation that patients with cancer not infrequently complicate fulminant hepatitis into consideration (36), the high incidence of the liver cell necrosis in our experiment is also noteworthy.

As we postulated in the previous experiment (33), that the Shwartzman mechanism might be involved in the pathogenesis of the fulminant hepatitis in cancer patients. Target organs of the Shwartzman reaction in a case depend on the combination of the mode of preparative state and provocation. And in some peculiar cases the foci are concentrated in a limited organ or tissue such as liver, kidney or tumor tissue. And when a focus of the reaction is aimed at the liver in a cancer patient, it would reveal as a fulminant hepatitis, which is also interpreted as a univisceral or single organ Shwartzman reaction (37, 38). Thus, even though the preparative state appears to be a generalized one the outcome could be a univisceral type of the Shwartzman reaction.

Therefore it would be helpful in treatments and researches to keep in mind that a patient with cancer could be in a preparative state to the Shwartzman reaction and trigger mechanism such as endotoxin is able to precipitate the reaction in the generalized form or in the visceral form.

5 Conclusion

1) We produced the generalized Shwartzman reaction in carcinoma-bearing rabbits with single intravenous injection of endotoxin, which coincide with the hemorrhagic necrosis of the tumor.

2) These observations suggest that, in some carcinoma-bearing hosts, not only the tumor tissue but also the hosts themselves are in a preparative state to the Shwartzman reaction.

3) The participation of the Shwartzman reaction as a pathogenetic mechanism of the disseminated intravascular coagulation, the fulminant hepatitis and other organ changes in patients with cancer is suggested.

4) This concept would be helpful to the clinicians in treatment and researches on patients with cancer.

6 References

(1) G. Shwartzman, Proc. Soc. Exp. Biol. Med. 15 (1928) 560-561.

(2) G. Sanarelli, Ann. Int. Pasteur 38 (1924) 11-72.

(3) K. Apitz, J. Immunol. 29 (1935) 255-266.

(4) R. A. Good, L. Thomas, J. Exp. Med. 96 (1952) 625-641.

(5) R. A. Good, L. Thomas, J. Exp. Med. 97 (1953) 871-887.

(6) S. S. Shapiro, D. G. McKay, J. Exp. Med. 107 (1958) 377-381.

(7) L. Thomas, R. A. Good, Proc. Soc. Exp. Med. 76 (1951) 604-608.

(8) J. G. Latour, J. B. Prejean, W. Margaretten, Am. J. Path. 65 (1971) 189-202.

(9) L. J. Old, E. A. Boyse, Harvey Lect. 67 (1973) 273-315.

(10) A. B. Hagedorn, E. J. W. Bowie, Mayo Clin. Proc. 49 (1974) 647-653.

(11) M. A. Amundsen, J. A. Spittell, Jr., J. H. Thompson, Jr., A. O. Charles, Jr., Ann. Int. Med. 58 (1963) 608-616.

(12) S. P. Miller, J. Sanchez-Avalas, T. Stefanski, L. Zuckerman, Cancer 20 (1967) 1452-1465.

(13) D. G. McKay, Disseminated intravascular coagulation. An intermediary mechanism of disease., Harper & Row, Publishers, New York 1965.

(14) Y. Ito, Proc. Soc. Exp. Biol. Med. 127 (1968) 1106-1111.

(15) G. Shwartzman, Phenomenon of local tissue reactivity and its immunological pathological and clinical significance, Paul B. Hoeber Inc., New York 1937.

(16) H. C. Nauts, G. A. Fowler, F. H. Bogatko, Acta Med. Scand. 145 (1953) 3-103.

(17) E. E. Ribi, D. L. Granger, K. C. Milner, S. M. Strain, J. Nat. Cancer Inst. 55 (1975) 1253-1257.

(18) I. Parr, E. Wheeler, P. Alexander, Br. J. Cancer 27 (1973) 370-389.

(19) A. Nowotony, Bacteriol. Rev. 33 (1969) 72-98.

(20) N. C. J. Sun, E. J. W. Bowie, F. J. Kazmier, L. R. Elveback, C. A. Owen, Mayo Clin. Proc. 49 (1974) 636-641.

(21) G. Mootse, D. Agostino, E. E. Cliffton, J. Nat. Cancer Inst. 35 (1965) 567-572.

(22) K. Abe, H. Fujimoto, E. Endo, y. Fukuoka, J. Khato, H. Sato, Tohoku J. Exp. Med. 117 (1975) 343-350.

(23) A. Poggi, N. Polentarutti, M. B. Donanti, G. de Gaetano, S. Garattini, Cancer Res. 37 (1977) 272-277.

(24) S. L. Robbins, M. Angel, Basic Pathology, 2nd Edt., W. B. Saunders, Philadelphia, Toronto 1976, p.363.

(25) P. F. Hjort, S. I. Rapaport, Ann. Rev. Med. 16 (1965) 135-168.

(26) E. A. Carswell, L. J. Old, R. L. Kassel, S. Green, N. Fiore, B. Williamson, Proc. Nat. Acad. Sci. USA 72 (1975) 3666-3670.

(27) G. F. Pineo, E. Regoeczi, M. W. C. Hatton, E. Rogoeczi, Ann. New York Acad. Sci. 230 (1974) 262-270.

(28) D. Deykin, New Engl. J. Med. 283 (1970) 636-644.

(29) J. P. Nolan, Hepatology 1 (1981) 458-465.

(30) T. G. Antikatzides, T. M. Saba, J. Reticuloendothel. Soc. 22 (1977) 1-12.

(31) N. K. Salky, N. R. DiLuzio, A. G. Levin, H. S. Goldsmith, J. Lab. Clin. Med. 70 (1967) 393-403.

(32) E. Pearlstein, L. I. Gold, A. Garcia-Pardo, Mol. Cellul. Biochem. 29 (1980) 103-128.

(33) N. Aoki, J. Shiga, W. Mori, Proceedings of the 1st conference of the clinical endotoxin research society, Yodo-sha, Tokyo 1979, p.57. (in Japanese)

(34) K. Apitz, Virchow Arch. Path. Anat. Phys. 293 (1934) 1-33.

(35) G. Sakuma, J. J. Reticuloend. Soc. 6 (1967) 92-120. (in Japanese)

(36) W. Mori, Gastroenterol. Jpn. 7 (1972) 46-47.

(37) W. Mori, Histopathology 5 (1981) 113-126.

(38) W. Mori, N. Aoki, J. Shiga, Am. J. Path. 103 (1981) 31-38.

Tumor Regression Induced by Endotoxin Combined with Trehalose Dimycolate

Kazue Fukushi, Hiroshi Asano and Jin-ichi Sasaki

Department of Bacteriology, Hirosaki University School of Medicine, Hirosaki 036, Japan

Abstract

Endotoxic lipopolysaccharide (LPS) or glycolipid (Gl) extracted from Enterobacteriaceae and trehalose dimycolate (cord factor) extracted from mycobacteria both possess adjuvant activity. Synergistic antitumor activity of these two kinds of adjuvants has been shown against guinea pig syngeneic line-10 hepatocellular carcinoma. Because of toxicity of LPS and Gl to the host, it is important to reduce the toxicity or detoxify them altogether in order to use them to control human cancer. In this paper it is reported that the glycolipid extracted from Salmonella minnesota R595 (Re mutant) deacylated with 2% hydroxylamine was rendered essentially nontoxic but retained tumor-regressive potency when combined with trehalose dimycolate of mycobacteria.

1 Introduction

Ribi et al. (1) reported that when certain preparations of endotoxin were combined with trehalose dimycolate of mycobacteria and oil droplets and injected into established malignant line-10 tumors, a high rate of cures and systemic tumor immunity developed. The most powerful endotoxin adjuvants were phenol-water or chloroform-methanol extracts from Re (heptoseless)-mutant, gram-negative bacteria. Further experiments demonstrated that the toxic glycolipid could be rendered essentially nontoxic and nonpyrogenic without loss of tumor regression potency by mild acid treatment (2).

We report in this paper that the endotoxic glycolipid of Salmonella minnesota Re mutant was selectively detoxified by hydroxylaminolysis and retained tumor-regressive potency when it was injected into tumors in combination with trehalose dimycolate of mycobacteria. This antitumor effect was not manifested by injection of the detoxified glycolipid alone.

2 Materials and methods

Endotoxic glycolipid was extracted from the cell walls of a heptoseless mutant of Salmonella minnesota 1167 R595 (chemotype Re) by the phenol-chloroform-petroleum ether method (3). The glycolipid was analyzed chemically and examined with an electron microscope.

Deacylation of the glycolipid was conducted by hydroxylaminolysis (2% NH_2OH in 4% NaOH in ethanol, 63°C, 3 min.) as described by Snyder and Stephens (4). An outline of the hydroxylaminolysis is shown in Figure 1. After hydroxylaminolysis fatty acids were analyzed as methyl esters in Shimadzu Gas Chromatograph GC-4A.

Trehalose dimycolate (cord factor) was extracted from Mycobacterium bovis (BCG) cells and purified as described by Noll and Bloch (5).

Antitumor activity was measured by the method of Ribi et al. (6) using tumor line-10, hepatocellular carcinoma, transplanted to syngeneic strain-2 guinea pigs. Donryu rats bearing methylcholanthrene-induced tumor and BALB/c mice implanted with methylcholanthrene-induced fibrosarcoma were also used in the experiment.

All experimental therapeutic agents were associated with droplets of light mineral oil (Violess U-6, Maruzen Oil Co., Tokyo), dispersed in Tween 80-PBS and injected directly into transplanted line-10 tumor 5 to 7 days after injection. The emulsion in Tween-PBS was prepared with a final concentration of 1% mineral oil and doses of 400 µg of Re glycolipid and 100 µg of trehalose dimycolate in 0.4 ml, the amount injected into each tumor. Diameter of the tumor was measured daily and the cure rate was determined after disappearance of tumor. A principle of our immunotherapy using microbial components is drawn in Figure 2.

After the intradermal tumor had regressed, living line-10 tumor cells were injected into other skin sites of these animals to observe for a reaction of delayed cutaneous hypersensitivity.

Endotoxicity of these therapeutic materials were measured by Limulus Amebocyte Lysate (LAL) test using Limulus-Test Wako (Wako Pure Chemical Industries, Tokyo), chick embryo lethality test as

```
Glycolipid
    ↓
Alkaline hydroxylamine
    │ (2% NH₂OH in 4% NaOH in ethanol)
    ↓
63°C, 3 min.
    ↓
Dialysis
    ↓
Lyophilization
    ↓
Extraction with CHCl₃
    ↓
Lyophilization
```

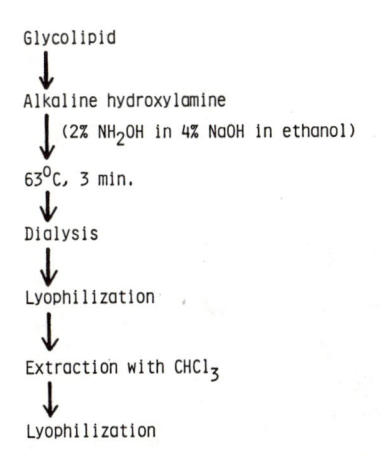

Fig.1. Hydroxylaminolysis of *Salmonella minnesota* glycolipid.

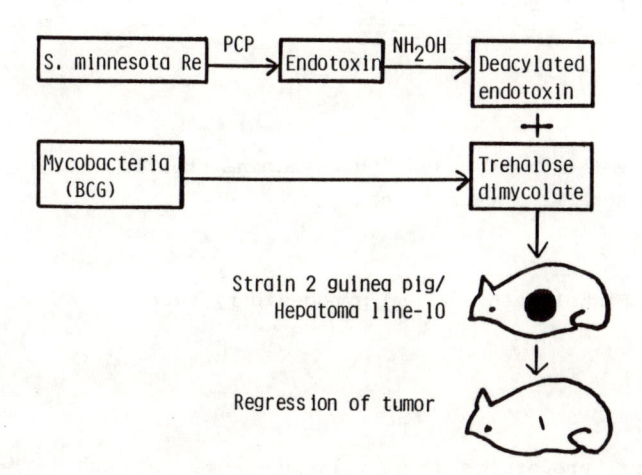

Fig. 2. A principle of immunotherapy of guinea pigs with line-10 tumor.

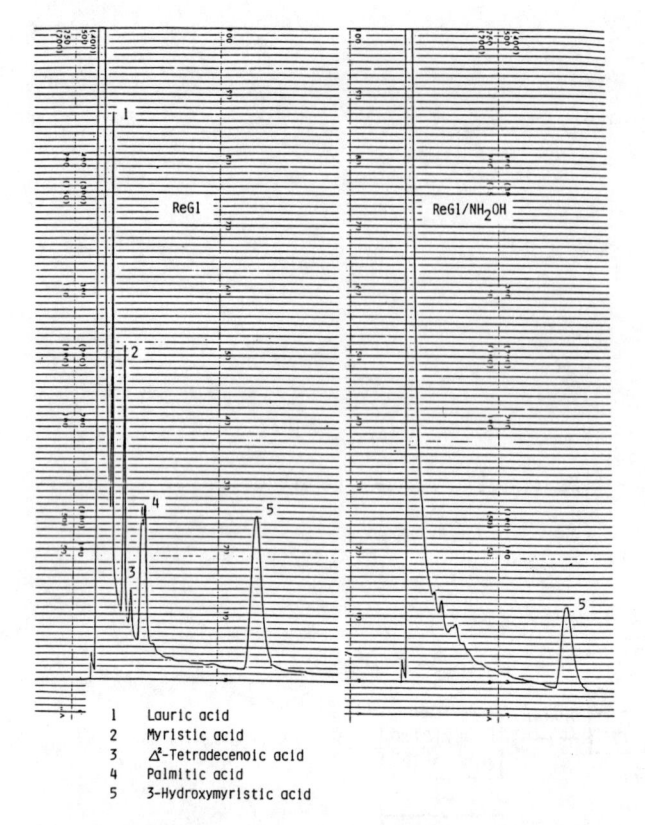

Fig. 3. Gas chromatogram of untreated and hydroxylamine treated Salmonella minnesota Re mutant glycolipid.

1 Lauric acid
2 Myristic acid
3 Δ^2-Tetradecenoic acid
4 Palmitic acid
5 3-Hydroxymyristic acid

described by Milner and Finkelstein (7) and pyrogenicity test.

3 Results

Chemical and biological properties of glycolipid —— Glycolipid extracted from cell walls of Salmonella minnesota R595 was composed of lauric, myristic, Δ^2-tetradecenoic, palmitic, and hydroxymyristic acids as fatty acid moiety (Fig. 3). The data were similar to those obtained in other laboratories (6). The extract was lethal to chick embryo at high levels (CELD$_{50}$=0.05 µg to 0.1 µg), highly pyrogenic (MPD=0.22 µg/rabbit), highly active in LAL test (5 X 10^{-5} µg/ml) and significantly effective in regression of line-10 tumor in strain-2 guinea pigs (24 cured out of 35 animals, cure rate 69%) (Table 1). Thus the toxicity of this extract paralleled its ability to regress

Table 1. Toxicity and antitumor activity of S. minnesota R595 endotoxin

Test		ReG1	Deacylated ReG1
Limulus	µg/ml	5×10^{-5}	5×10^{-2}
Chick embryo lethality (CELD$_{50}$)	µg/egg	0.1 - 0.05	>50
Pyrogenicity (MPD)	µg/rabbit	0.22	89
Tumor regression of guinea pigs No.cured animal/No. treated animal		24/35 (69%)	22/70 (31%)

tumors.

Chemical and biological properties of glycolipid deacylated with alkaline hydroxylamine —— Figure 3 shows that hydroxylaminolysis of Re glycolipid liberated almost all normal fatty acids (lauric, myristic, Δ^2-tetradecenoic and palmitic acids), but 3-hydroxymyristic acid was retained. These data are similar to those reported by Lüderitz et al. (8) and Amano et al. (9). Treatment of Re glycolipid with hydroxylamine destroyed all activities of LAL gelation (5×10^{-2} µg/ml), chick embryo lethality (CELD$_{50}$ >50 µg), and pyrogenicity (MPD=89 µg), whereas the deacylated glycolipid retained the ability to regress line-10 tumors in strain-2 guinea pigs (22 cured out of 70 animals, cure rate 31%) (Table 1). Even though the deacylated glycolipid was nearly 1,000 times less toxic than the untreated glycolipid, the tumor-regressive potency was not significantly reduced.

Relationship among LAL, chick embryo lethality and pyrogenicity —— A simplified schematic diagram (Fig. 4) shows the relationship among potencies of LAL, chick embryo lethality and pyrogenicity of both Re glycolipids (untreated ReG1 and deacylated ReG1). There is a remarkable parallelism among all potencies tested in the present experiment.

Efficacy of Re glycolipids in regression of line-10 tumor in strain-2 guinea pigs —— Survival and cure rate of guinea pigs with line-10 tumor are shown in Figures 5 and 6, which represent the data obtained by a single injection of the glycolipid. All animals treated

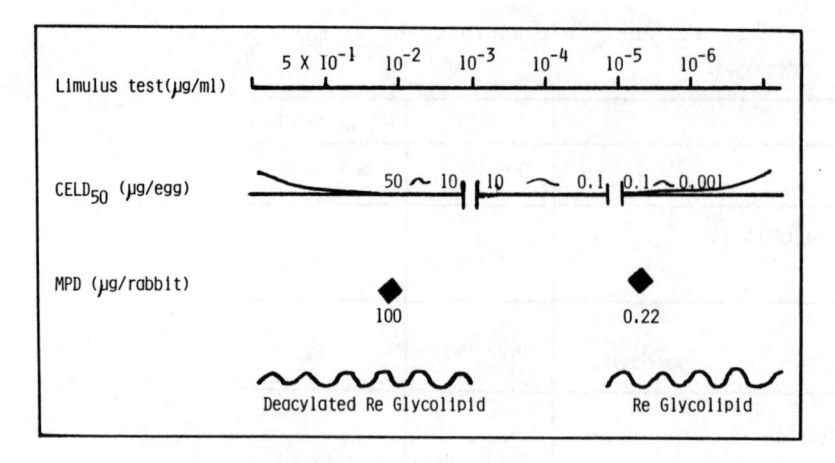

Fig. 4. Relationship among Limulus test, chick embryo lethality and pyrogenicity of the untreated and deacylated endotoxic glycolipids.

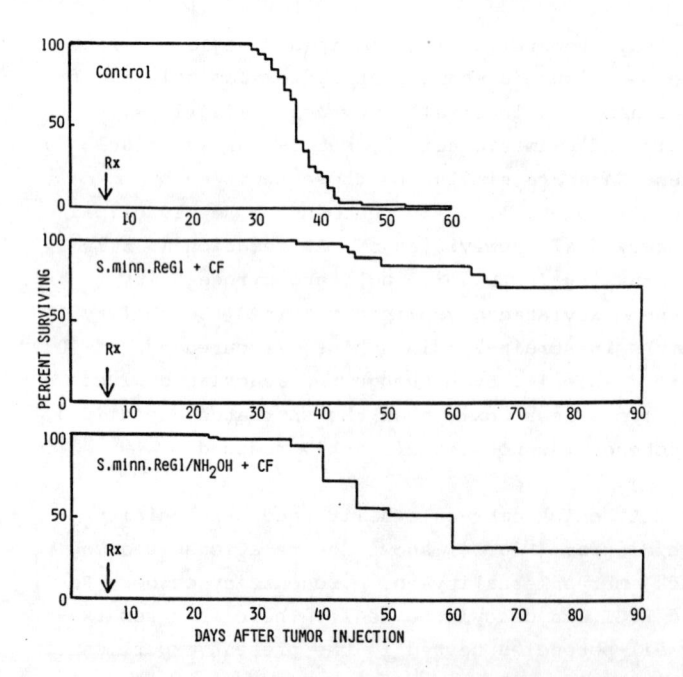

Fig. 5. Survival time and survival rate obtained by a single injection of a combination of endotoxic glycolipid and trehalose dimycolate.

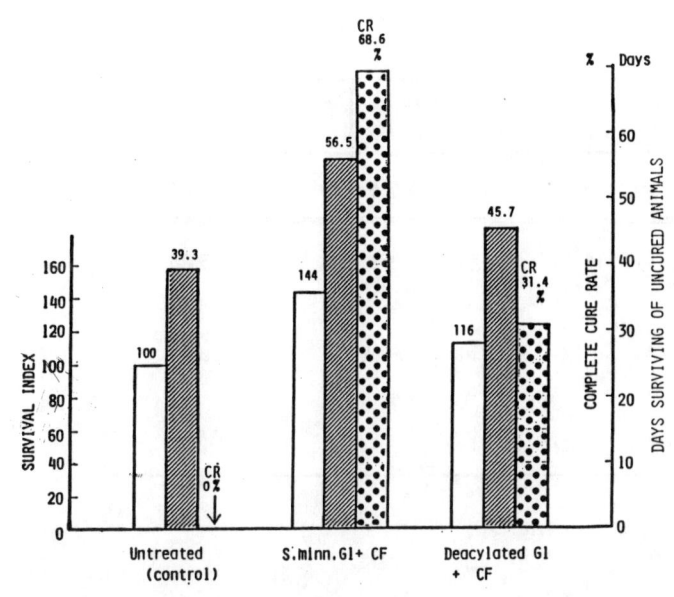

Fig. 6. Complete cure rate, survival time and survival index obtained by a single injection. ▦ cure rate, ▨ survival time, ☐ survival index.

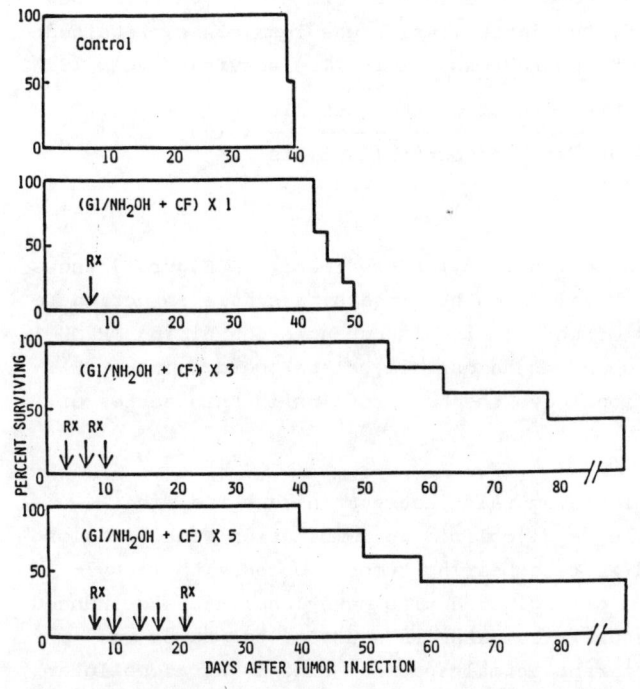

Fig. 7. Survival time and survival rate obtained by multiple sequential injections of a combination of adjuvants.

Table 2. Tumor regression in guinea pigs, rats and mice by intratumoral treatment

Animal	Guinea pigs Strain 2	Rats Donryu	Mice BALB/c
Tumor	Hepatoma line-10	MC induced	Fibrosarcoma Meth-A
ReGl + TDM*	24/35 (69%)	5/20 (25%)	9/24 (38%)
Deacylated ReGl + TDM*	22/70 (31%)	ND	2/8 (25%)
Untreated	0/33 (0%)	0/2 (0%)	0/32 (0%)

*TDM, Trehalose dimycolate.

with oil-Tween-PBS, the controls in this experiment, died within 50 days after injection of tumor cells. Though the combined treatment of Re glycolipid with trehalose dimycolate was more effective than the deacylated Re glycolipid, the latter, which was less toxic, retained significantly high ability to regress tumors. The survival index (SI)

$$SI = \frac{\text{Days surviving of treated animals}}{\text{Days surviving of control animals}} \times 100$$

is shown in Figure 6.

Multiple injections of glycolipid for treatment —— Figure 7 shows a comparison between survival rates obtained by a single injection to 5 sequential injections of the combined substances containing NH_2OH-deacylated Re glycolipid and trehalose dimycolate (cord factor; CF). Three sequential injections gave the best results in this series of experiments.

Tumor regression by intratumoral treatment in rats and mice —— Therapeutic effect of the Re glycolipid in combination with trehalose dimycolate was observed in rats bearing tumor induced with methylcholanthrene and in mice transplanted with methylcholanthrene-induced fibrosarcoma. Table 2 shows a comparative study of the tumor-regressive potency in different animal species. Line-10 hepatocellular carcinoma implanted in strain-2 guinea pigs regressed most significantly to a complete cure. It seems that there is some difference in susceptibility of animal species.

Rejection of rechallenge with tumor cells —— After the intradermal tumor had regressed, all cured guinea pigs showed positive skin reaction to a lethal dose of line-10 tumor cells and rejected challenge with tumor cells. Similar injection into non-immunized guinea pigs elicited no skin reactions. This appears to indicate that growth and regression of line-10 tumor cells in the skin leads to the acquisition of systemic immunity.

4 Discussion

Ribi et al. (1) reported the consistent high efficacy of combination of the endotoxic glycolipid from Re mutant Salmonella and trehalose dimycolate from mycobacteria in causing regression of line-10 tumors in strain-2 guinea pigs, elimination of lymph node metastasis and resistance to rechallenge with line-10 tumor cells. They showed in further studies that the toxicity of the endotoxin may be separable from tumor-regressive activity.

Tanamoto et al. (10) reported that LPS obtained from protein-LPS complex of autolysates of Pseudomonas aeruginosa and LPS of Salmonella and E. coli inhibited experimental tumor development in ascites and induced interferon activity in vitro. The tumor inhibitory activity was retained in the LPS de-O-acylated by 0.1 N NaOH in ethanol, but diminished by complete de-O-acylation by NH_2OH.

In the present paper, it was demonstrated that Salmonella minnesota Re glycolipid deacylated by alkaline hydroxylamine caused regression of a significant proportion of line-10 tumors, without endotoxicity. We assume that the fatty acids linked to amino groups of the glucosamine disaccharide residue of the lipid A are essential for tumor-regression activity.

Recently, Amano et al. (9) reported that the majority of the glycolipids of the Re mutant of Salmonella typhimurium were rendered nontoxic but retained antitumor activity when hydrolyzed with boiling 0.1 N hydrochloric acid, which split KDO and glycosidic phosphate from the glycolipid molecules. They concluded that glycosidic bound phosphate and at least a portion of the fatty acids of the lipid A moiety are essential for toxicity, but that this phosphate is not an essential structural feature for tumor-regressive activity.

Our studies do not elucidate the role of KDO as a component of glycolipid in the tumor-regressive activity.

Acknowledgments. The authors wish to thank Prof. Dr. O. Lüderitz of the Max-Planck-Institut für Immunologie, Freiburg, for kindly providing the mutant strains of Salmonella, and Dr. S. Sato of the Mitsubishi Chemical Industries Ltd., Tokyo, for his experiments on chick embryo lethality and pyrogenicity.

5 References

(1) E. Ribi, D. L. Granger, K. C. Milner and S. M. Strain, J. Nat. Cancer Inst. 55 (1975) 1253-1257.

(2) E. Ribi, K. Amano, J. Cantrell, S. Schwartzman, R. Parker and K. Takayama, Cancer Immunol. Immunother. 12 (1982) 91-96.

(3) C. Galanos, O. Lüderitz and O. Westphal, Eur. J. Biochem. 9 (1969) 245-249.

(4) F. Snyder and N. Stephens, Biochim. Biophys. Acta, 34 (1959) 244-245.

(5) H. Noll and H. Bloch, J. Biol. Chem. 214 (1955) 251-265.

(6) T. J. Meyer, E. E. Ribi, I. Azuma and B. Zbar, J. Nat. Cancer Inst. 52 (1974) 103-111.

(7) K. C. Milner and R. A. Finkelstein, J. Infect. Dis. 116 (1966) 529-536.

(8) O. Lüderitz, C. Galanos, V. Lehmann, H. Mayer, E. T. Rietschel and J. Weckesser, Naturwiss. 65 (1978) 578-585.

(9) K. Amano, E. Ribi and J. L. Cantrell, J. Biochem. 93 (1983) 1391-1399.

(10) K. Tanamoto, C. Abe, J. Y. Homma and Y. Kojima, Eur. J. Biochem. 97 (1979) 623-629.

Hemorrhagic Tumor Necrosis Induced by Endotoxin

Nikolaus Freudenberg, Kensuke Joh*, Chris Galanos, Marina A. Freudenberg,
and Otto Westphal

Pathologisches Institut der Universität Freiburg, Albertstr. 19,
7800 Freiburg, Federal Republic of Germany
Pathologisches Institut der Jikei-Medizinischen Hochschule Tokyo, Japan
Max-Planck-Institut für Immunbiologie, 7800 Freiburg, Federal Republic of Germany

ABSTRACT

Endotoxin-induced necrosis of the Methylcholantren (Meth A) mouse tumour has been investigated using morphological, immunohistochemical and enzyme histochemical methods. The earliest alteration in the tumour occurred already 30 minutes after lipopolysaccharide (LPS) injection. Endotoxin could be detected immunohistochemically in interstitial macrophages of the tumour, but never in neoplastic cells. This was followed by local granulocytosis, increase of acid phosphatase activity and hyperemia in the tumourous tissue. Hemorrhagic necrosis of the tumour was complete 4 hours after endotoxin administration. Rejection of the necrotic tumour was complete by the end of the 2nd week after treatment. The hemorrhagic tumour necrosis may be compared with a localized Shwartzman-like reaction induced by endotoxin, and probably involving endogenous mediators such as the tumour necrotizing factor.

1 INTRODUCTION

It is well known that endotoxins from the outer cell-wall of Gram-negative bacteria can both inhibit the induction of a tumour and lead to its regression when present. The regression of a tumour due to bacterial infection in a human being has already been reported by Busch (1) and Coley (2) in the last century. These original observations initiated the investigation on the role of bacterial toxins particularly of Gram-negative bacterial endotoxin in the inhibition of tumour development. After parenteral administration of endotoxin hemorrhagic necrosis occurs in solid tumours.

*Stipendiat des Deutschen Akademischen Austauschdienstes (DAAD)

The pathomechanism of this necrosis is still unclear, however, a number of mediators such as the tumour necrotizing factor (TNF, 3) have been claimed to participate in the above necrosis.

Although much information is available on the LPS-induced tumour necrosis there exists practically no morphological characterization of the above phenomenon. In the limited number of morphological studies carried out in earlier years on the hemorrhagic tumour necrosis by bacterial toxins a role of the LPS may be assumed. In such studies ill-defined bacterial extracts, such as bacterial filtrates or mixed bacterial toxins were employed and the necrosis observed may not be regarded as a pure LPS-induced effect. In the present study we investigated the alterations in the Meth A mouse tumour brought about by pure standardized endotoxin preparations using morphological, immunohistochemical and enzyme histochemical methods.

2. METHODOLOGY

The Meth A tumour cells were cultivated intraperitoneally in CBF_1-mice. 5×10^5 tumour cells were injected intradermally into the abdominal wall of BALB/c-mice. Under these conditions a solid tumour develops which is visible already 3 days after transplantation.

On day 8 after transplantation the tumour bearing mice received 20 µg of Salmonella abortus equi S-form LPS injected intravenously.

The necrosis of the tumour was followed for 14 days after LPS-administration using direct observation, histology, transmission electron microscopy, indirect immunoperoxidase staining in order to localize the endotoxin, and demonstration of the acid phosphatase activity in the tumour.

3 RESULTS

Morphology of intradermally growing Meth A tumour

By the 8th day after the transplantation the Meth A tumour is about 8 mm in diameter and replaces the entire thickness of the subcutaneous tissue and the underlying abdominal musculature as far as the parietal layer of the peritoneum. Up to the 4th or 5th week after transplantation the tumour grows extensively and brings about massive ulceration of the skin (fig. 1). By the 6th week after transplantation at the latest the mouse dies.

Histologically one sees a malignant, poorly differentiated neoplasm of which the histogenesis is not clearly determined (fig. 2).

The morphological appearance of a malignant tumour under the light microscope is confirmed by electron microscopy (moderate nuclear polymorphy with increase in heterochromatin and also striking malignant nucleoli). The tumour is composed of closely packed cells between which a narrow interstitial space can be seen. Intercellular connections such as desmosomes cannot be demonstrated. Against a finely granulated, moderately electron dense background single bundless of collagen fibres are seen lying between the cells.

Tumor Necrosis Induced by Endotoxin

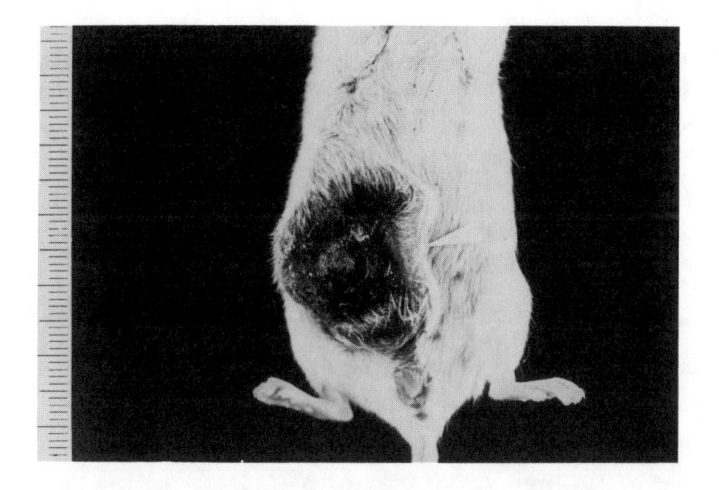

Fig. 1   Meth A tumour, 4 weeks after intradermal transplantation of tumour cells (from (17)

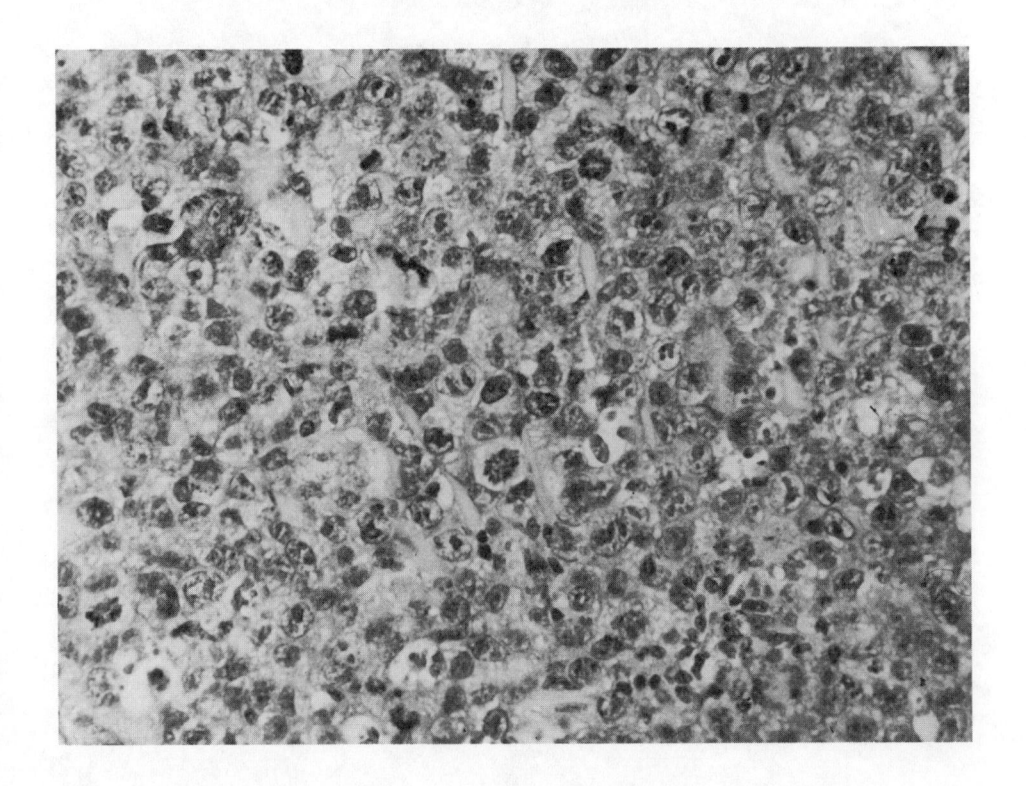

Fig. 2   Histology of the Meth A mouse tumour, 8 days after transplantation (from (17)

Time-course of the endotoxin induced tumour necrosis

Already 30 minutes after LPS-administration endotoxin could be demonstrated immunohistochemically in the tumour (fig. 3). It was internalized by mononuclear cells in the intima of capillaries in the tumour. Between 60 minutes and 24 hours following endotoxin administration the LPS was found in phagocytes, not only of the vessel walls but also of the interstitial space of the tumour.

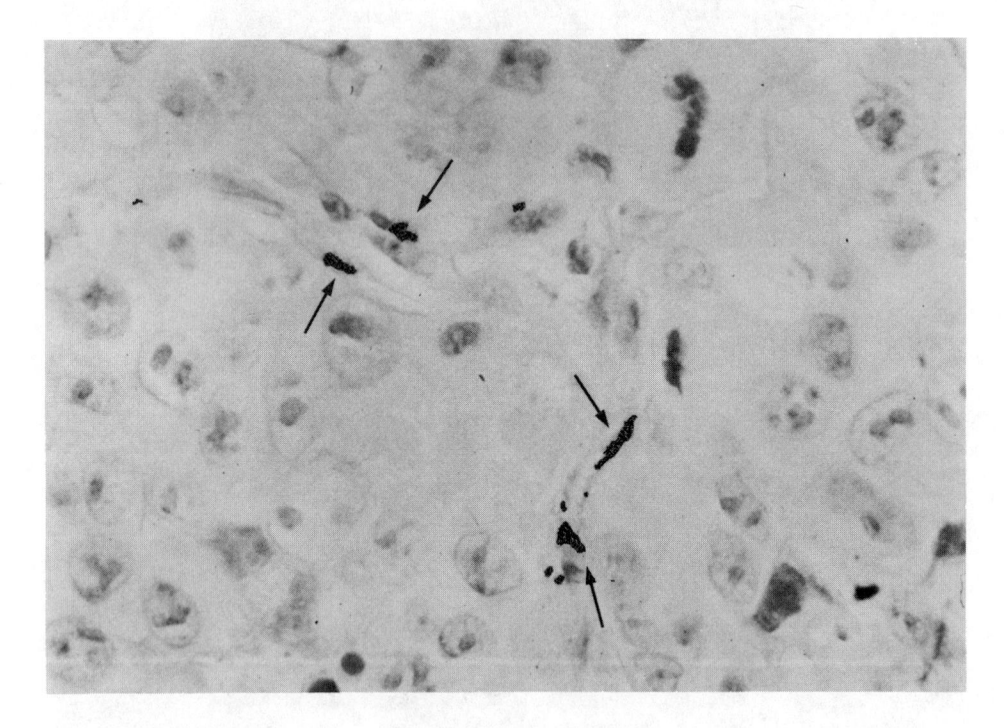

Fig. 3   Immunohistochemical demonstration of endotoxin in perivascular macrophages (→) in the tumour, 30 minutes after LPS administration (from (18)

The first ultrastructural changes occurred 90 minutes after injection of LPS. The tumour cells were forced apart by an increase in the interstitial oedema fluid, and the separated cells showed a tendency to become circular. Single leukocytes can be seen here and there in the tumour. Two hours after LPS injection the necrosis is much advanced (fig. 4). Among single cells with pyknotic nuclei other cells can be seen with strongly degenerative changes in the cytoplasm mostly consisting of the appearance of vacuoles, and the intercellular spaces are now packed with detritus.

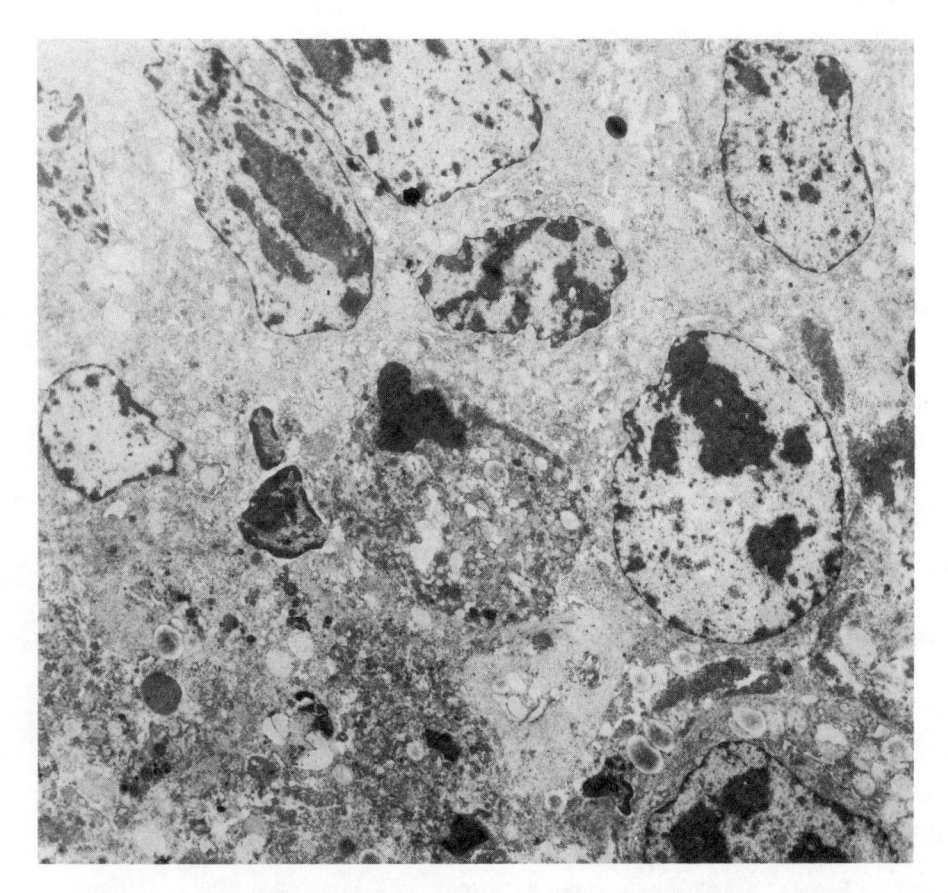

Fig. 4   Ultrastructure of the Meth A tumour, 2 hours after LPS-treatment, showing marked degenerative alterations

Up to 60 minutes after endotoxin treatment under the light microscope a granulocytosis and a hyperemia of the vessels running through the tumour were striking.

Three hours following LPS injection an increased number of extracellular matrix vesicles was recognized under the electron microscope. This alteration was accompanied by a marked increase of the activity of acid phosphatase in numerous tumour cells as well as in interstitial macrophages in the tumour.

Four hours after endotoxin injection even with the naked eye an extensive central hemorrhage can be recognized, leaving only a narrow intact rim of neoplastic tissue at the periphery.

Histologically the individual tumour cells are widely dissociated and show degenerative changes (fig. 5).

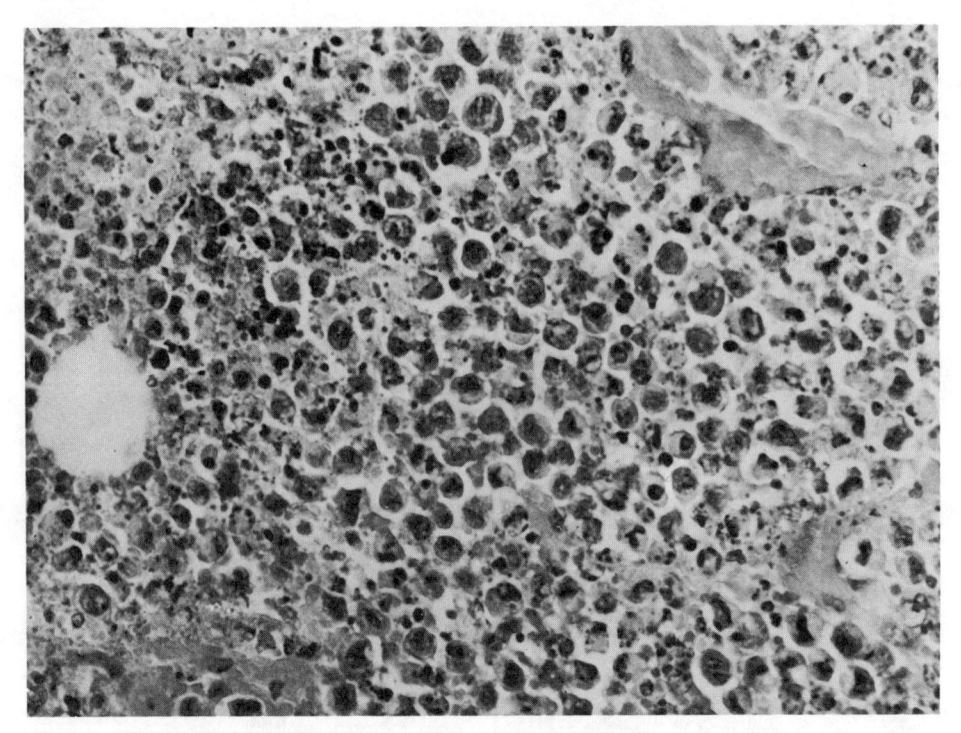

Fig. 5   Histology of the hemorrhagic necrosis of the Meth A mouse tumour, 4 hours following endotoxin treatment (from (17))

At day 4 after endotoxin treatment necrosis is far advanced and the greater part of the tumour has been rejected.

Fourteen days after beginning of LPS treatment the original cutaneous defect has been reduced to a mere dempla, beneath which a deficiency of the abdominal musculature remains. Round about the rejected and necrotic tumour newly developed skin with a delicate epidermis and a corium poor in collagen fibres can be seen.

3 CONCLUSIONS

The hemorrhagic tumour necrosis, induced by relatively non toxic amounts of endotoxin, is an impressive example of a useful property of LPS.

At present, one may only speculate on the mechanism of the endotoxin-induced tumour necrosis. Our own immunohistochemical observation that endotoxin, although if appeared in the tumour, internalized in macrophages, was however never detectable on or in the tumour cells themselves is in agreement with the findings of other investigators (7,8,9), which could not demonstrate direct cytotoxic effects of LPS on tumour cells. Till now a direct damaging effect of endotoxin on cells was shown only on macrophages in in-vitro experiments (10).

The early activation of lysosomes in interstitial leukocytes and in tumour cells, 3 hours after LPS application, parallels with the accumulation of endotoxin in the tumour. These findings may suggest that endotoxin induces activation of lysosomes and liberation of their enzymes (11). Whereby in the case of tumour lysosomes the activation may be indirect since endotoxin was not found associated with these cells. Together with the increase of cellular and extracellular lysosomal activity we observed a marked degeneration of the interstitium and of tumour cells. Ultrastructurally, the tumour necrosis was advanced, two hours before macroscopically visible alterations became evident by an increasing accumulation of an interstitial detritus, which is probably due to the catabolic effect of lysosomes at this time of the experiment. The source of the numerous lysosomes should be seen amongst macrophages and tumour cells above all in granulocytes, which invaded the tumour in an increasing number up to 60 minutes after LPS injection.

The excessive hyperemia, which can be found as an early effect of LPS administration in the tumour, must be interpreted as a circulatory disturbance during the degenerative and inflammatory process in the neoplastic tissue following endotoxin application.

We know now that a "tumour necrotizing factor" (TNF, 3) plays a part in the generation of tumour necrosis. This factor has been demonstrated in the serum of mice infected with BCG and to which endotoxin was administered (3). TNF produces a massive hemorrhagic necrotizing inflammation in solid tumour, a response which Shwartzman and Michailovsky (12) interpreted as a localized Shwartzman reaction. If this be so, the presence of the tumour itself corresponds to the first injection of the classical experiment, and it is on the tumour that some unidentified substance presumable acts in senzitation of the vascular endothelium. Stetson (13) demonstrated that solid tumours treated with endotoxin reacted with a marked but short-lived vascular disturbance. Our own observations on the severe degree of hyperemia in the vessels of the tumour after LPS administration confirms this finding. In spite, however, of the dramatic destruction of the central tissue of the tumour by hemorrhagic necrosis, it is generally believed that only in a relatively small number of cases does the remainder of the neoplastic tissue completely regress (14). So far as the size of the tumour represents a decisive factor for ultimate complete regression after the hemorrhagic necrosis we agree with Shear (15), Donelly at al. (6), Parr et al. (16) and Berendt et al. (14). Only in the case of an 8-day-old tumour of 8 mm in diameter were we able to bring about completely satisfactory result with the use of LPS. In contrast to the hemorrhagic necrosis, the tumour regression induced by endotoxin was shown to be dependent upon a T-cell mediated anti-tumour immunity that is only produced against immunogenic tumours (Berendt et al. (14). Also in this case the size of the tumour was shown to be critical for an effective anti-tumour immunity. Since the Meth A tumour investigated here is reckoned among the immunogenic neoplasms (14), in our experiments, in addition to the hemorrhagic necrosis a T-cell mediated regression may contribute to the complete tumour destruction induced by endotoxin.

4 REFERENCES

(1) Busch, Klin. Wschr. 3 (1866) 245
(2) W.B. Coley, Bull. Johns Hopkins Hosp. 65 (1896) 157
(3) E.A. Carswell, L.J. Old, R.L. Kassel, S. Green, N. Fiore, B. Williamson, Proc. Nat. Acad. Sci. USA 72 (1975) 3666-3670
(4) H.B. Andervont, Am. J. Cancer 27 (1936) 77-83
(5) I.C. Diller, Cancer Research 7 (1947) 605-626
(6) A.J. Donelly, H.F. Havas, M.E. Groesbeck, Cancer Research 18 (1958) 149-154
(7) C. Yang, A. Novotny, Infect. Immun. 9 (1974) 95-100
(8) M.J. Berendt, P.H. Saluk, Infect. Immun. 14 (1976) 965
(9) D. Tripodi, L. Hollenbeck, W. Pollack, Int. Arch. Allergy Appl. Immunol. 35 (1970) 575
(10) R.V. Maier, J.C. Mathison, R.J. Ulevitch in J.A. Majde, R.J. Person (Eds) Pathophysiological effects of endotoxins at the cellular level, Allan, R. Liss Inc. New York 1981 pp. 133-155
(11) S.G. Bradley in J.A. Majde, R.J. Person (Eds) Pathophysiological effects of endotoxins at the cellular level, Allan R. Liss Inc. New York 1981 pp. 3-14
(12) G. Shwartzman, N. Michailovsky, Proc. Soc. Exper. Biol. Med. 29 (1932) 737-741
(13) J. Stetson, Bull. N.Y. Acad. Sci. 101 (1961) 80
(14) M.J. Berendt, G.F. Mezrow, P.H. Saluk, Infect. Immun. 21 (1978) 1033-1035
(15) M.J. Shear, J. Natl. Cancer Inst. 4 (1943) 81
(16) I. Parr, E. Wheeler, P. Alexander, Br. J. Cancer 27 (1973) 370
(17) N. Freudenberg et al., in preparation
(18) K. Joh. et al., in preparation

Endotoxin Target Cells and Endogenous Mediators: The Role of Macrophages and Arachidonic Acid Metabolites

Ulrich Schade, Thomas Lüderitz, and F nst Th. Rietschel

Forschungsinstitut Borstel, D-2061 Borstel, FRG

1 INTRODUCTION

Lipopolysaccharides (LPS), the endotoxins of gramnegative bacteria, are known to induce in humans and experimental animals a variety of acute pathophysiological reactions which resemble manifestations of gramnegative sepsis. These effects include fever, neutropenia and thrombocytopenia, hypotension, disseminated intravascular coagulation (DIC), abnormal glucose levels and metabolic changes like respiratory acidosis and metabolic alkalosis. If administered in higher doses LPS causes irreversible shock and death. These reactions are largely due to the action of endogenous mediators which are produced and released by host cells in response to LPS activation. In this sense, LPS represents a toxin only because the host reacts to it with the formation of secondary "toxic mediators". In the present paper some current views on the nature of LPS-sensitive host cells and the mediator role of arachidonic acid metabolites, secreted by such cells, will be discussed.

2 ARACHIDONIC ACID METABOLITES AS MEDIATORS OF ENDOTOXIN ACTION

Among the various mediators postulated to participate in LPS action, special attention has been paid to arachidonic acid metabolites, in particular to the biologically highly active prostaglandins (PG's) and thromboxane A_2 (TxA_2). This is because a number of endotoxic phenomena can be suppressed by inhibitors of PG synthesis, and in a variety of studies the beneficial effect of these drugs was demonstrated in several shock models (1, 2, 3, 4).

Studies in dogs (5, 6, 7), cats (8, 9), cows (10), sheep (11), and horses (12) suggest that different nonsteroidal antiinflammatory drugs (NSAID's) such as aspirin, indomethacin or meclophenamate, which inhibit the cyclooxygenase, act mainly on the initial phase of the animal's response to endotoxin. The delayed phase of endotoxin shock was beneficially influenced by NSAID's only in dogs (13) and baboons (14) (in a LD 70 model). Contradictory data were reported for the action of these inhibitors in the late shock phase in cats, rats, and sheep (8, 15, 17). From these findings it appears that PG's may play an active role in immediate responses to endotoxin. This is underlined by the finding that polyphloretin phosphate, an antagonist of $PGF_{2\alpha}$ prevents the rise in pulmonary artery pressure. In addition, the production of $PGF_{2\alpha}$ has been found to be increased in bovine lungs during the early phase of shock (9).

More recently, thromboxane A_2 (TxA_2), a potent vasoconstrictor with high platelet aggregating activity, has been associated with early endotoxin-induced pulmonary phenomena (12, 31). The involvement of TxA_2 in endotoxicosis is suggested by the finding that treatment of rats with imidazol (an inhibitor of Tx synthesis) and 13-azaprostanoic acid (an antagonist of Tx action) provided effective protection against the lethal effects of LPS (16). However, in studies with baboons the Tx biosynthesis inhibitors imidazol and OKY-1581 did not influence the survival of animals although TxA_2 synthesis was effectively suppressed (17). Thus, the involvement of TxA_2, predominantly in the early events of shock, appears likely, its participation in the lethal (late) phase of endotoxaemia remains contradictory.

The finding that arachidonic acid given together with LPS effectively diminishes endotoxin lethality shows that eicosanoids do not, exclusively, have a detrimental action (18). The beneficial effect of arachidonic acid has been postulated to originate from endogenously formed prostacyclin (PGI_2), a very potent inhibitor of platelet aggregation and a strong vasodilator (19). Thus, when PGI_2 was infused concomitantly with LPS it increased survival, but an enhanced fall in systemic blood pressure was observed (20). Interestingly, PGI_2 has been found to be produced during the late phase of endotoxin shock and has been deemed responsible for the fall in systemic blood pressure. Endotoxemic dogs treated with PGI_2 reacted with a prompt increase in cardiac index, a late increase in mean arterial pressure after discontinuation of PGI_2 infusion and an increase in circulating platelets (no effect on circulating leukocyte levels was seen) (21). These data suggested an influence of PGI_2 on the number of circulating platelets, and its positive effects in endotoxemic states could be based on an inhibition of platelet aggregation. However, when LPS-induced platelet aggregation was tested in canine platelet rich plasma, neither NSAID's nor PGI_2 had any demonstrable, anti-aggregatory effect (20). Therefore, although PGI_2 is very effective in the treatment of experimental endotoxemia, its exact mode of action remains to be elucidated (the possible involvement of leukotrienes in endotoxin shock is briefly discussed below).

3 TARGET CELLS FOR LPS

LPS, after injection, is known to interact with a variety of host systems. The nature of injurious humoral or cellular mediation systems remains, however, so far only poorly defined. Among the very early host reactions to LPS application are thrombocytopenia and granulo cytopenia. It was demonstrated that 0.2-2 min after injection of radiolabelled LPS most of the radioactivity was associated with platelets and smaller amounts with monocytes, and neutrophils (while no radioactivity was found in the lymphocyte fraction (23)). If LPS was injected as a complex with high density lipoprotein (HDL), however, no significant binding to platelets occurred. Nevertheless, despite the prevention of LPS binding to platelets the animals went into endotoxic shock (22). This could also be observed when animals were, prior to LPS injection, treated with cobra venom factor, whereby complement (C')-dependent association of LPS with platelets is inhibited. These results provide evidence that the initial C'-dependent responses of platelets to LPS are not required for development of shock (24, 28). Further, binding of LPS to thrombocytes does not appear to play an essential role in LPS-induced hypotension and DIC.

Protein-free preparations of LPS from different strains of E. coli and Salmonella sp. were ineffective in initiating several in vitro responses in polymorphonucleated neutrophils (PMN). As assessed by a number of parameters, including chemiluminescence, superoxide anion generation, iodination and release of granular constituents, endotoxin failed to stimulate these cells (41). These data showing the complete lack of direct activation of PMN's in vitro by LPS are in contrast to their response to LPS in vivo. The role of PMN's in endotoxic events, therefore, remains so far unclear. However, it can be concluded from these findings that, although LPS has very profound effects on the number of platelets and granulocytes in vivo, interaction of these cells with LPS does not seem to be essential for the development of endotoxic shock.

The results of numerous studies show that the most likely candidate as primary target cells for endotoxic actions are macrophages. This is underlined by the fact that radioactive LPS applied intravenously to experimental animals accumulated mainly in the phagocytic vacuoles of macrophages in liver and spleen (23). The coagulopathies of Gramnegative sepsis are most probably mediated by a macrophage product, i.e. by a complex of phospholipids and protein, the tissue factor or procoagulating factor, which promotes association of factor VII with calcium (25). Moreover, macrophages produce a number of hormone-like proteins (monokines) which act on lymphoid and other cells (for review see 26).

4 MACROPHAGES: A SOURCE OF LPS-INDUCED ARACHIDONIC ACID METABOLITES

Macrophages not only produce numerous biologically highly active proteins but they also represent a major source of PG's which are released in response to a variety of different stimuli. PG synthesis is observed in macrophage cultures in vitro during phagocytosis of inflammatory particles (zymosan), opsonized red blood cells or heat killed bacteria. The trigger for the activation of arachidonate metabolism appears to be correlated with the attachment phase of phagocytosis (27). However, PG synthesis can also be stimulated by soluble compounds such as phorbolmyristate acetate, ConA, muramyldipeptide, and notably, LPS. Thus, we and others have shown that mouse peritoneal macrophages or cultures of bone marrow derived (pure) macrophages, incubated with LPS, release substantial amounts of PGE$_2$ and PGF$_{2\alpha}$ (29, 30).

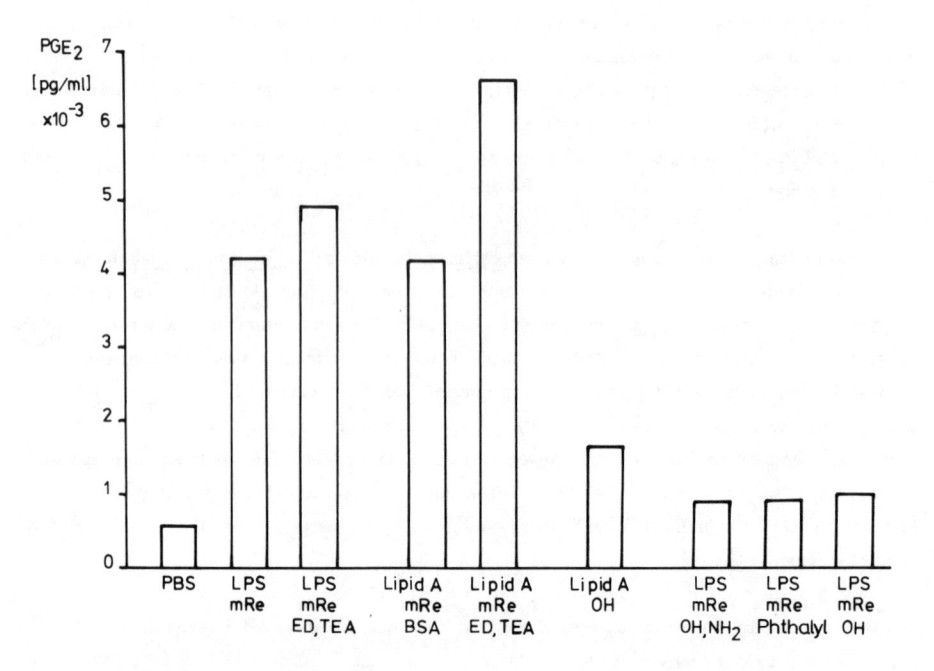

Fig. 1: Induction of Prostaglandin E$_2$ - Release from Mouse Peritoneal Macrophages by LPS, free Lipid A and nontoxic Derivatives
The amount of PGE$_2$ was estimated by radioimmunoassay. (ED, electrodialized; TEA, triethylammonium salt; BSA, complexed to bovine serum albumine; OH, treated with alkali (0.2 N NaOH, 100°C, 1 h); NH$_2$, treated with hydrazine (100°C, 10 h); phthalyl, treated with phthalic acid anhydride).

The hypothesis that LPS induced PG synthesis from macrophages is involved in the genesis of shock is based on the correlation of this cellular activity and the endotoxic properties of LPS. The active principle of LPS, responsible for the majority of endotoxic effects observed in vivo, is located in the lipid A component (32). Certain chemical modifications of lipid A result in a significant attenuation of its biological activity, including pyrogenicity and lethality. Thus, it is known that O-deacylated LPS (or free lipid A) and preparations which are devoid of both, ester and amide-bound fatty acids exhibit low or no endotoxicity in vivo. Such materials were also found to be of low activity in inducing PG synthesis in mouse peritoneal macrophages in vitro. As shown in Fig. 1, alkali-treated free lipid A or LPS, as well as hydrazine-treated or phthalylated preparations were not active. PG synthesis was induced, however, to a significant extent by untreated LPS or free lipid A. Thus, S-form LPS from Salmonella minnesota (Sms) and LPS from a deep rough (Re) mutant of Salmonella minnesota as well as the triethylammonium salt of free lipid A and free lipid A complexed to bovine serum albumine, increased the levels of PGE_2 in macrophage cultures (Fig. 1). These experiments show that the lipid A component represents the active moiety of LPS responsible for the induction of PG synthesis in macrophages and they further indicate that the endotoxic activity of lipid A in vivo is paralleled by its capacity to induce PG synthesis in vitro.

Further evidence for the involvement of macrophage-derived PG's in endotoxicosis emerges from the following findings.

i. The mouse strain C3H/HeJ has been recognized as being largely resistant to typical endotoxin activities including lethal toxicity (33). As was shown in our and other laboratories, peritoneal macrophages of this mouse strain could not be stimulated by LPS to secrete enhanced levels of PGE_2 and $PGF_{2\alpha}$. In contrast, cells from genetically related LPS responder mouse strains C3H/HeN or C3H/SwSn released, as expected, significant amounts of PGE_2 and $PGF_{2\alpha}$ when treated with LPS (34, 35). Thus, the genetically determined host resistance to LPS correlates with the unresponsiveness of host macrophages to respond to LPS stimulation with PG production.

ii. It is well documented that a single dose or repeated injections of sublethal amounts of LPS render experimental animals and humans less responsive to a second LPS challenge. This phenomenon is generally referred to as endotoxin tolerance (33). Tolerance can be induced to most of the activities associated with LPS, including pyrogenicity and lethality. When macrophages of mice rendered tolerant to LPS lethality (as judged by LD 50 measurements) were incubated with endotoxin, no increased PG production was observed. This cellular refractoriness could not be overcome by higher LPS doses or variations in culture conditions of macrophages. It is noteworthy that macrophages from tolerant mice incubated with zymosan released significant amounts of PG's (35).This findings shows that the capacity of "tolerant macrophages" to produce PG's was intact, and it further indicates that the observed unresponsiveness to endotoxin was LPS-specific. Therefore, also in this model, the susceptibility of mice to LPS lethality in vivo agrees well with the capacity of their macrophages to produce PGE_2 and $PGF_{2\alpha}$ after incubation with LPS in vitro.

Table 1

Arachidonic Acid Metabolites released from Peritoneal Macrophages (Mouse) after Stimulation with Lipopolysaccharide and Zymosan
(Metabolites were, after derivatisation, estimated by combined gas-liquid chromatography/mass spectrometry)

Stimulus	PGE_2	TxB_2	6-keto-$PGF_{1\alpha}$
	pg/ml		
Saline	60 ± 14	b.d.[1]	60 ± 12
LPS[2]	880 ± 28	b.d.	70 ± 15
Zymosan[3]	160 ± 12	1 120 ± 56	624 ± 72

[1] below detection limit   [2] 24 h, 37°C, 10 µg/ml   [3] 1 h, 37°C, 500 µg/ml

As was pointed out, among the arachidonate metabolites, TxA_2 and PGI_2 seem to play a dominating role in LPS activities. In order to test the possibility that macrophages also represent the cellular source for PGI_2 and TxA_2, mouse peritoneal macrophages were incubated with LPS in vitro, and the stable derivatives of TxA_2 and PGI_2, i.e. TxB_2 and 6-keto-$PGF_{1\alpha}$ were determined by combined gas-liquid chromatography/mass spectrometry. For comparison cells were also incubated with zymosan. Table 1 shows that the amounts of TxB_2 detected in supernatants of LPS-stimulated macrophages was below the limit of detection. The PGI_2 derivative 6-keto-$PGF_{1\alpha}$ could be detected, but the levels in the supernatants of stimulated cells were similar to those of controls. In contrast, LPS-treated cells released, as expected, high amounts of PGE_2. When cells were incubated with zymosan, the production of PGE_2 was very low, whereas TxB_2 and 6-keto-$PGF_{1\alpha}$ were detected in large amounts. These results indicate that peritoneal macrophages are not able to produce TxA_2 and PGI_2 to a significant extent in response to LPS under the experimental conditions applied, although they possess the potential to synthesize these metabolites. From these data it may be concluded that cells other than peritoneal macrophages are responsible for the LPS-induced production of TxA_2 and PGI_2 which are observed in the circulation of mammals in LPS shock states. An alternative hypothesis that LPS-induced (serum) factors change the pattern of PG's produced by macrophages in vivo is presently under investigation in our laboratory.

More recently it was found that mononuclear phagocytes have a set of lipoxygenases at their disposal to convert arachidonate to hydroxylated derivatives, which include the leukotrienes (LT's) (36). These metabolites represent highly active mediators of inflammatory and hypersensitivity reactions (37). It has been shown that inhibitors of the lipoxygenase pathway as well as antagonists of LTC_4 are able to prevent, to a large degree, endotoxic shock in D-galactosamine-sensitized mice (38), indicating that lipoxygenase products may play a significant role in endotoxicosis.

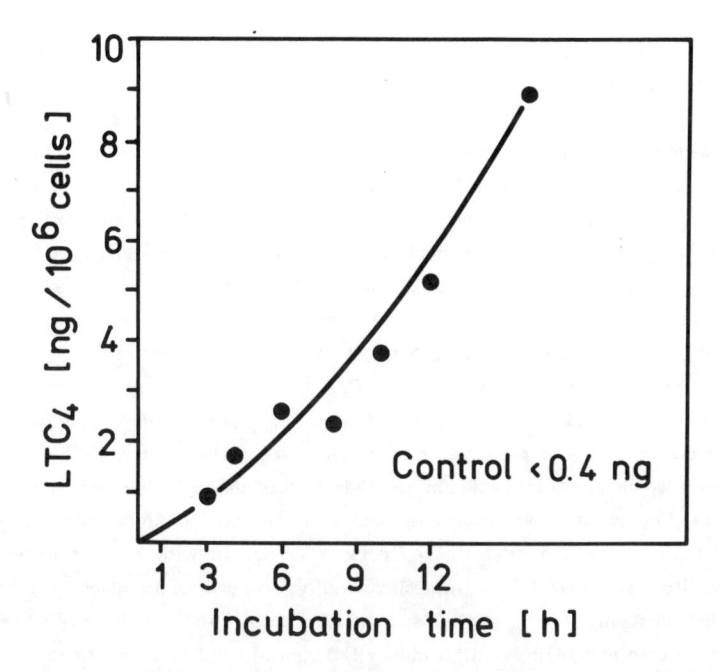

Fig. 2: Amount of Leukotriene C_4 detectable in Supernatants of Mouse Peritoneal Macrophages incubated for different Time Periods with LPS

The amount of LTC_4, eluted by reversed phase HPLC, was estimated by integration of the UV trace at 280 nm in comparison with synthetic LTC_4.

We have recently shown (39) that peritoneal macrophages of mice, incubated with LPS (10 µg/ml, 16 h, 37°C, 5×10^6 cells/ml), produced, in a dose and time dependent manner, substantial amounts of LTC_4 (approximately 9 ng/10^6 cells). The time dependence of LTC_4 formation is shown in Fig. 2.

These results provide evidence that macrophages are the possible source for lipoxygenase products possessing significance in endotoxin shock. However, it remains to be elucidated whether the role of lipoxygenase products is restricted to the above test system of sensitized mice or whether it applies to endotoxin shock states in general.

5 CONCLUDING REMARKS

In the present manuscript some findings are described which support the concept that arachidonic acid metabolites are of importance in the mediation of acute and delayed endotoxin effects. These metabolites include prostaglandins, thromboxane A_2 and leukotrienes (LT's). Macrophages are capable of producing both cyclooxygenase and lipoxygenase products when treated with bacterial endotoxin in vitro, and it is likely that this cell type also represents a major source of arachidonic acid metabolites in endotoxic reactions in vivo. Little is known, however, on the mechanisms by which these metabolites exert their action in endotoxicosis. Since several LPS activities (including fever, abortion, increased vascular permeability and constriction of airway smooth muscles) are inducible by prostaglandins or leukotrienes, these may represent true mediators which act on susceptible target cells, organs or body sites which are distinct from the reticuloendothelial system. On the other hand, it has been demonstrated that the LPS-induced release of collagenase from macrophages is regulated by PGE_2 (40). Additionally, it was suggested that LPS-induced PG synthesis in phagocytes is under the control of LTC_4 (43). Thus, PG's and LT's may, at the site of production, function as regulators of the formation (and/or release) of other macrophage products. It, therefore, is possible that arachidonic acid metabolites promote, in macrophages, the synthesis of secondary mediators which are responsible for the induction of pathophysiological, i.e. endotoxic reactions. The finding that cyclooxygenase and lipoxygenase inhibitors provide protection against endotoxin shock and other LPS-effects is in accord with both concepts. Today, effective inhibitors of PG and LT biosynthesis as well as antagonists are at our disposal. Since such drugs indeed suppress some of the detrimental endotoxin effects, and since irreversible (septic) shock due to endotoxin is still a clinically important but therapeutically unresolved problem, future intensified research in this field appears to be indicated and justified.

References

(1) J.R. Fletcher, P.W. Ramwell, Br. J. Pharmacol. 61 (1977) 175-181.

(2) J.R. Parratt, R.M. Sturgess, Br. J. Pharmacol. 60 (1977) 209-219.

(3) E.F. Smith, J.H. Tabas, A.M. Lefer, Prostagland. Med. 4 (1980) 215-225.

(4) J.R. Fletcher, P.W. Ramwell, Adv. Shock Res. 4 (1980) 103-111.

(5) J.R. Culp, E.G. Erdös, L.B. Hinshaw, D.D. Holmes, Proc. Soc. Exp. Biol. Med. 137 (1971) 219-223.

(6) E.G. Erdös, Biochem. Pharmacol. Suppl. 17 (1968) 283-291.

(7) E.G. Erdös, L.B. Hinshaw, C.C. Gill, Proc. Soc. Exp. Biol. Med. 125 (1967) 916-919.

(8) C.V. Greenway, V.S. Murphy, Brit, J. Pharmacol. 43 (1971) 259-269.

(9) J.R. Parratt, R.M. Sturgess, Brit. J. Pharmacol. 58 (1976) 547-551.

(10) F.L. Anderson, T.J. Tsagaris, W. Jubiz, H. Kuida, Amer. J. Physiol. 228 (1975) 1479-1482.

(11) R.C. Hall, R.L. Hodge, R. Irvine, F. Catic, J.M. Middleton, J. Trauma 16 (1976) 968-973.

(12) E.R. Jacobs, M.E. Soulsby, R.C. Bone, F.J. Wilson, F.C. Hiller, J. Clin. Invest. 70 (1982) 536-541.

(13) F.L. Anderson, W. Jubiz, T.J. Tsagaris, H. Kuida, Amer. J. Physiol. 228 (1975) 410-414.

(14) J.R. Fletcher, P.W. Ramwell, Adv. Prost. Thromb. Res. 7 (1980) 821-828.

(15) M.J. Reichgott, K. Engelman, Circ. Shock 2 (1975) 215-219.

(16) J.A. Cook, W.C. Wise, P.V. Halushka, J. Clin. Invest. 65 (1980) 227-230.

(17) L.C. Casey, J.R. Fletcher, M. Zmudka, P.W. Ramwell, J. Pharmacol. Exp. Ther. 222 (1982) 441-446.

(18) J.T. Flynn, J. Pharmacol. Exp. Ther. 206 (1978) 555-566.

(19) B.J.R. Wittle, S. Moncada, J.R. Vane, Prostaglandins 16 (1978) 373-388.

(20) J.R. Fletcher, P.W. Ramwell, Circ. Shock 7 (1980) 299-308.

(21) M.M. Krausz, T. Utsunomya, G. Feuerstein, J.H.N. Wolfe, D. Shepro, H.V. Hechtman, J. Clin. Invest. 67 (1981) 1117-1125.

(22) J.C. Mathison, R.J. Ulevitch, J. Immunol. 123 (1979) 2133-2143.

(23) J.C. Mathison, R.J. Ulevitch, J. Immunol. 126 (1981) 1575-1580.

(24) J.C. Mathison, R.J. Ulevitch, J.R. Fletcher, C.G. Cochrane, Am J. Path. 101 (1980) 245-263.

(25) D.C. Morrison, R.J. Ulevitch, Am. J. Path. 93 (1978) 526-617.

(26) R. van Furth, in: Mononuclear Phagocytes, Functional Aspects, Part 2 (R. van Furth ed.), Nijhoff, Boston (1980) 1271-1417.

(27) W. Hsueh, F. Gonzalez-Crussi, E. Hanneman, Nature 283 (1980) 80-82.

(28) M.A. Freudenberg, C. Galanos, Infect. Immun. 19 (1978) 875-882.

(29) U. Schade, H. Fischer, M.L. Lohmann-Matthes, B. Peskar, E.Th. Rietschel, 4. Europ. Immunol. Meeting Budapest (1978), Abstract.

(30) J.I. Kurland, R. Bockman, J. Exp. Med. 147 (1978) 952-957.

(31) J.R. Fletcher, in: Adv. Prostag. Thromb. Res. (B. Samuelsson, P.W. Ramwell and R. Paoletti eds.), Raven Press, New York (1980) 843-849.

(32) E.Th. Rietschel, C. Galanos, O. Lüderitz, O. Westphal, in: Immunopharmacology and the Regulation of Leukocyte Function (D.R. Webb ed.), M. Dekker, New York and Basel (1982) 183-229.

(33) C.H. Johnston, S. Greissman, in: Pathophysiology of Endotoxin (L. Hinshaw ed.), Elsevier /North Holland, Biomedical Press, in press.

(34) L.M. Wahl, D.L. Rosenstreich, L.M. Glode, A.L. Sandberg, S.E. Mergenhagen, Infect. Immun. 23 (1979) 8-13.

(35) U. Schade, E.Th. Rietschel, in: Natural Toxins (D. Eaker and T. Wadström eds.), Pergamon Press, Oxford, New York (1980) 271-277.

(36) C.A. Rouzer, W.A. Scott, Z.A. Cohn, P. Blackburn, J.M. Manning, Proc. Natl. Acad. Sci. U.S.A. 77 (1980) 4928-4932.

(37) B. Samuelsson, S. Hammarström, R. Murphy, P. Borgeat, Allergy 35 (1979) 375-381.

(38) W. Hagmann, D. Keppler, Naturwissenschaft 69 (1982) 594-595.

(39) Th. Lüderitz, E.Th. Rietschel, U. Schade, Ann. Immunol. Hung. 23 (1983) in press.

(40) R. Bhatnagar, U. Schade, E.Th. Rietschel, K. Decker, Eur. J. Biochem. 125 (1982) 125-130.

(41) D.C. Morrison, M.E. Wilson, S. Raziuddin, S.J. Betz, B.J. Curry, Z. Oades, P. Mundenbeck, in: Microbiology-1980 (D. Schlessinger ed.), Am. Soc. Microbiol., Washington DC (1980) 30--35.

(42) J. Rokach, Y. Girard, Y. Guindon, J.G. Atkinson, M. Larue, R.N. Young, P. Masson, G. Holme, Tetrahed. Lett. 21 (1980) 1485-1488.

(43) N. Feuerstein, J.A. Bash, J.N. Woody, P.W. Ramwell, Biochem. Biophys. Res. Comm. 100 (1981) 1085-1090.

ACKNOWLEDGEMENTS

We thank B. Müller, S. Schwechten and A. Roth for excellent technical assistance, M. Lohs and B. Köhler for preparing illustrations and photographs, and R. Hinz and G. Dettki for typing this manuscript. Synthetic LTC_4 was kindly donated by Dr. J. Rokach (42).
Part of the work described in this paper was supported by a grant from the Stiftung Volkswagenwerk.

Recovery of Immune Response in Immunodeficient Mice after Administration of Lipopolysaccharide: LPS-Induced Non-Specific Resistance in Beige Mice with Chèdiak-Higashi Syndrome

Masayasu Nakano, Kazuyasu Onozuka, Tatsuo Saito-Taki, Nagahiro Minato*

Department of Microbiology and Department of Clinical Immunology*,
Jichi Medical School, Tochigi 329-04, Japan

Abstract

Bacterial lipopolysaccharide (LPS)-induced non-specific resistance in beige (bg/bg) mutant mice with Chediak-Higashi syndrome (CHS) against salmonella or staphylococcal infection was studied. The beige mice were more susceptible than the original C57BL/6 ($+/+$) mice or heterozygous ($bg/+$) mice derived from this strain to the infection with a small number of virulent organisms of *Salmonella enteritidis* No.11 strain or *S. typhimurium* LT2 strain. Bactericidal ability of beige mice for these salmonella or the *Staphylococcus aureus* 248βH strain were obviously weaker than that of control $+/+$ or $bg/+$ mice. When LPS was injected into the mice 1 day before challenge, the beige mice as well as control mice gained some resistance to the infection and showed an increase of bactericidal activity. The effect of LPS was dose-dependent. When LPS was given several days before infection or simultaneously, LPS-induced bactericidal activity in beige mice became obscure. The effect of lipid A-bovine serum albumin (BSA) complex was less than that of LPS, and alkali-treated detoxified LPS had a very weak effect. Furthermore, natural killer (NK) cell activity in the spleen cell population and interferon activity in the sera of beige mice were heightened by LPS-injection.

1 Introduction

The Chèdiak-Higashi syndrome (CHS) in man is rare autosomal recessive disorder characterized by pale skin that sunburn easily, pale or blue-gray hair color, decreased uveal pigment, photophobia, frequent pyrogenic infections and the presence of giant, irregularity shaped lysosomes in most granual containing cells (1). A similar disorder has been known in minks, cattle, killer whales, cats and mice (2). Beige mutation in mice is a readily available small animal model of CHS. The mutation arose spontaneously as a recessive coat color mutant located at the centromeric end of chromosome 13 (3). This mutation in the homozygous condition results in increased susceptibility to infection (3, 4), decreased chemotaxis and bactericidal capability of granulocytes (5), and marked deficiency in natural killer (NK) cell activity (6). The basic defect in CHS is thought to be cellular, especially in microtubles, and it may cause abnormal motility and fusion of lysosomes in phagocytic cells (2).

As bacterial lipopolysaccharide (LPS) is a potent immunomodulator (7), it may have some ability to improve the immunodeficiency in beige mice. We have examined some effects of LPS *in vivo* in beige mice. In the present paper, we report improvements of protectivity and bactericidal activity against salmonella or staphylococcal infection and significant augmentation of NK cell activity in these mice by LPS.

2 Materials and Methods

<u>Animals</u>. The beige (*bg/bg*) mice, a mutant from the C57BL/6 (+/+) strain, were originally obtained from the National Institutes of Health, Bethesda, U.S.A., and have been maintained in our laboratory. As controls, we used C57BL/6 mice or heterozygous F1 hybrid (*bg/+*) mice by mating female beige mice with male C57BL/6 mice. These mice were supplied from our own animal colony and used at 8 to 10 weeks of age.

<u>Bacterial lipopolysaccharide (LPS) and other immunomodulators</u>. LPS was extracted from *Salmonella typhimurium* LT2 by the hot-phenol-water procedure as described by Westphal (8) and then purified by repeated centrifugation at 100,000 x g for 60 min. Lipid A was prepared from the LPS by hydrolysis with 1% acetic acid at 100°C for 4 hr, and then the lipid A was complexed (9) by the method of Galanos *et al.*(10) with bovine serum albumin (BSA). Alkali-detoxified LPS was prepared from the LPS by hydrolysis with 0.25N NaOH at 56°C for 1 hr according to the method of Neter *et al.*(11). These materials were dissolved in physiological saline and injected intraperitoneally (ip) into mice (0.2ml/mouse). Synthetic *N*-acetyl-muramyl-*L*-alanyl-*D*-isoglutamine (MDP) and its derivative *N*-acetyl-muramyl-*L*-alanyl-*D*-isoglutamyl-*L*-stearoyl-

lysine (L18-MDP) were gifts from Daiichi Seiyaku Company, Tokyo. MDP and L18-MDP were dissolved in 0.02M phosphate buffered saline (pH 7.6) and injected subcutaneously (sc) into mice (0.2ml/mouse).

Bacterial strains. Virulent *Salmonella enteritidis* No.11 strain, *Salmonella typhimurium* LT2 strain and *Staphylococcus aureus* 248βH strain were used. For infection, these strains were cultivated at 37°C overnight on nutrient agar and then suspended in physiological saline.

Infection. Mice were challenged ip with the bacteria (0.2ml/mouse). Viable numbers of infecting organisms were determined by quantitative cultivation. Mortality of mice and mean survival days were estimated on the twenty-first day after the challenge.

Quantitative cultivation of inoculated organisms. For the determination of bactericidal activity in mice, viable numbers of the infecting organisms in the peritoneal cavity, spleen and kidney (in the case of 248βH-infection) were counted at 24 hr after the infection. The peritoneal cavity was repeatedly washed out with 5 ml aliquots of saline. The spleen and kidney were emulsified with 5 ml each of saline in homogenizers. The washings and homogenates were serially diluted, and the dilutions were mixed with melted nutrient agar to prepare a pour plate cultivation for salmonella organisms or spread on the surface of nutrient agar plates to cultivate staphylococcal organisms. Colony counts were made after 48 hr incubation.

Natural killer (NK) cell activity. NK cell activity in the spleen cell population was assessed by a short term (4 hr)-^{51}Cr-release assay *in vitro* with RL male lymphoma cells as target cells (12).

Interferon assay. Interferon activity in the sera obtained from mice was determined by microcytopathic effect-inhibition-assay with L929F cells against encephalomyocarditis virus (12).

3 Results

Susceptibility of beige mice against salmonella-infection and protective effect of LPS

To estimate the susceptibility of beige mice, these mice and ones from their parentage strain, C57BL/6 mice, were challenged with *S. typhimurium* LT2 organisms. As shown in Table 1, the beige mice seemed to be more susceptible than the C57BL/6 mice only under very limited conditions. When the mice were infected with 4×10^6 organisms, all mice, regardless of whether they were beige or C57BL/6 mice, were killed within several days after the infection. However, when the mice were infected with 4×10^5 organisms, there seemed to be some difference in the mortality rate between beige and C57BL/6 mice. Furthermore,

TABLE I
Mortality of beige (bg/bg) and C57BL/6 (+/+) mice after infection with *Salmonella typhimurium* LT2

Number of infecting organisms	Strain	No. of deaths / No. of mice	Mortality rate (%)	Mean survival day ± S.D.
4×10^6 ip	bg/bg	5 / 5	100	3.6 ± 0.9
	+/+	5 / 5	100	4.0 ± 0
4×10^5 ip	bg/bg	6 / 6	100	4.5 ± 0.5
	+/+	3 / 6	50	5.6 ± 1.6
7.2×10^4 ip	bg/bg	1 / 5	20	6
	+/+	1 / 5	20	4

TABLE II
Protective effect of LPS on bg/bg or bg/+ mice infected with *Salmonella enteritidis* No.11 strain

Mice	No. of infecting organisms	Treatment	No. of deaths / No. of mice	Mortality rate (%)	Mean survival day ± S.D.
bg/+	10^3 ip	LPS[a]	8 / 8	100	4.8 ± 1.5
		none	8 / 8	100	6.6 ± 0.5
	10^2 ip	LPS	8 / 8	100	6.5 ± 2.0
		none	7 / 8	88	6.8 ± 0.9
	10^1 ip	LPS	0 / 8	0	
		none	3 / 8	38	8.0 ± 2.6
bg/bg	10^2 ip	LPS	6 / 8	75	7.2 ± 0.5
		none	7 / 8	88	9.0 ± 3.5
	10^1 ip	LPS	0 / 5	0	
		none	5 / 5	100	9.0 ± 5.0

[a] 20 μg, ip, 24 hr before the infection.

if we used more less number of organisms for the infection, the majority of mice of both strains were survived, and the difference of mortality rate was again obscure.

The data suggests an increased susceptibility of beige mice to *S. enteritidis* No.11 strain because No.11 organisms are highly virulent to

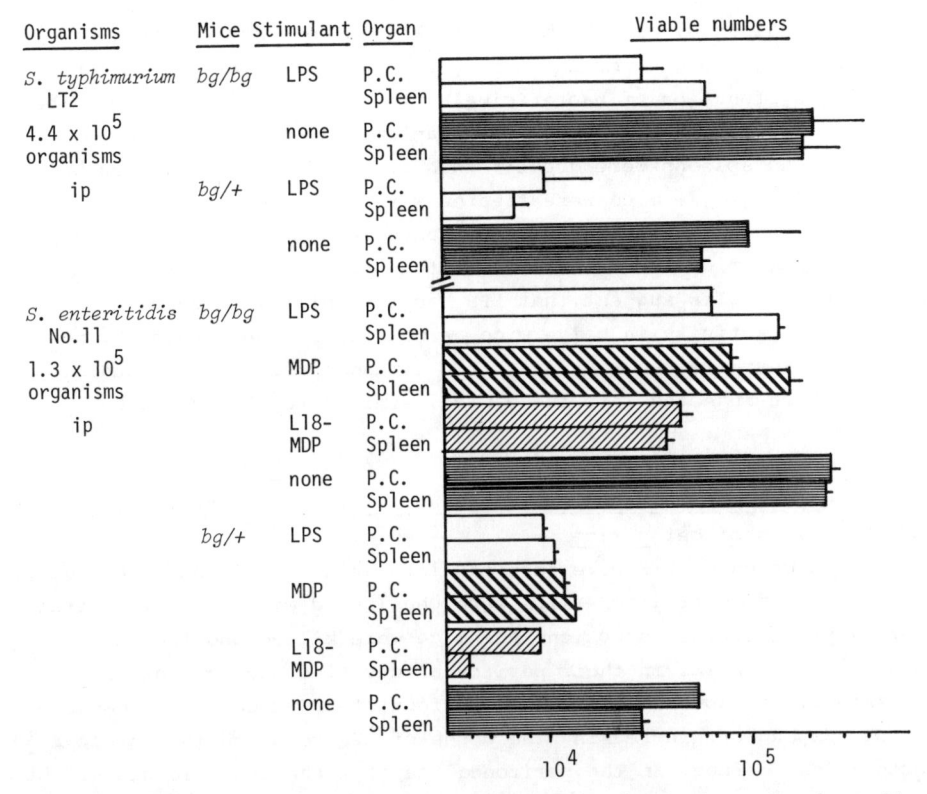

Figure 1. Effect of LPS, MDP or L18-MDP on the infection of salmonella in beige (bg/bg) and control (bg/+) mice. LPS (20µg), MDP (100µg) or L18-MDP (100µg) was injected 24 hr before the infection. Viable numbers of organisms in the peritoneal cavity (P.C.) and spleen were determined by quantitative cultivation at 24 hr after the infection. Results are expressed as the arithmetic mean of 3 mice ± the standard error.

mice. Only ca. 10 organisms are enough to kill all the beige mice and 3 out of 8 heterozygous mice (Table 2). However, these mice, either bg/bg or bg/+, that had been injected with LPS 24 hr before the infection could survive the infection with ca. 10 organisms of the No.11 strain.

Inhibitory effects of LPS, MDP and L18-MDP on multiplication of infecting organisms in the peritoneal cavity and spleen in mice

Some groups of bg/bg or Bg/+ mice were injected with LPS, MDP or L18-MDP. Twenty-four hr later, these mice and other untreated control groups of mice were challenged with LT2 or No.11 organisms, and 24 hr after the challenge, all the mice were killed and examined for the number of viable organisms in their peritoneal cavities and spleens (Fig. 1). Dysfunction of bactericidal activity in bg/bg mice could be obviously seen, as the recovery of organisms in their peritoneal cavities and spleens were greater than those of $bg/+$ mice. When beige mice as well as $bg/+$ mice were treated with LPS before the infection, these mice showed an augmentation of bactericidal ability. The numbers in LPS-treated mice were clearly less than those in untreated control mice. These results suggest that LPS has the ability to enhance bactericidal activity in beige mice as well as in phenotypically normal $bg/+$ mice. Furthermore, when synthetic immunomodulators, MDP and L18-MDP, had been injected into these mice, similar results could be obtained.

Dose-effect of LPS on bactericidal activity in the peritoneal cavity and the spleen of beige mice

Various doses of LPS were injected into beige mice, and 24 hr later, these mice and control $bg/+$ mice were challenged with the No.11 strain. Then 24 hr after the infection, all mice were killed and the viable numbers of organisms in their peritoneal cavities and spleens were determined. As shown in Fig. 2, the effect of LPS-treatment seemed to be dose-dependent: increasing the doses of LPS resulted in a greater bactericidal effect. In the peritoneal cavity, the bactericidal effect induced by LPS was prominent, but its effect was uncertain in the spleen.

Effect of time between LPS-injection and bactericidal activity

Groups of beige mice were injected with 20µg of LPS on day 7, 5, 3, 1 or 0 before the challenge with the No.11 strain. Twenty-four hr after the infection, numbers of viable organisms in the peritoneal cavity and the spleen were examined. As shown in Fig. 3, the most obvious effect was seen in the mice that had been injected with LPS at 24 hr before the challenge.

Effect of lipid A-BSA complex and alkali-detoxified LPS

To examine the active part of LPS which consists chemically of lipid A and polysaccharides, comparative studies were carried out by using the lipid A-BSA complex and alkali-detoxified LPS. The lipid A-BSA complex is still biologically active as an endotoxin (13), while the alkali-detoxified LPS has lost the majority of the lipid molecules originally present in LPS and has quite a low toxic. Induction of an antibacterial ability against strain No.11 in beige and control mice by these substances were examined. Groups of mice were injected with

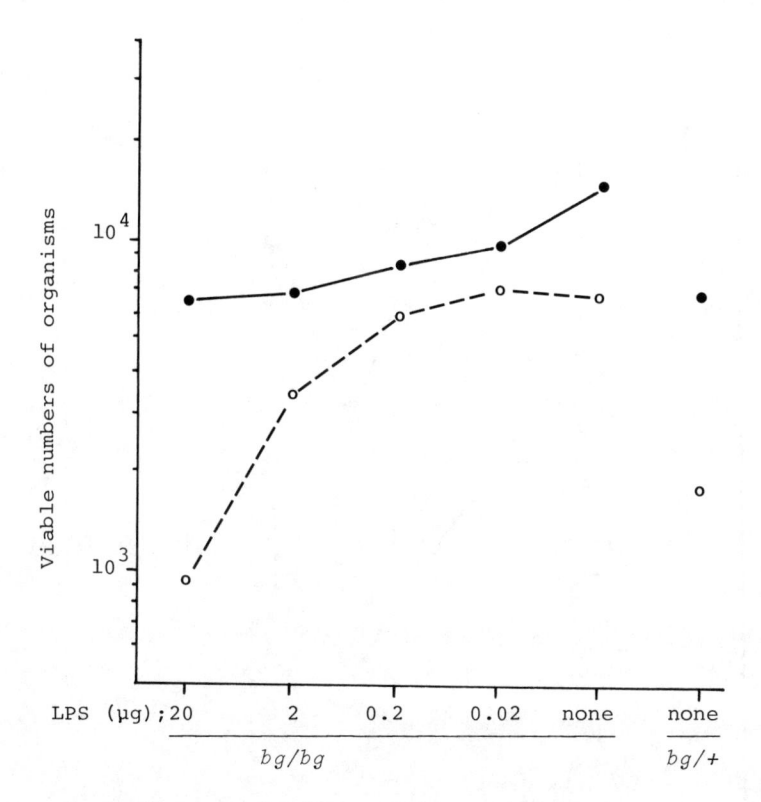

Figure 2. Antibacterial effect of LPS on beige mice. LPS was ip injected 24 hr before infection by S. enteritidis No.11 (2.0 x 10^5 viable organisms, ip), and the viable numbers of infecting organisms in the peritoneal cavity (o----o) and spleen (●——●) were determined by quantitative cultivation at 24 hr after the infection. Each point represent the mean value of 3 mice.

one of the substances 24 hr before challenge with No.11 organisms. Viable numbers of infecting organisms in the peritoneal cavities and the spleens of these mice were examined at 24 hr after the challenge. As shown in Fig. 4, the bactericidal activity induced by lipid A-BSA complex was stronger than that by alkali-detoxified LPS, but the original LPS showed the strongest effect among the substances examined.

Effect of LPS in beige mice against infection of S. aureus 248βH

Salmonella enteritidis and *S. typhimurium* are intracellular pathogens, and they cause in mice a systemic infection, murine salmonellosis. On the other hand *Staphylococcus aureus* is thought to be an extracellular pathogen, and it is difficult for this organism to multiply in phago-

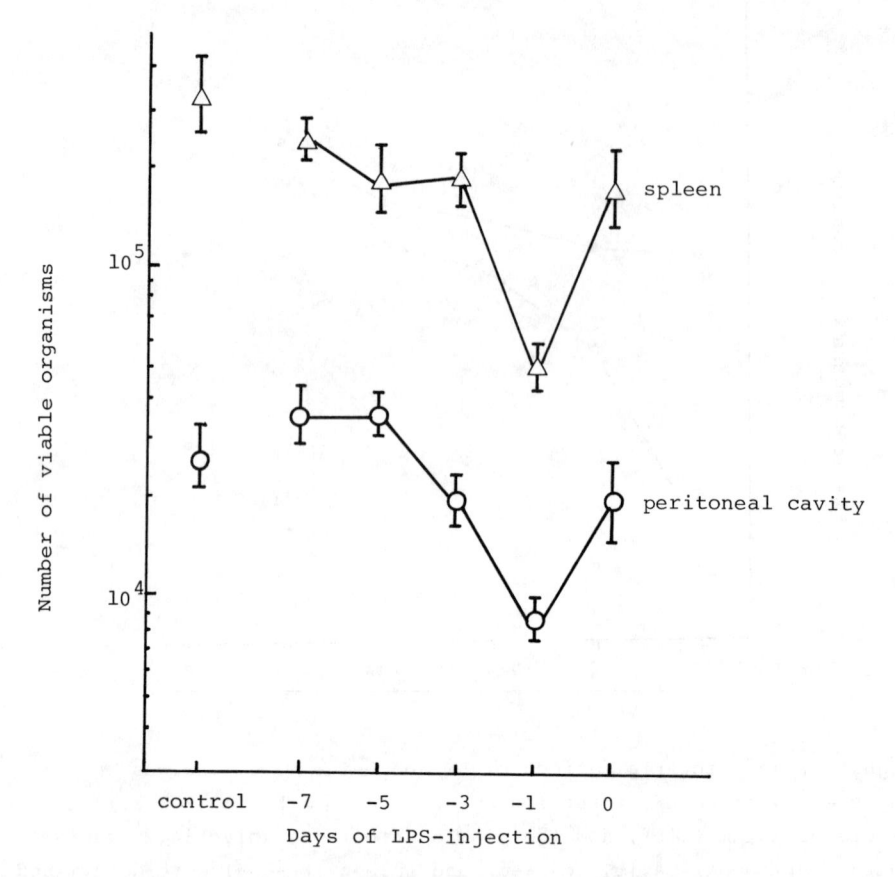

Figure 3. Effect of time between LPS-injection and bactericidal activity in beige mice. Beige mice were injected with 20μg of LPS at varying times as indicated and challenged with *S. enteritidis* No.11 strain (4.4 x 10^5 organisms) on day 0. Twenty-four hr after the infection, viable numbers of infecting organisms in the peritoneal cavity and spleen were examined. Values represent the means ± standard errors of 3 mice.

cytic cells of hosts. It seemed worthwhile to examine whether or not beige mice have an elevation their bactericidal ability against this extracellular pathogen after the treatment with LPS. Groups of mice, either beige or *bg/+* mice, were injected with LPS, and 24 hr later, these mice and untreated control mice were challenged with the 248βH strain of *S. aureus*. The bacterial numbers in their peritoneal cavity, spleen and kidney were determined at 24 hr after the infection. As

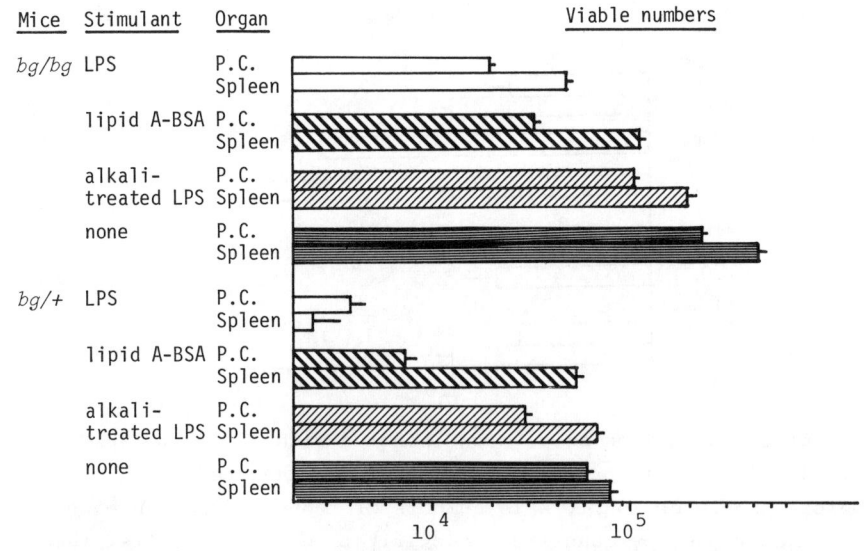

Mice	Stimulant	Organ	Viable numbers
bg/bg	LPS	P.C. Spleen	
	lipid A-BSA	P.C. Spleen	
	alkali-treated LPS	P.C. Spleen	
	none	P.C. Spleen	
bg/+	LPS	P.C. Spleen	
	lipid A-BSA	P.C. Spleen	
	alkali-treated LPS	P.C. Spleen	
	none	P.C. Spleen	

10^4 10^5

Figure 4. Effect of LPS, lipid A-BSA or alkali-detoxified LPS on beige mice infected with *S. enteritidis* No.11 strain (3.6 x 10^5 viable organisms, ip). These stimulants were ip injected (20μg/mouse) 24 hr before the infection. Viable numbers of the organisms in the peritoneal cavity (P.C.) and spleen were determined by quantitative cultivation at 24 hr after the infection. Values represent the means ± standard errors of 3 mice.

shown in Fig. 5, viable counts in LPS-treated mice, regardless of whether bg/bg or bg/+ mice, were greatly reduced. These results indicate that LPS can induce in beige mice some anti-bacterial ability against virulent *S. aureus* organisms.

Effect of LPS on NK cell activity in beige mice

Recently, it has been found that there is a selective impairment of Nk cell activity in beige mice and human patients with Chèdiak-Higashi syndrome (6,15). The NK cell system is believed to be an important effector of immune surveillance, tumor immunity and viral immunity. A further experiment was conducted to examine whether or not LPS can recover the defective ability of NK cells in beige mice. Mice were injected with LPS, and 24 hr later, the NK cell activity of their spleen cells was examined by a short-term ^{51}Cr-release assay *in vitro* with RL male lymphoma cells as the target cells. Interferon activity in the sera obtained from these mice was also determined.

As shown in Fig. 6, bg/+ mice without LPS-treatment showed slightly higher NK cell activity than untreated bg/bg mice, and no interferon

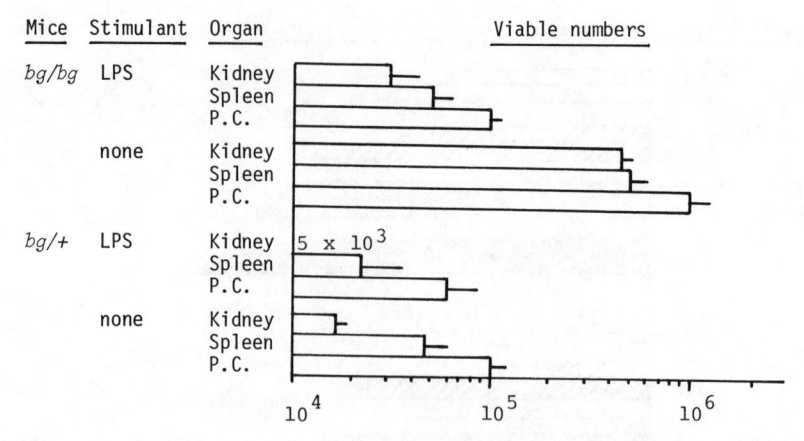

Mice	Stimulant	Organ
bg/bg	LPS	Kidney Spleen P.C.
	none	Kidney Spleen P.C.
bg/+	LPS	Kidney Spleen P.C.
	none	Kidney Spleen P.C.

Figure 5. Effect of LPS on beige mice infected with S. aureus 248βH strain (1.2×10^8 viable organisms, ip). LPS was injected (20µg/mouse) 24 hr before the infection. Viable numbers of the organisms in each organ were determined by quantitative cultivation at 24 hr after the infection. Values represent the means $\pm$ standard errors of 3 mice.

activity could be detected in the sera of both types of mice. After the LPS-treatment, NK cell activity in these mice increased significantly, though the increment of the activity in beige mice was rather weak than that of bg/+ mice. The interferon activity in the sera of LPS-treated mice was variable. Some sera had significant increases of the activity, but others did not.

4 Discussion

As increased susceptibility to infection has previously been documented in CHS human, minks, cattle (16) and mice (3,4). Lane and Murphy reported (3) that homozygous beige mice had a significantly higher incidence of pneumonitis than non-beige mice, and their bacteriological studies on the lungs of the pneumonitis demonstrated causal agents of a beta hemolytic streptococcus, Staphylococcus epidermidis, Escherichia coli and Pasteurella pneumotropica. A significant increase in mortality rate for the CHS mice was demonstrated by Elin et al. (4) with challenge organisms such as Candida albicans, E. coli, Klebsiella pneumoniae, S. aureus and Streptococcus pneumoniae. The present study demonstrates an increased susceptibility of CHS mice to challenge with S. enteritidis and S. typhimurium. The mechanism of the increased susceptibility to infection is probably related to impaired cellular function and is not humoral in

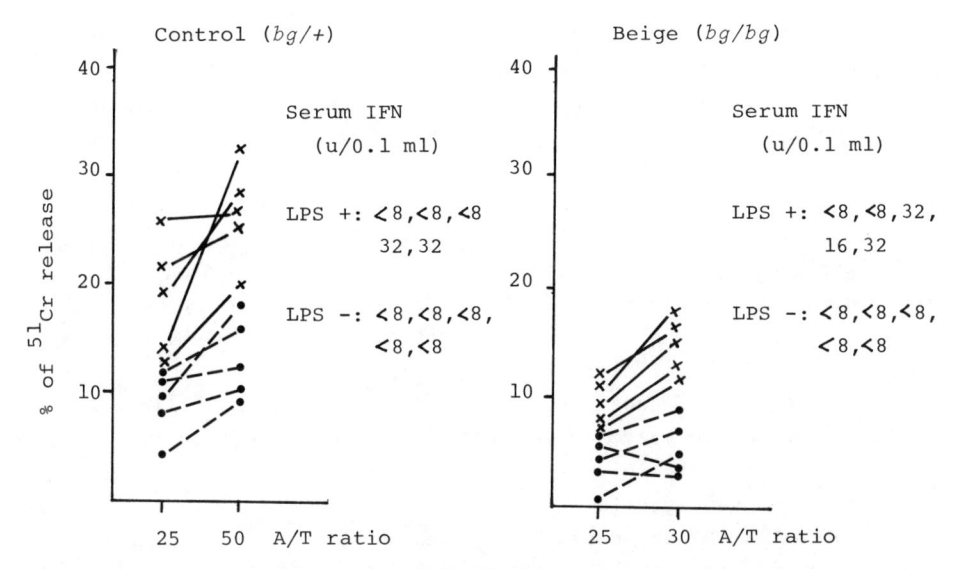

Figure 6. Effect of LPS on NK cell activity in the spleen cells of beige and control bg/+ mice. Mice were injected ip with 20μg of LPS, and 24 hr later, the NK cell activity of the spleens in mice was determined by short-term (4 hr) ^{51}Cr-release assay *in vitro* with RL male lymphoma cells as the target. Interferon (IFN) activity in the sera of these mice was also determined. x——x: LPS-treated cells, •——•: LPS-untreated control. Each point represents the mean from triplicate cultures, and a line connecting the points indicates that the results were obtained using the identical spleen cell population of the mouse.

origin (4,5).

LPS has some potencies to increase phagocytic ability in the reticuloendothelial system and to elevate the resistance to infection nonspecifically (17). Our results demonstrated that the previous treatment of beige mice with LPS could elevate their resistance to bacterial infection and increase their *in vitro* bactericidal ability. The viable numbers of salmonella organisms as well as staphylococcal organisms were significantly reduced in the peritoneal cavities or the LPS-treated beige mice. Salmonella organisms are thought to be a facultative intracellular pathogen (18), but staphylococcus organisms are presumably in the category of extracellular parasites (19). The LPS-activated macrophages in beige mice must have gained an increased capability of killing both of these intra- and extracellular parasites as those of normal mice have. The effect of LPS is temporary and dose-dependent. When LPS has been injected only 24 hr before infection, its certain effect can be seen. However, the injection of LPS simultaneously or several days before challenge does not produce an obvious effect.

LPS consists of a specific polysaccharide covalently linked to lipid A. The lipid A is presumed to be an endotoxic principle, but it can hardly show its activity as it is insoluble in water. When lipid A is solubilized by complexing with BSA, the solubilized complex can gain endotoxic potency (20). The major part of the lipid portion in LPS is removed by alkali-treatment, and the lipid-deprived LPS becomes low in toxicity (14). The findings that the effect of the original LPS was stronger than those of either the lipid A-BSA complex or alkali-detoxified LPS (Fig. 4) suggest that both portions of endotoxic lipid A and nontoxic carrier polysaccharide may be related to the potency to induce bactericidal activity in these mice. Synthetic MDP and its derivative are known to have a prominent protective effect against infection (21). The MDP and L18-MDP are capable of inducing bactericidal activity in beige mice, but the effect of LPS seems to be not inferior to that of MDP or L18-MDP (Fig. 1).

Beige mutation in the homozygous condition results in the marked deficiency of NK cell activity *in vitro* (6). We demonstrated that significant augmentation of NK cell activity *in vitro* can be achieved by an injection of LPS *in vivo*, but the effect in beige mice is far less than that in normal mice (Fig. 6). The cytotoxic activity by NK cells can be stimulated in beige mice by an administration of interferon (6). LPS is capable of producing interferon in mice, but it unlikely that the augmentation of NK cell activity is *via* interferon since we could not observe a direct relation between interferon activity in the serum and NK cell activity in the spleen cell population of mice treated by LPS. Therefore, we are convinced that LPS can directly activate NK cells in beige and normal mice.

The exact mechanism for the correction of the beige defect by LPS is presently unclear. Recently, data has indicated that it is conceivable that elevating the levels of cyclic guanosine 3',5'-monophosphate (cGMP) in CHS NK cells in man promotes microtubule assembly, which in turn leads to lysosomal enzyme release with resultant target lysis (15). It should be necessary to study whether LPS elevates cGMP levels in phagocytic cells or NK cells in beige mice or not.

References

(1) R. S. Blume, S. M. Wolff, Medicine 51 (1972) 247-280.

(2) J. E. Brandt, R. T. Swank, E. K. Novak, Immunologic defects in laboratory animals (edited by Gershwin, M. E., and Merchant, B.), Plenum Press 1981, pp. 99-117.

(3) P. W. Lane, E. D. Murphy, Genetics 72 (1972) 451-460.

(4) R. J. Elin, J. B. Edelin, S. M. Wolff, Infect. Immun. 10 (1974) 88-91.

(5) J. I. Gallin, J. S. Bujak, E. Pattern, S. M. Wolff, Blood 43 (1974) 201-206.

(6) B. N. Beck, C. S. Henney, Cell. Immunol. 61 (1981) 343-352.

(7) D. C. Morrison, J. L. Ryan, Adv. Immunol. 28 (1979) 293-450.

(8) O. Westphal, O. Lüderitz, F. Bister, Naturforsch. 76 (1952) 148-155.

(9) M. Nakano, T. Saito, H. Asou, Jpn. J. Microbiol. 19 (1975) 403-406.

(10) C. Galanos, E. T. Rietschel, O. Lüderitz, O. Westphal, Y. B. Kim, D. W. Watson, Eur. J. Biochem. 31 (1972) 230-233.

(11) E. Neter, O. Westphal, O. Lüderitz, E. A. Gorzynski, E. Eichenberger, J. Immunol. 76 (1956) 377-385.

(12) N. Minato, L. Reid, H. Cantor, P. Lengyel, B. R. Bloom, J. Exp. Med. 152 (1980) 124-137.

(13) C. Galanos, E. T. Rietschel, O. Lüderitz, O. Westphal, Eur. J. Biochem. 19 (1974) 143-152.

(14) S. Britton, Immunology 16 (1969) 513-526.

(15) P. Katz, J. C. Roder, A. M. Zaytoun, R. B. Herberman, A. S. Fauci, J. Immunol. 129 (1982) 297-302.

(16) G. A. Padgett, C. W. Reiquam, J. B. Henson, J. R. Gorham, J. Pathol. Bacteriol. 95 (1980) 509-522.

(17) M. Landy, W. Braun, Bacterial Endotoxin, Rutgers University Press, 1964.

(18) F. M. Collins, Bacteriol. Rev. 38 (1974) 371-402.

(19) E. Suter, Bacteriol. Rev. 20 (1956) 94-132.

(20) O. Lüderitz, C. Galanos, V. Lehmann, M. Nurminen, E. T. Rietschel, G. Rosenfelder, M. Simmon, O. Westphal, J. Infect. Dis. 128s (1973) 17-29.

(21) Y. Osada, M. Mitsuyama, K. Matsumoto, T. Une, T. Otani, H. Ogawa, K. Nomoto, Infect. Immun. 37 (1982) 1285-1288.

Distribution, Degradation and Elimination of Intravenously Applied Lipopolysaccharide in Rats

Marina A. Freudenberg, Bernhard Kleine, Nikolaus Freudenberg*, Chris Galanos

Max-Planck-Institut für Immunbiologie, D-7800 Freiburg, Germany
*Pathologisches Institut der Albert Ludwigs Universität, D-7800 Freiburg, Germany

Abstract

Lipopolysaccharide (LPS) after its intravenous administration in rats undergoes fast complex formation with high density lipoprotein (HDL) in plasma. This interaction prevents LPS from attaching to red blood cells, and lowers the rate of uptake of LPS by macrophages. The main part of LPS is removed from the circulation by phagocytic cells, in the case of rough form LPS eventually also by hepatocytes. In the liver LPS passes from Kupffer cells further to hepatocytes. Intracellularly LPS undergoes partial deacylation, whereby ester bound as well amide bound fatty acids are lost. The elimination of LPS from the body is relatively slow and takes place over weeks. The altered LPS material is excreted mainly through the gut, a smaller part appears also in urine. Passage of LPS carrying macrophages through the respiratory tract was observed.

Introduction

A great deal of information exists on the beneficial and damaging properties of bacterial LPS, however, the fate of the LPS molecule after entering the animal or human organism is not completely understood. Our knowledge of how the host may handle LPS is very unsatisfactory, and important questions such as whether resistance towards endotoxin, hereditary or acquired, has something to do with a better or faster detoxification of the LPS molecule remain unanswered. In the present communication more recent results on the distribution, degradation and elimination of intravenously applied LPS in the rat are reported.
Abbreviations: LPS: lipopolysaccharide; HDL: high density lipoprotein; R form: rough form; S form: smooth form.

In most of the experiments the complete smooth (S) form LPS of S.abortus equi was used. In some studies we also used the LPS from S.minnesota R 595, which is the most defective rough (R) form LPS, however, biologically fully active (1). To avoid misinterpretations due to presence of bacterial contaminants such as nucleic acids and proteins, LPS of high purity was used. The exact way of the preparation was described earlier (2,3).

Blood clearance of LPS

After intravenous injection of ^{14}C-labeled LPS preparations in rats, the R form LPS was cleared from the circulation much faster (half life time = 30 min) than the S form (half life time = 7,5 h) (4). This is in agreement with earlier observations showing that in general R form is cleared faster than S form LPS (5,6). In both cases the rate of clearance is unusually long (Fig. 1) compared with data of other laboratories. The high purity and the very low aggregation state of the injected preparations as well as the absence of anti LPS antibodies in the animals employed may be factors accounting for the above slow removal of LPS from the blood. The rate of clearance is not dose dependent because 75 and 200 µg of LPS were cleared at the same rate as 5 mg. Three minutes after injection all circulating LPS was found in plasma and no radioactivity was associated with erythrocytes or leucocytes.

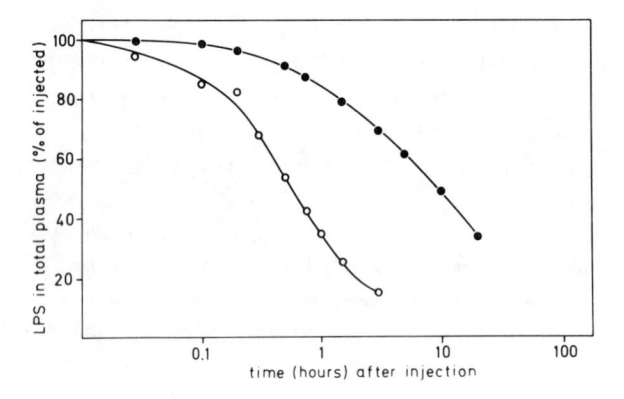

Fig. 1: Clearance of ^{14}C-LPS of S.abortus equi (o) and S.minnesota R 595 (o) from the blood.

Interaction of LPS with plasma high density lipoprotein

In the circulation both LPS (S and R form) bind to plasma high density lipoprotein (HDL) forming a stable complex. Evidence for this was obtained when plasma samples of LPS treated rats were analyzed by crossed immunoelectrophoresis using antibodies to rat plasma proteins and to LPS (4).
The injection of LPS induced an early change in the precipitation pattern of high density lipoprotein (Fig. 2), patterns of other plasma proteins remained unchanged. With S form LPS a

part of the HDL precipitate shifted towards the cathode and formed a shoulder on the original HDL peak. With R form LPS all HDL was shifted towards the anode. The degree of alterations was dependent on the dose of injected LPS. The alteration of HDL precipitation peak in plasma lasted as long as the LPS circulated, in the case of rough form LPS up to 6 hours after injection, in case of smooth form up to 48 hours.

S. abortus equi

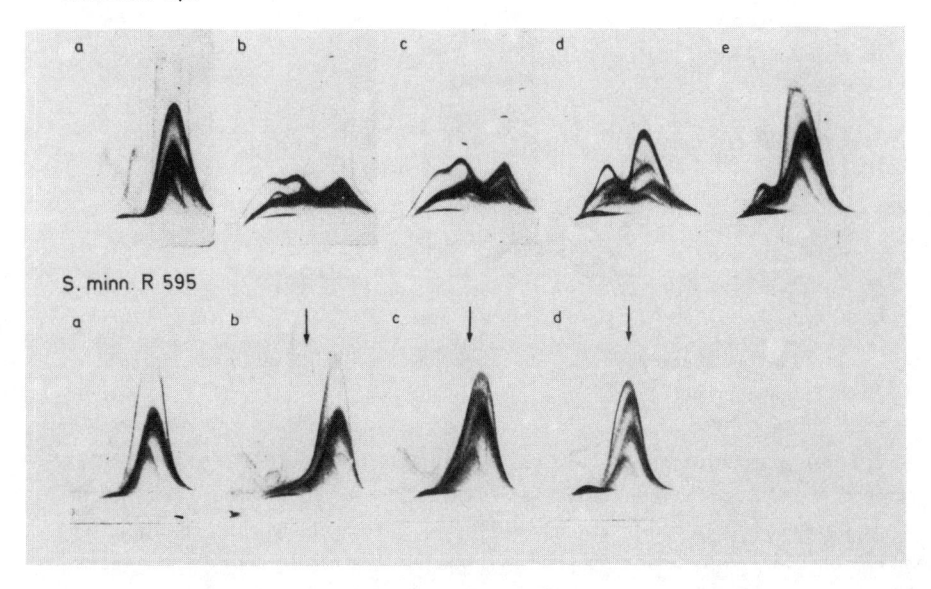

S. minn. R 595

Fig. 2: Time dependent changes in rat HDL induced by 5 mg LPS i.v.: a) before treatment, b) 3 min., c) 6 h, d) 24 h and e) 48 h after injection. In case of R form LPS arrows indicate original position of HDL.

In Fig. 3 evidence is shown that the alteration of the HDL precipitation pattern was due to complex formation between HDL and LPS. Plasma of a rat injected with LPS was subjected to crossed immunoelectrophoresis with anti LPS serum in the intermediate gel and anti HDL serum in the upper gel. The anti LPS serum caused a retardation of the precipitate which was particularly evident with the shifted part of the lipoprotein. No such retardation was seen when instead of anti LPS serum normal serum was included in the intermediate gel. This shows that the shifted part of HDL contains LPS. Conversely, the LPS precipitation peak obtained in crossed immunelectrophoresis with anti LPS serum was totally retarded when anti HDL serum was added into the intermediate gel. From this experiment it may also be concluded that appart from HDL there was no other major protein in plasma carrying LPS.

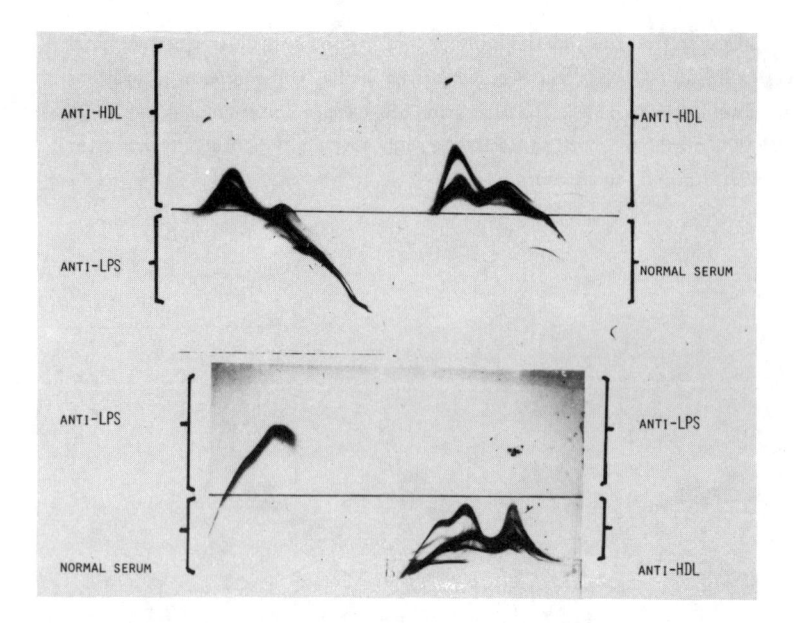

Fig. 3:   Crossed immunoelectrophoresis with intermediate gel of rat plasma 1 h after 5 mg LPS of S.abortus equi i.v.

The very fast association of LPS with rat HDL raised the question of the biological significance of the LPS - HDL interaction. As mentioned before following administration of ^{14}C-LPS we were not able to detect any radioactivity bound to red blood cells, despite of the fact that LPS bind very easily to erythrocytes in vitro in the absence of serum. For this reason we investigated whether HDL inhibits the attachment of LPS to erythrocytes in vitro (7) by the passive hemolysis test. Erythrocytes were incubated with S form LPS in the presence of graded amounts of isolated rat HDL, washed and used in a hemolysis test with anti LPS antiserum and complement. The curve of Fig. 4 illustrates that HDL inhibited in a dose dependent manner the attachment of LPS onto erythrocytes and the subsequent hemolysis by anti LPS antiserum. In vivo this phenomenon would be of biological importance because the binding to HDL would inhibit the indiscriminate interaction of LPS with cells and tissues thus ensuring that the cellular uptake of LPS is effected as thougroughly as possible only by cells carrying specific recognition sites.

It is generally accepted that the cells of RES (reticuloendolethal system) mainly macrophages participate in the removal of LPS. The role of HDL in the uptake of LPS of S.abortus equi by macrophages was tested in vitro (Fig. 5). Bone marrow derived mouse macrophages were cultivated with different amounts of free and HDL bound ^{14}C labeled LPS for 24 hours. The results obtained indicate that the binding of LPS to mouse HDL results in significantly lower LPS uptake. The decrease of uptake was also observed when instead of mouse serum, sera of numerous different species were used as source of HDL.

The slow internalization of LPS - HDL complexes in comparison to free LPS seen in vitro is in good agreement with in vivo results of Mathison and Ulevitch (8) who showed that binding of LPS to HDL results in the decrease of the rate of LPS clearance.

INHIBITION OF FIXATION OF LPS TO ERYTHROCYTES BY PURIFIED RAT HDL.

(ATTACHMENT OF LPS TO ERYTHROCYTES WAS TESTED BY THE PASSIVE HEMOLYSIS TEST)

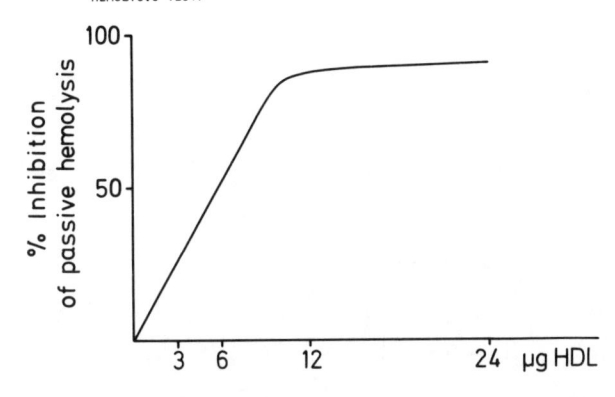

S. ABORTUS EQUI LPS (10 µG) WAS INCUBATED WITH 0.2 ML ERYTHROCYTES IN THE PRESENCE OF DIFFERENT AMOUNTS OF HDL, AND THE CELLS TESTED WITH ANTI-S. ABORTUS EQUI SERUM AND COMPLEMENT.

Fig. 4

Cellular uptake of LPS

As mentioned above most of the circulating LPS is cleared by cells of the reticuloendothelial system mainly in the liver and spleen.

The percentage of radioactive LPS trapped in here is independent of the amount of the injected dose. Highest amounts of ^{14}C, ^{3}H labeled LPS of S.abortus equi bound in the liver were observed on day 3 after injection, being in average 40% of injected material. LPS originated radioactivity persists here for a rather long time, up to weeks.

Cells participating in the uptake and processing of LPS in the liver were visualized by immunohistochemical methods using anti LPS antibodies coupled with horseradish peroxidase (9). Two to 7 hours after injection the S form LPS of S.abortus equi is first detectable in Kupffer cells and granulocytes. It was often observed that LPS positive granulocytes are internalized by liver macrophages (Fig. 6).

On day 3 after injection a redistribution of LPS from Kupffer cells to hepatocytes occurs. The R form LPS of S.minnesota R 595 is detectable already 30 min. after injection in Kupffer cells, granulocytes and hepatocytes suggesting a direct uptake of the LPS also by hepatocytes.

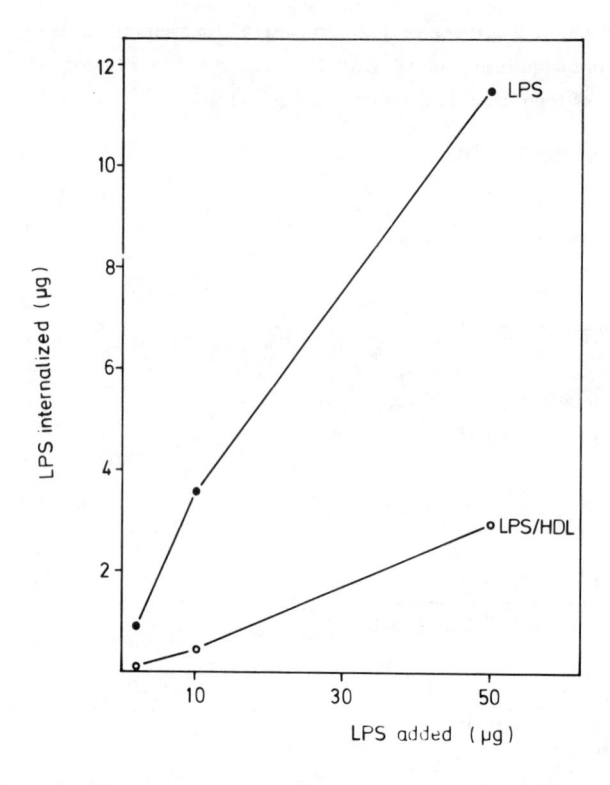

Fig. 5: Uptake of free (o) and HDL bound (o) ^{14}C-LPS by mouse macrophages (10^7 cells/24 h).

In vivo degradation and excretion of LPS

To study whether changes in the LPS molecule occurred in vivo, ^{14}C and ^{3}H labeled S form LPS was used. The preparation was labeled with ^{3}H exclusively in fatty acids and 93% of ^{14}C was found in the sugar part (10). The use of LPS labeled with both ^{14}C and ^{3}H is useful because from measuring the ratio of ^{14}C to ^{3}H one can tell whether changes have occurred in LPS.

Experiments with this preparation revealed (11) that on the basis of measured ^{14}C, the LPS internalized in the liver contains less tritium than expected. The ratio between ^{14}C and ^{3}H which in the original LPS is 26 increased on day 1 after injection in average to 32 on day 3 to 34, on day 14 to 38 and reached finally on day 35 a ratio of 45. Because in the LPS preparation used only fatty acids were labeled with tritium, the loss of ^{3}H-activity in the liver indicates a partial loss of fatty acids. The loss of tritium relative to ^{14}C was observed also in spleen and other organs.

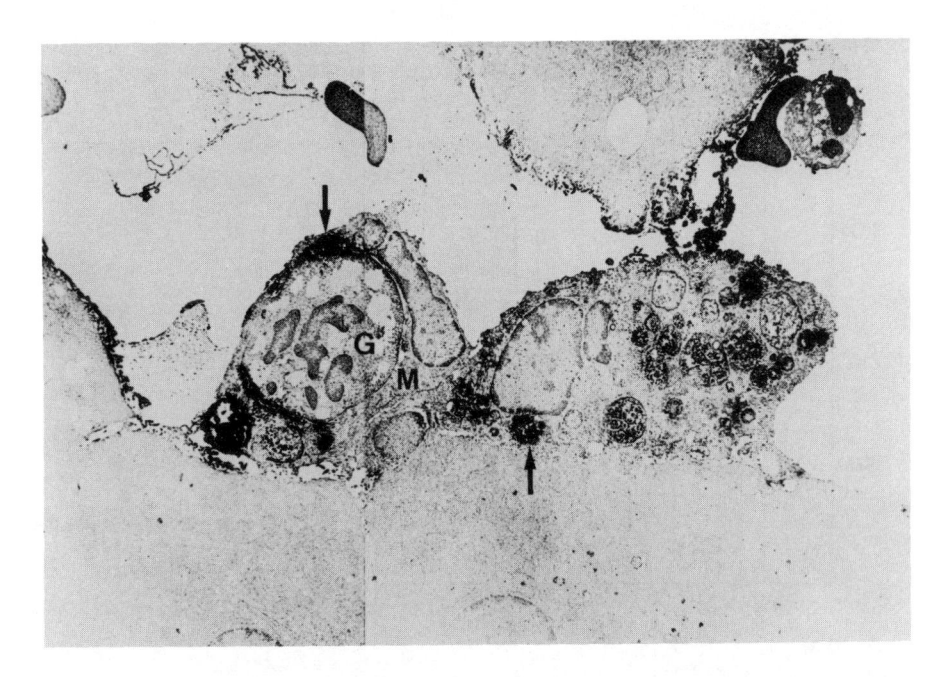

Fig. 6: Ultrastructural immunohistochemical demonstration of endotoxin in the liver, 4 hours after LPS administration. Electron dense staining (indicated by arrows) represents lipopolysaccharide. G = endotoxin-coated granulocyte internalized by a liver macrophage (M). Photomounting. Non contrasted ultrathin section. Primary magnification: × 5 700. (N. Freudenberg et al., in preparation)

Mild alkali treatment of original LPS with 0.25 N sodium hydroxide for 90 min. at 56°C leads to a complete cleavage of ester bound fatty acids and a corresponding loss of ^{3}H activity. The remaining tritium represents exclusively amide bound fatty acids. Alkali treatment of liver samples obtained from rats injected with LPS revealed that the observed loss of tritium includes ester bound as well as amide bound fatty acids (Fig. 7).

The present data suggest very strongly that LPS in vivo (rats) undergoes degradation in the lipid A component. It was well documented in the past that the presence of fatty acids in the LPS molecule is necessary for the expression of its endotoxic activities. Further work must be done to decide if the partial deacylation of LPS observed in rats leads also to detoxification.

The excretion of injected LPS was followed in faeces and urine with double labeled S form LPS (10). It became obvious that the excretion of the LPS proceeds mainly through the gut (Fig. 8). Four to 5 times more ^{14}C radioactivity was collected in faeces than in urine. The excretion after injection of 200 μg LPS proceeded gradually and reached on day 14, on the average, 32% in faeces and 7% in urine.

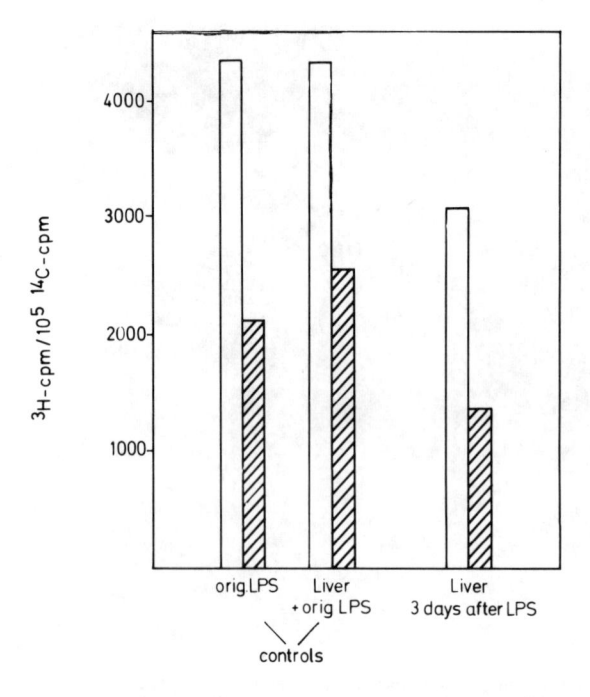

Fig. 7: Fatty acids remaining in in vivo processed ^{3}H, ^{14}C-LPS of S.abortus equi after cleavage of ester bound fatty acids with alkali (0.25 n NaOH/56°C/90 min.). ^{3}H cpm before ▯ and after ▨ cleavage

LPS derived material found in faeces showed the same loss in tritium as the material detected in the liver. It was not possible to measure the ratio of ^{14}C to ^{3}H in urine because of the low content of radioactivity, however, the material found in urine was different from the administered LPS, because half of it was dialysable against water.

Regarding the route of elimination of LPS it should be mentioned that using immunohistochemistry it was found that LPS positive macrophages migrate into the lungs and leave the body as alveolar and bronchiolar macrophages (9). One could speculate as to the consequences the accumulation of such cells may have for the lung and if there is a connection to the development of the shock lung syndrom seen in patients.

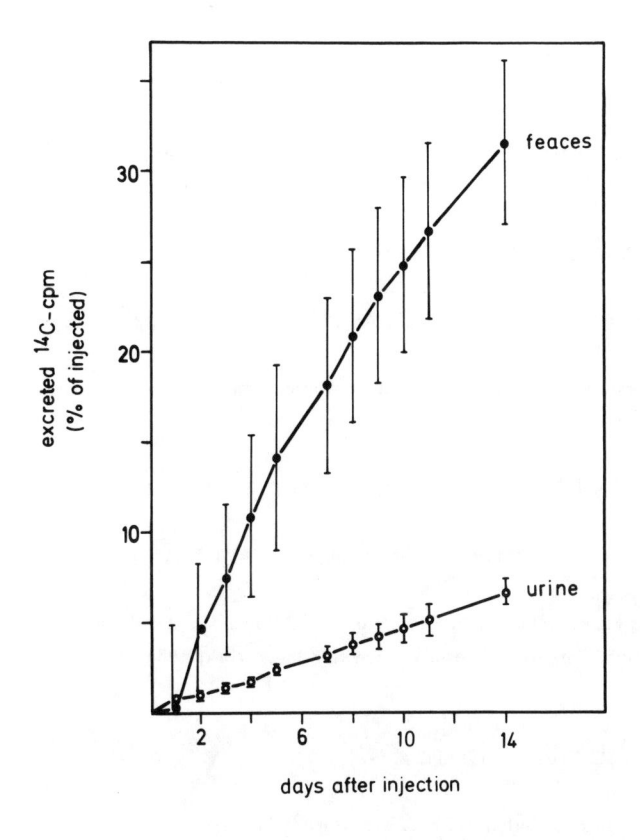

Fig. 8: Excretion of ^{14}C activity in feaces and urine after injection of ^{3}H, ^{14}C-LPS of S. abortus equi (200 µg i.v.) in rats.

References

1. O. Lüderitz, M.A. Freudenberg, C. Galanos, V. Lehmann, E.Th. Rietschel, D.H. Shaw. In F. Bronner, A. Kleinzeller (Ed.) Current topics in membranes and transport. Vol. 17, Academic Press, New York, London, 1982, pp. 79-151.

2. C. Galanos, O. Lüderitz and O. Westphal, Eur. J. Biochem. 9 (1969) 245-249.

3. C. Galanos, O. Lüderitz, O. Westphal, Zbl. Bakt. Hyg.,I. Abt. Orig. A 243 (1979), 226-249.

4. M.A. Freudenberg, T.C. Bøg-Hansen, U. Backs, C. Galanos. Infect. Immun. 28 (1980) 373-380.

5. L. Chedid, F. Parant, M. Parant and F. Boyer, Ann. N.Y. Acad. Sci. 133 (1966) 712-726.

6. J. Hofman, V. Dlabac, J. Hyg. Epidemiol. Microbiol. Immunol. 18 (1974) 447-453.

7. M.A. Freudenberg, T.C. Bøg-Hansen, U. Back, E. Jirillo, C. Galanos. In: D. Eaker, T. Wadström (Ed.), Natural Toxins, Pergamon Press, Oxford and New York, 1980, pp. 349-354.

8. J.C. Mathison, R.J. Ulevitch, Fed. Proc. 39 (1980) 674.

9. M.A. Freudenberg, N. Freudenberg, C. Galanos, Br. J. exp. Path. 63 (1982) 56-65.

10. B. Kleine, M.A. Freudenberg, C. Galanos, in preparation.

11. M.A. Freudenberg, C. Galanos, in preparation.

Endotoxemia: Clinical Aspects

Clinical Aspects of Endotoxemia

Masaru Ishiyama, Chiyuki Watanabe, Shoetsu Tamakuma

First Department of Surgery, University of Tokyo, Faculty of Medicine, 7-3-1, Hongo, Bunkyo-ku, Tokyo, Japan, 113

Abstract

Endotoxin in 818 blood samples of 478 surgical patients over 10-year period were tested by Limulus test. Positive cases were 222. Causative diseases of endotoxemia were divided in two groups, septic and non-septic. Major causes were septic lesions such as acute peritonitis or acute suppurative cholangitis and minor causes were other septic lesions or non-septic life-threatening state such as shock or liver dysfunction. Common symptoms in endotoxemia were septic and shock symptoms. Mortality rate of endotoxemia cases was 42.3 percent. Common laboratory findings were leukocytosis, thrombocytopenia, hypocalcemia, hypophosphatemia. Hypophosphatemia in endotoxemia was a specific and important finding. Reduction of serum complement activity and opsonic activity were closely related to the incidence of endotoxemia.

1 Introduction

Minimal amounts of endotoxin in the blood stream can be detected by Limulus test. This technique first proposed by Bang and Levin has been developed and applied to assay endotoxin in patient's blood sample by Levin and Fine independently. Levin compared Limulus test to blood culture in septic patients as a clinical examination[1]. Caridis and Fine reported endotoxemias not only in septic cases but also in other critical cases[2]. Since then Limulus test positive cases have been reported, for instance, in various diseases by Fossard and Kokkar[3] in liver cirrhosis by Liehr et al. (4), and in Reye's syndrome by Cooperstock et al. (5).

The source of endotoxin in patient's circulating blood is believed to be either septic focus or intestinal flora. The defense mechanism may consist of intercepting barrier against invading endotoxin such as intestinal mucosa, and clearing function of circulating endotoxin by liver. When the defense mechanism breaks down, endotoxin may spill over into the blood stream. Septic lesion, a life-threatening state appearing as shock, liver cirrhosis or liver dysfunction, or bowel diseases such as ileus and ulcerative colitis, may involve failure of the defense mechanism and be a causative disease of endotoxemia.

The present study was undertaken to analyze incidence and prognosis of endotoxemias among surgical patients and to clarify the pathophysiology of clinical endotoxemia.

2 Materials and Methods

For 10 years from 1972 to 1982, 818 samples from 478 cases were tested for Limulus test. The test basically followed Levin's original technique, using Limulus amoebocyte lysate Pre-Gel (Teikokuzoki Pharmaceutical Co. Japan). The one modification was the use of platelet rich plasma instead of conventionally centrifugalized supernate, therby increasing recovery of endotoxin from blood sample by two or three times. For the endotoxin positive cases, clinical symptoms and laboratory data such as leukocyte and thrombocyte counts, serum transaminases, blood urea nitrogen, serum calcium and inorganic phosphate, were evaluated. Serum complement hemolytic activity was determined by Mayer's method. Serum opsonic activity was determined according to Hirsch and Strauss' method and formula using live E.coli O75 (6).

3 Results

Three hundred twenty five blood samples from 222 cases were Limulus test positive. The causative diseases of these 222 cases are shown in Table 1. Acute peritonitis due to gastro-intestinal perforation or post operative anastomosis dehiscence and acute suppurative cholangitis were the main causative diseases, accounting for 63 percent of total endotoxemias. If other septic lesions such as pyothorax, pneumonia, and urinary tract infection are included, septic lesions were responsible for 81.5 percent of the cases of endotoxemia. While non-septic causes account for 18 percent, they were mainly lesions

Septic	Cases	Died
peritonitis due to gastro-intestinal perforation	44	18
disruption of anastomosis	34	17
cholangitis	43	16
direct cholangiography	14	0
liver abscess	5	2
pyothorax	7	4
wound infection	3	0
other septic lesion	24	12
leukemia and sepsis	3	3
liver cirrhosis and sepsis	4	2
Non-septic		
hepatic coma	14	14
liver cirrhosis	17	1
other non-septic lesion	10	5
total	222	95

mortality rate 42.3 percent

1972-1982 First Dept. Surg. Univ. Tokyo

Table 1 Causes of endotoxemia

related to liver dysfunction. Other non-septic causes included terminal cancer, ileus, severe hemorrhagic shock, and adrenal insufficiency. The mortality rate of patients with endotoxemia was 42.3 percent.

About 85 percent of the patints with endotoxemia showed sweating, tachypnea, tachycardia, and fever. Oliguria, hypotension, mental disturbance, cold extremities, and chillness were observed in 50-65 percent of endotoxin positive patients. Upper and/or lower gastro-intestinal bleeding occurred in 15 percent. Leukocytosis and thrombocytopenia were significantly related to endotoxemia. S-GOT, S-GPT, and blood urea nitrogen levels tended to be higher in endotoxin positive cases but were not ststistically significant. Hypocalcemia and hypophosphatemia were seen among these with endotoxemia. These laboratory data are shown in Figs. 1-3. Serum complement hemolytic activity (CH50) and opsonic index were depleted in endotoxemia, as shown in Figs. 4 and 5.

4 Discussion

The mortality rate of endotoxemic patients was 42.3 percent, clear-

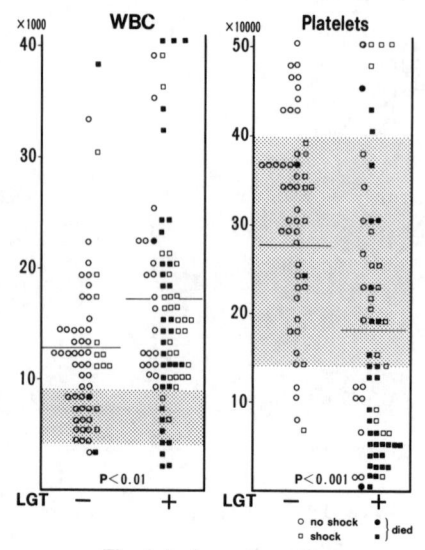

Fig.1 Laboratory Data

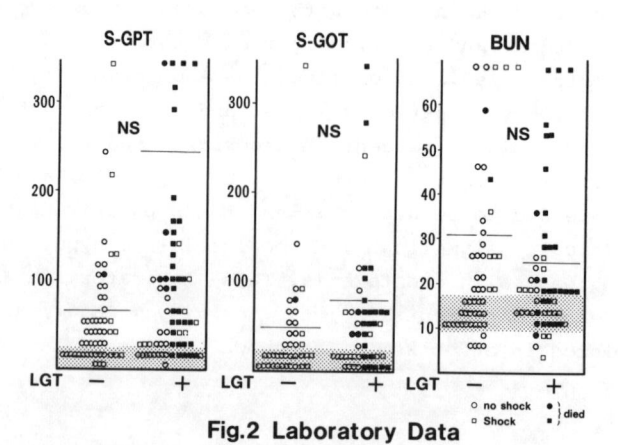

Fig.2 Laboratory Data

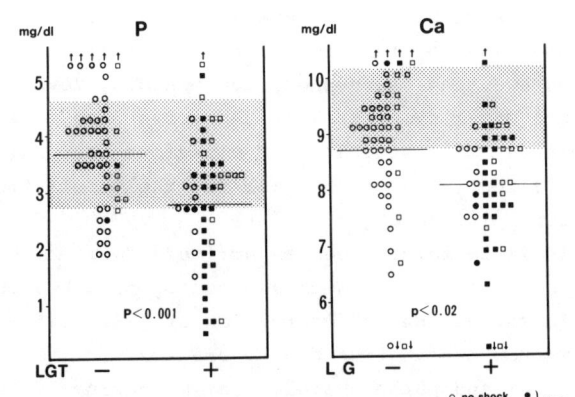

Fig.3 Laboratory Data ○ no shock ●) died
○ Shock ■)

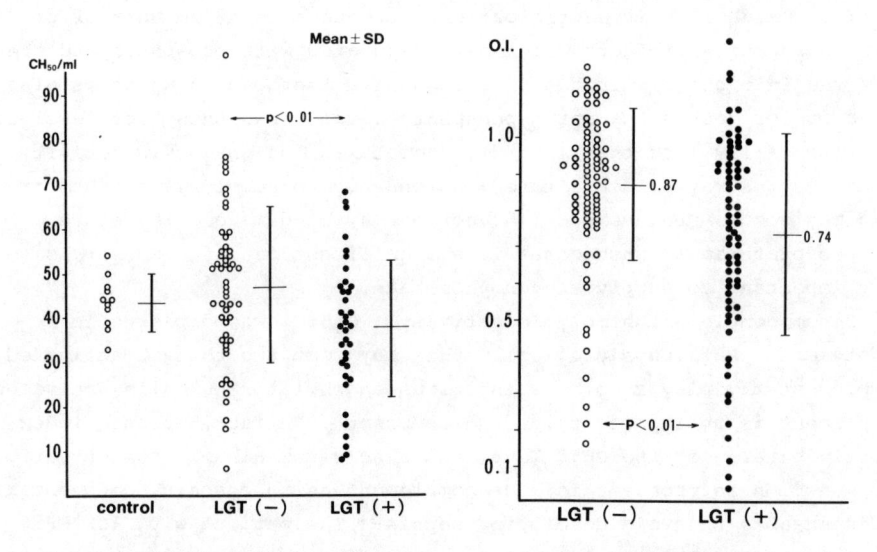

Fig.4 Serum CH$_{50}$ level

Fig.5 Serum opsonic index

ly higher than endotoxin negative cases, suggesting that endotoxin was a lethal factor. The clinical symptoms which various patients with endotoxemia presented are divided in two groups; 1) those with septic symptoms such as sweating, tachypnea, tachycardia, fever and chills, and 2) those with shock symptoms such as hypotension, oliguria, coldness of extremities, or disturbance of consciousness. This seems to suggest that a patient with endotoxemia is likely to suffer septic shock and develop multiple organ failure.

Leukocytosis is certainly seen as active infection, but among the fatal cases there were several with leukocytopenia. Leukocytopenia and thrombocytopenia may be due to destruction or aggregation of blood cells or bone marrow dysfunction due to endotoxin.

The etiological and pathophysiological mechanism of hypocalcemia and hypophosphatemia is not precisely elucidated. Hypocalcemia is probably due to hypoalbuminemia, which is frequently seen in such septic patients. Serum calcium and albumin levels parallel each other closely in septic patients. Specific hypophosphatemia in sepsis was reported by Riedler in 1969 (7). Factors for reducing serum inorganic phosphate described in papers are decreased intake and increased excretion of phosphate, and disturbance of glucose metabolism. Large amounts of glucose injection also a reducing factor which tends to take extracellular phosphate into intracellular space. Most septic patients are fasting. Catabolism in septic patients produces large amounts of protein degradation products which may conjugated with phosphate and then excreted into urine. Glucose injection as a therapeutic agent is also a factor for decreasing serum phosphate level. When phosphate level is low, especially less than 1 mg/dl, chemotaxis and phagocyic activity of leukocytes may be debilitated, a tendency to hemolysis and hemorrhagia may occur, and every cell function may be damaged. Therefore, hypophosphatemia in such cases is a significant finding, and it is very important to supply enough phosphate.

Serum complement hemolytic activity (CH50) was depleted in endotoxemia. Pathophysiologically this may mean two things. Activated complement may play a role as injurious anaphylatoxin, while activated complement is an important opsonin component. In fact, opsonic index closely paralleled the CH50 level and also serum albumin concentration. The question is problematic, but complement as an opsonin component is indispensable to avoid developing sepsis. The patient with low CH50 level is susceptible to infection and is likely to develop endotoxemia.

5 References

(1) J. Levin, et al., Ann. Intern. Med. 76 (1972) 1-7.

(2) D. T. Caridis, J. Fine, Lancet 1 (1972) 1381-1385.

(3) D. P. Fossar, V. V. Kokkar, Brit. J. Surg. 61 (1974) 798-804.

(4) H. Liehr, et al., Lancet 1 (1975) 810-811.

(5) M. S. Cooperstock, et al., Lancet 1 (1975) 1272-1274.

(6) J. G. Hirsch, B. Strauss, J. Immunol. 92 (1964) 145-154.

(7) G. F. Riedler, et al., Brit. Med. J. 1 (1969) 753-756.

Factors and Mechanisms of Clinical Endotoxemia

Masahiko Onda, Kenji Adachi and Akiro Shirota

First Department of Surgery, Nippon Medical School,
1-1-5 Sendagi, Bunkyo-ku, Tokyo 113, Japan

Abstract

   Limulus gelation tests for endotoxemia in the blood of 237 patients with various digestive diseases revealed unexpectedly higher positive results in cases of esophageal varices than in other diseases with underlying infectious backgrounds. More detailed studies revealed higher incidence of portal endotoxemia than systemic endotoxemia in these cases. The results of our studies indicate a strong possibility that endotoxemia in this type of liver disease could be endogenous, originating with the intestinal microflora, released through the portal blood and developing into systemic endotoxemia because of injured hepatic detoxification function. To elucidate this type of endotoxemia, experiments were conducted by inducing fatty livers in germfree and conventional rats. The results of these experiments suggest that while intestinal microflora have some protective role in fatty liver formation, endotoxin originating from intestinal microflora also play a role in the deterioration of the liver detoxification activity.

1 Introduction

   Despite the judicious use of antibiotics, mortality due to gram

---

Abbreviations: LGT: Limulus gelation test; C-CA: Cholesterol cholic acid.

negative sepsis is still very high, and this type of sepsis remains a substantial cause of morbidity and mortality in hospitalized patients. The authors are interested in intestinal microflora, particularly gram-negative bacteria, which play an important role in the pathophysiology of severe sepsis in digestive disorders such as intestinal obstruction, peritonitis, biliary tract infection, and others. As a result of accumulated findings in clinical and experimental studies on intestinal microflora, we believe that endotoxin is closely related to the cause of death and refractory shock in severe septic digestive disorders. However, recent Japanese studies reported high positive rates of endotoxemia in liver cirrhosis with no infectious background, calling attention to the relationship between the pathophysiology of liver disorders and endogenous endotoxemia originating from the intestines. Our clinical series of surveys also revealed that unexpectedly high positive Limulus gelation test (LGT) results in esophageal varices connected with liver cirrhosis may be due to the decreased clearance activity of the hepatic reticuloendothelial system. To study the mechanism and pathophysiology of endotoxemia in liver disorders, we produced hepatic dysfunction experimentally in germfree and conventional rats through fatty liver induction. The following discussion is based on our clinical and experimental results and explores the effects and mechanisms of endogenous endotoxemia in chronic hepatic disorders.

2 Materials and methods

Detection of endotoxin: Two hundred and thirty-seven patients with digestive disorders who were hospitalized in our hospital were investigated. Venous blood samples were obtained from each patient from a site prepared with iodine and alcohol. Plasma was prepared from blood anticoagulated with heparin (75 units/ml of blood) for determination of endotoxin. In some cases of esophageal varices, blood samples were obtained from portal and hepatic veins simultaneously, by catheterization of the portal vein through percutaneous transhepatic puncture and catheterization of the hepatic vein through the cephalic vein. Endotoxin was measured using the LGT as previously described (1). Tests were performed immediately after or within 12 hours of sampling. The heparinized plasma was diluted 1:3 with a pyrogen free saline solution and heated to 100°C for 10 minutes to extract plasma inhibitors in the LGT (2). The endotoxin assay was carried out in a water bath at 37°C, and the incubation mixture for the assay consisted of 0.1 ml of Lymulus lysate (Pregel, Teikoku Zoki Co., Tokyo, Japan) and 0.1 ml of the material tested. The presence of increased viscosity or solid gel formation when the tube was inverted after incubation represented a positive test for endotoxin.

Experimental induction of liver disorders: Male germfree rats (120-140 gm body weight) of the Sprague-Dawley strain (Japan CLEA Co. Tokyo, Japan) bred in our laboratory were used in this series of experiments. Liver dysfunction was induced by feeding germfree and conventional rats a fatty diet. CL-2 chow (Japan CLEA Co. Tokyo, Japan) was used as basal diet and CL-2 chow combined with 1% w/w cholesterol and 0.5% w/w cholic acid was used as the experimental diet (C-CA diet) for the fatty liver induction. Two groups of 8-week-old germfree and conventional rats were fed the C-CA diet and the basal diet, respectively. After 12 weeks of feeding for the germfree rats and 12 and 24 weeks for the conventional rats, the animals were sacrificed and examined. Blood samples for detection of endotoxin were obtained through the puncture of aorta and portal veins after laparotomy. Accidental contamination was checked through feces specimens taken every week of the experimental period. All germfree rats were maintained successfully to the end of the experiment.

The cyclic-AMP response to glucagon was measured in germfree and conventional rats for evaluation of the compensatory reserve of the liver. After sampling blood, glucagon (10 µg/kg, Novo, Denmark) was injected intravenously to fasted rats. Blood samples were collected every 10 minutes for an hour after the injection for the assay of cyclic-AMP. The assay was performed by the radioimmunoassay described previously (3). Peak value/basal value (PB ratio) was calculated for the evaluation.

3 Results and discussion

The study involved the detection of endotoxin by LGT in the systemic blood of 237 patients with various digestive disorders. Eighty three of the 237 patients (35%) showed positive LGT in this survey. Positive tests were found in 49.2% of the cases of esophageal varices, 48.2% of acute peritonitis, 15.6% of intestinal obstruction and 15.2% of acute cholangitis (Table 1). It is worth noting that the positive rate of endotoxemia was equally high in esophageal varices as in acute peritonitis due to the perforative hollow viscera, and that it was higher than that in diseases caused by underlying bacterial infections such as intestinal obstruction or acute cholangitis. Many recent Japanese studies (4,5) have reported a higher rate of endotoxemia in liver cirrhosis and discussed this phenomenon enthusiastically. However, the effect and pathophysiology are not known. It is difficult to understand

Table 1. Positive systemic LGT in 237 patients of various diseases.

Disease	No. of cases	Positive LGT (%)
Esophageal varices	59	29 (49.2)
Peritonitis	81	39 (48.1)
Intestinal obstruction	64	10 (15.6)
Acute cholangitis	33	5 (15.2)
Total	237	83 (35.0)

why such a high rate of endotoxemia is detected in patients with esophageal varices who show no infectious background and no manifestation of septic shock.

To elucidate the higher rate of endotoxemia in esophageal varices, the following studies were performed clinically and experimentally.

In 36 out of 59 patients with esophageal varices, LGTs of both systemic and portal blood were performed simultaneously. A comparison of LGTs between systemic and portal blood in these patients is shown in Figure 1. The tests showed a higher rate in portal blood than in systemic blood. Nineteen out of 28 positive portal LGT patients also showed positive LGT in the systemic blood sample. All 8 negative portal LGT patients showed negative LGT results in systemic blood samples. Systemic endotoxemia, therefore, may originate from an intestinal source, because all negative portal LGT patients may be interpreted to mean that

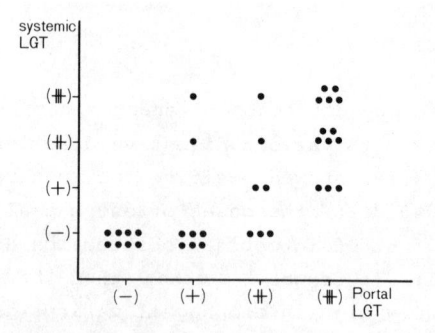

(+++) solid gel formation in LGT
(++) clearly increased viscosity in LGT
(+) increased viscosity in LGT

Fig. 1. Comparison of LGT between systemic and portal blood

the impaired detoxification ability of the liver combined with the effects of the development of collateral pathway from the portal system were responsible for the 19 cases of positive portal and systemic tests and that the efficient detoxification function of the liver was responsible for the 9 cases of positive portal tests with systemic negative tests.

For a more detailed analysis of the endotoxemia of esophageal varices as a result of liver cirrhosis, three blood samples were taken from the portal, hepatic and systemic veins simultaneously.

Table 2. Comparison of LGT of portal, hepatic venous and systemic blood in 23 patients of esophageal varices.

Type	Portal vein	Hepatic vein	Systemic artery	No. of cases
I	+	+	+	6
II	+	+	−	3
III	+	−	−	4
IV	+	−	+	6
V	−	−	−	4

The five types of endotoxemia classified from the results are shown in Table 2. In the six cases of Type I, all blood samples showed positive LGT, indicating the most common type of endotoxemia. In the three cases of Type II, only the systemic sample showed negative LGT. In the four cases of Type III, the portal blood sample showed positive LGT while hepatic and systemic samples showed negative LGT. In the six cases of Type IV, portal and systemic samples showed positive LGT, while hepatic samples showed negative LGT. The four cases of Type V showed negative LGT in all three samples. These clinical observations indicate that endotoxin in the portal blood may be detoxified mainly in the liver. The three Type II patients whose systemic samples showed negative LGT may indicate that the endotoxin coming into the systemic blood was diluted. The six Type IV patients with negative hepatic and positive systemic and portal tests clearly suggest the possibility that endotoxin in the portal vein flows directly into the systemic circulation through the extrahepatic collateral shunts as in esophageal varices.

Transabdominal esophageal transection with splenectomy and devascularization of shunts was performed on Type IV patients with endotoxemic esophageal varices. The preoperative and postoperative situations of these patients are illustrated in Figures 2 and 3. In all four patients, the LGT for systemic blood

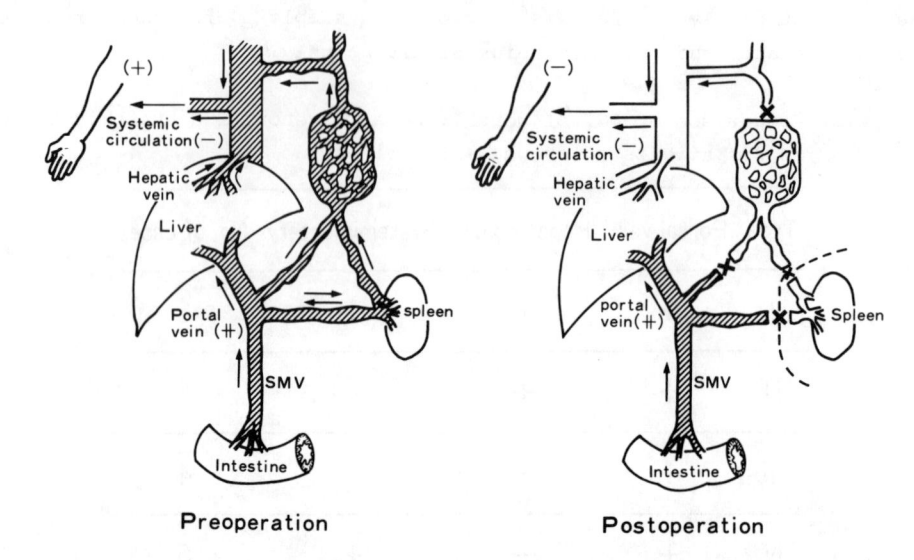

Figs. 2, 3. Pre- and postoperative Scheme of the type of endotoxemia of esophageal varices

changed from positive to negative postoperatively. The results of this operation indicate that the endotoxin in the systemic circulation was released through the portal shunts. Interception of the extrahepatic port-venous shunts resulted in negative LGT in the systemic blood as a result of the efficient detoxification function of the liver. These clinical results motivated us to elucidate the factors and mechanisms of endotoxemia in chronic liver diseases using germfree animals.

Eight-week-old germfree and conventional rats were fed a C-CA diet to produce fatty livers. Animals in the germfree group were killed after 12 weeks and in the conventional group after 12 and 24 weeks. A macroscopic view of three liver specimens of rats is shown in Figure 4. The right specimen is the liver before the experimental diet feeding; the middle after 24 weeks of C-CA diet feeding in conventional rats; and the left after 12 weeks of C-CA diet feeding in germfree rats. The specimens taken after the C-CA diet clearly show the findings of fatty liver, marked enlargement and milky whitish color.

Factors and Mechanisms of Endotoxemia

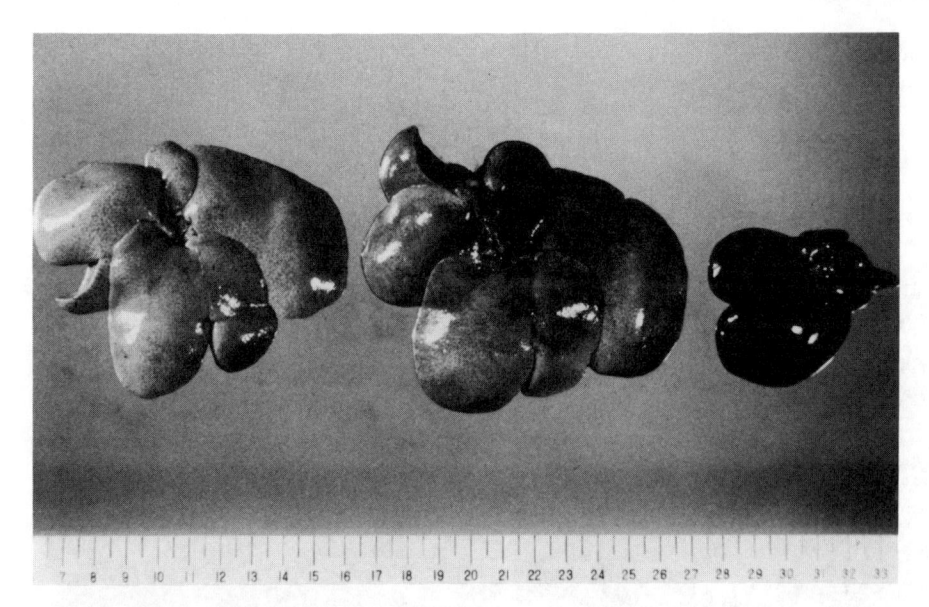

Fig. 4. Macroscopic views of the liver
  right: before experimental diet feeding
  middle: after 24 weeks of C-CA diet feeding in conventional rat
  left: after 12 weeks of C-CA diet feeding in germfree rat

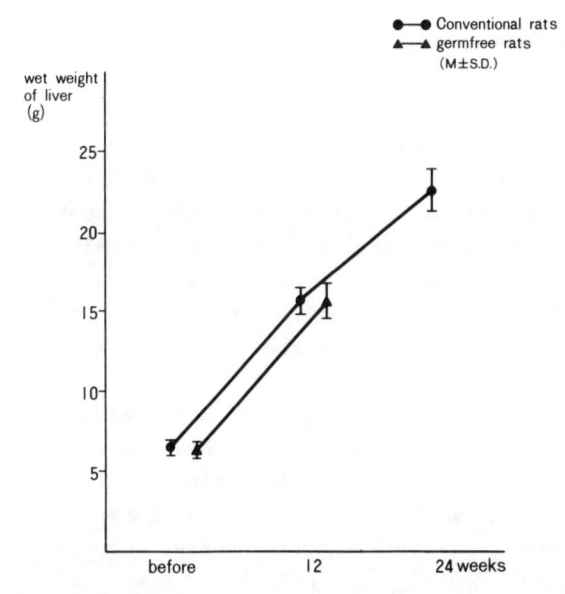

Fig. 5. Changes in weight of the liver of experimental fatty liver in rats

Endotoxemia: Clinical Aspects

The average weight of the liver was 6.6 grams in the conventional rats before the experimental diet, and linear weight increases after the C-CA diet were observed. After 12 weeks of the C-CA diet, the liver weight increased to 15.6 grams and by week 24, the liver weight was five times greater in conventional rats than at the start of the experiment (Figure 5). Figure 6 shows the histologic findings of the

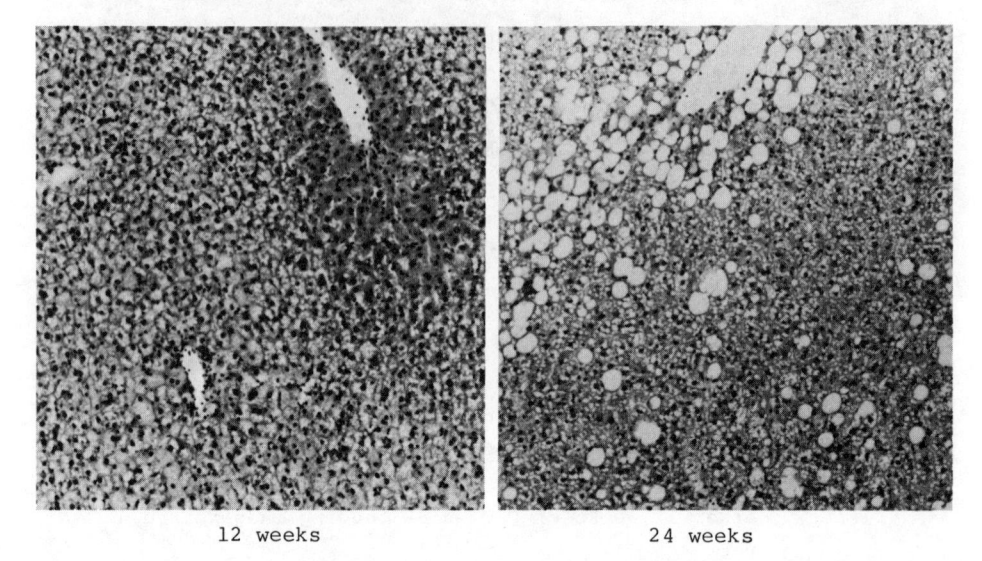

| 12 weeks | 24 weeks |

Fig. 6. Microscopic views of the liver of rats of the conventional group after 12 and 24 weeks of C-CA diet feeding (x20 Hematoxylin-Eosin stain)

conventional rat liver after the 12th and 24th week of the C-CA diet. The slide was stained with Hematoxylin-Eosin. The right side of the figure showed the liver of a rat sacrificed at the 24th week. Remarkable vaculoization was found at the peripheral area to the central lobulus and a swelling of the cytoplasm of the hepatocyte was also oberved. The left side of the figure shows the liver of the rat at 12 weeks. The histologic changes are milder than those at 24 weeks. Figure 7 shows the histologic findings of the germfree rat after 12th week of the C-CA diet. A more severe vacuolization was observed in this rat than in the conventional rat after 12 weeks. This vacuolization was identified as adipose deposition using a Suddan III stain. The results of endotoxin detection by LGT in germfree and conventional rats after the C-CA diet are shown in Table 3. As was expected, portal and systemic blood of germfree rats showed negative LGT before and after 12 weeks of experimentation. Before the C-CA diet feeding in conventional rats, 2 out of 10 rats showed positive LGT in the portal blood. However,

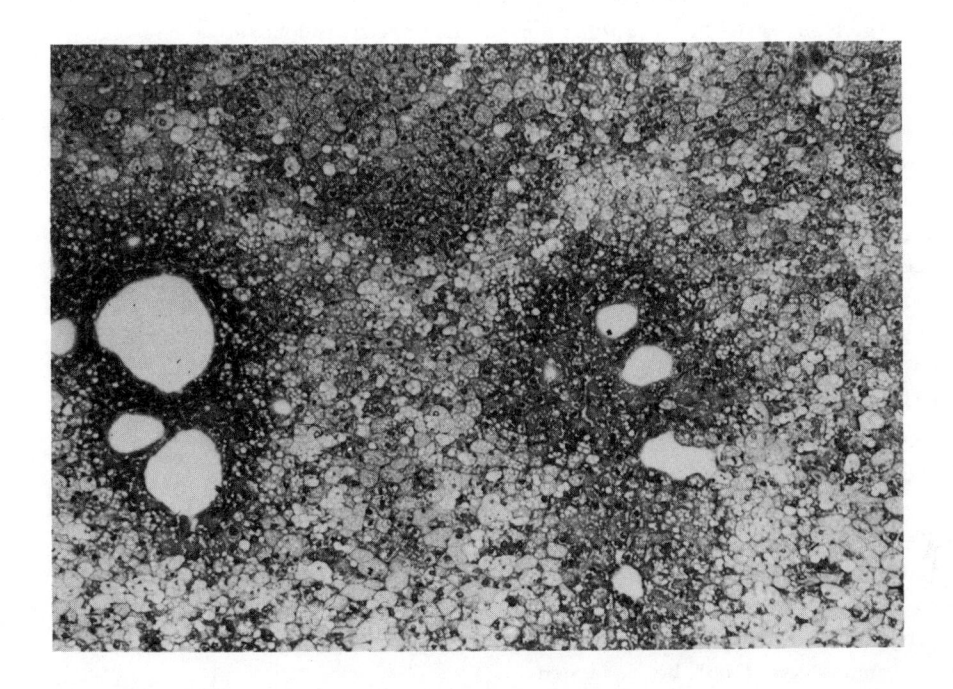

Fig. 7. Microscopic view of the rat after 12 weeks of C-CA diet feeding in germfree group. Hematoxylin-Eosin stain

Table 3. Detection of endotoxin in rat fed by fatty-liver-diet

Weeks after feeding	Germfree rats Portal blood	Germfree rats Systemic blood	Conventional rats Portal blood	Conventional rats Systemic blood
before	0/5 (0%)	0/5 (0%)	2/10 (20%)	0/10 (0%)
12 weeks	0/5 (0%)	0/5 (0%)	9/16 * (56%)	8/18 * (44%)
24 weeks			8/18 * (44%)	8/18 * (44%)

(* P<0.005)

none of the conventional rats showed positive LGT in the systemic blood. After 12 weeks of the C-CA diet, portal and systemic blood in conventional rats showed positive LGT in 9 out of 16 cases (56%) and 8 out of 18 cases (44%) respectively. After 24 weeks of the C-CA diet, 8 out of 18 rats (44%) showed positive LGT in portal and systemic blood. In this series of experiments we observed two rats in the conventional group which developed portal endotoxemia without systemic endotoxemia before feeding of C-CA diet. This type of portal endotoxemia has been reported perviously in clinical observations (6) and is believed to be the result of a disturbed intestinal mucosal barrier. Moreover, endotoxin released in the portal vein may be eliminated by hepatic reticuloendothelial system under normal conditions. In contrast, in the rats with chronic liver diseases induced by fatty diet feeding, equally high rates of endotoxemia were observed in both portal and systemic blood. It may be suggested that the higher rate of portal endotoxemia is caused by a disturbed intestinal mucosa which is affected by dietry cholic acid.

Moreover, it is reasonable to conclude that the impaired detoxification function of the liver could not eliminate the released endotoxin from the portal blood. Changes in glucagon induced plasma cyclic-AMP value are shown in Figure 8.

After 12 weeks of the C-CA diet, a remarkable decrease of P/B ratio occurred during the 12 and 24 weeks of the C-CA diet, particularly the LGT positive rats. These rats showed a significantly lower P/B ratio than LGT negative rats of the same period. The same tendency was observed at 24 weeks as at 12 weeks. However, statistical significance could not be obtained.

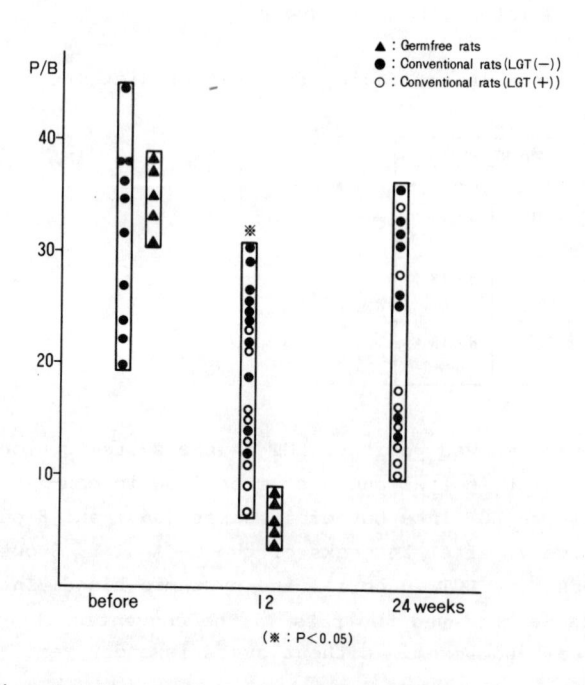

Fig. 8. Changes in glucagon induced plasma c-AMP value (P/B ratio)

From this series of experimental studies, it was confirmed that more severe fatty livers were induced in germfree animals than in conventional animals, using a fatty diet. It should be noted that germfree animals cannot metabolize cholesterol and cholic acid because of the absence of intestinal microflora. In contrast to the above mentioned protective effects of intestinal microflora, endotoxins originating from the microflora are one of the factors contributing to impairment of liver detoxification function. It should be pointed out that there is a possibility that endotoxin contributes to the detoxification function of the liver.

4 References

(1) J.Levin and F.B.Bang, Annals of Internal Medicine. 76 (1972) 1-7.

(2) M.S.Cooperstock, R.P.Tucker, and J.V.Baublis, The Lancet, June 7 (1975) 1972-1274.

(3) A.E.Broadus, N.I.Kaminsky, R.C.Northcutt, J.G.Hardman, E.W. Sutherland and G.W.Liddle, J. Clin. Invest. 49 (1970) 2222-2236.

(4) K.Tarao, K.So, T.Moroi, T.Ikeuchi, T.Suyama, O.Endo and K.Fukushima, Gastroenterology. 73 (1977) 539-542.

(5) M.Iwasaki, I.Maruyama, N.Ikejiri, M.Abe, T.Maeyama, E.Osada, H.Abe and H.Tanigawa, Japanese J. Gastroenterol. 76 (1979) 239-248.

(6) A.I.Jacob, P.K.Goldberg, N.Bloom, G.A.Degenshein, and P.J.Kozinn, Gastroenterology. 72 (1977) 1268-1270.

Clinical Aspects of Endotoxemia in Liver Diseases

Kyuichi Tanikawa, Masataka Iwasaki

The Second Department of Medicine, Kurume University School of Medicine, Kurume-shi, 830 Japan

Abstract

   The role of endotoxin seems to be important in the pathogenesis of some liver injuries and in the induction of further hepatocytic damages or extrahepatic manifestations in liver diseases. Firstly, possible roles of endotoxin in the pathogenesis of various liver diseases were discussed. Secondly, clinical significance of endotoxemia and its treatment were studied. Approximately half of our patients with liver cirrhosis showed positive for endotoxemia, which was more frequently found in patients with ascites, advanced stage of esophageal varix and low hepatic blood flow. Thus, cirrhotic patients with long-term endotoxemia showed poor prognosis. More than half of patients with fulminant hepatitis showed positive for endotoxemia, which may be one of causative factors for serious extrahepatic manifestations in this disease. Thus, proper treatments for endotoxemia are mandatory. Our experiences revealed that long-term administration of lactulose or short-term oral administration of polymyxin B was of choice in treatment of endotoxemia.

1 Introduction

   This study is mainly focused on two aspects of endotoxin in liver diseases. One is possible roles of endotoxin in the pathogenesis of liver injuries. The other is clinical significance of endotoxemia and its treatment. The first is more speculative or hypothetical, however

, is very challenging subject to elucidate. The second is more practical problem in managing patients with liver diseases.

2 Role of endotoxin in the pathogenesis of liver diseases

For the past there are many interesting experimental studies reported on the role of enteric endotoxin in the development of liver injuries. In acute carbon tetrachloride liver injury in rats, the liver damage is significantly reduced by oral administration of polymyxin B (1) or endotoxin tolelant state (2). In D-galactosamine induced liver injury, it is known that colectomy or combined with small bowel resection performed prior to galactosamine administration protects almost completely against hepatic necrosis or shows no serum transaminase elevation (3). It is also known that the development of cirrhosis or fibrosis on a choline deficient diet is inhibited by chronic oral administration of neomycin (4). These studies suggest that intestinal factors, especially endotoxin, is necessary for development of the hepatic lesions. Massive hepatic necrosis by Schwarzman reaction, beautifully demonstrated recently by Mori and his associates (5), is more direct proof of endotoxin in the pathogenesis of fulminant hepatic failure. Similar mechanisms may be involved in human liver diseases. Possible mechanisms for the pathogenesis of liver injuries in association with endotoxin is illustrated in Figure 1.

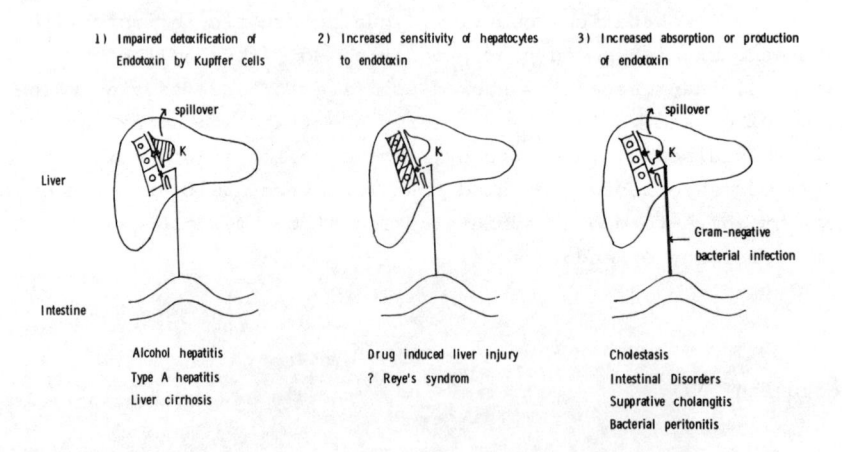

Fig.1. Possible mechanisms of the liver injury in association with edotoxin.

Endotoxin, originated mainly from the intestinal bacterial flora is detoxified by the Kupffer cell in the liver. Thus, when the Kupffer cell is primarily damaged, detoxification of endotoxin is impaired, resulting in hepatocytic damage by non-detoxified endotoxin.
In alcohol induced liver injuries and acute type A viral hepatitis, this mechanism seems to be involved in the development of hepatocytic damages. Alcohol is known to impair the function of the Kupffer cell. Even a single oral bolus of ethanol delays a removal of endotoxin from the circulation (6). Clinically a transient endotoxemia is frequently seen in alcoholic liver injuries (7). Liver scintigram taken in patients with alcoholic hepatitis shows a depressed ability of the Kupffer cell to take up radioactive colloid particles. However, it returns to normal after stop of drinking. It is, thus, expected that the hepatocyte is exposed to a large amount of endotoxin in alcoholics because of depressed function of the Kupffer cell and this leads to further hepatocytic damages.

In type A hepatitis, hepatitis A virus is mainly localized in hepatocytes and Kupffer cells near the portal tract where is prominent in tissue damages. Under electron microscope the Kupffer cell is remarkably damaged with numerous lipofuscin-like granules in the cytoplasm which contain hepatitis A virus, and hepatocytes, very close to those Kupffer cells, appear to be also markedly altered (8). From these morphological evidences it is speculated that hepatocytes in type A hepatitis are exposed to a large amount of endotoxin because of damaged Kupffer cells and also to mediators such as lysosomal enzymes released from the damaged Kupffer cells.

The second mechanism is more speculative. In experimental liver injuries, induced by carbon tetrachloride or D-galactosamine, the following may be considered as a mechanism of hepatocytic damages ; pretreated hepatocytes by those agents become very susceptible to even a small amount of endotoxin, resulting in further hepatocytic damages. Similar mechanisms may occur in various drug induced liver injuries in man. In Reye's syndrom in which by one report (9), endotoxemia is observed in all cases, our study by electron microscopy demonstrates characteristic mitochondrial changes in the hepatocyte which are very similar to those of endotoxin induced liver injuries.

The third is an increased absorption or production of endotoxin, which causes hepatocytic alterations. An increased absorption of endotoxin from the intestine could occur in cholestasis or intestinal disorders. In severe gram-negative bacterial infections such as suppurative cholangitis or bacterial peritonitis, a large amount of endotoxin are released from the lesions and entered to the general circulation, resulting in hepatocytic damages.

Thus, those mentioned are very suggestive that endotoxin plays at least some role in the pathogenesis of their liver injuries and we must elucidate its role more extensively in future.

3 Clinical significance of endotoxemia in liver diseases

Endotoxin is mainly detoxified by the Kupffer cell in the liver. Thus, spillover of endotoxin to the general circulation could occur in fulminant hepatitis or liver cirrhosis because of depressed function of the Kupffer cell or portal-systemic shunting. Frequency of endotoxemia detected by Limulus test in various liver diseases is shown in Figure 2. Endotoxemia is most frequently found in fulminant hepatitis, the most serious disease among various liver injuries. In cirrhosis or cirrhosis with hepatocellular carcinoma, endotoxemia is also found in half of cases. Hepatic blood flow index measured by radioactive Tc phytate is low in cases with endotoxemia in liver cirrhosis. Endotoxemia is very often detected under 0.124 of hepatic blood flow index and over 34% of BSP or ICG retension tests (Fig. 3). Liver scintigram study in cirrhosis shows that endotoxemia is frequently noted in remarkably atrophied liver (Table 1) and is also often found in patients with ascites than without ascites (Fig. 4). These studies indicate that endotoxin is a common complication in advanced stage of cirrhosis.

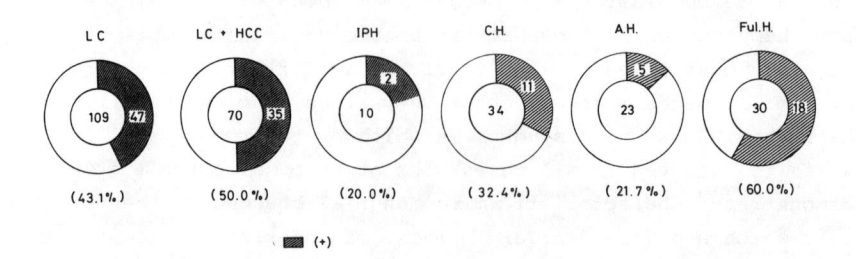

Fig. 2. Frequency of endotoxemia in various liver diseases.
LC=liver cirrhosis; HCC=hepatocellular carcinoma; IPH= idiopathic portal hypertension; CH=chronic hepatitis; AH=acute hepatitis; Ful.H=fulminant hepatitis.

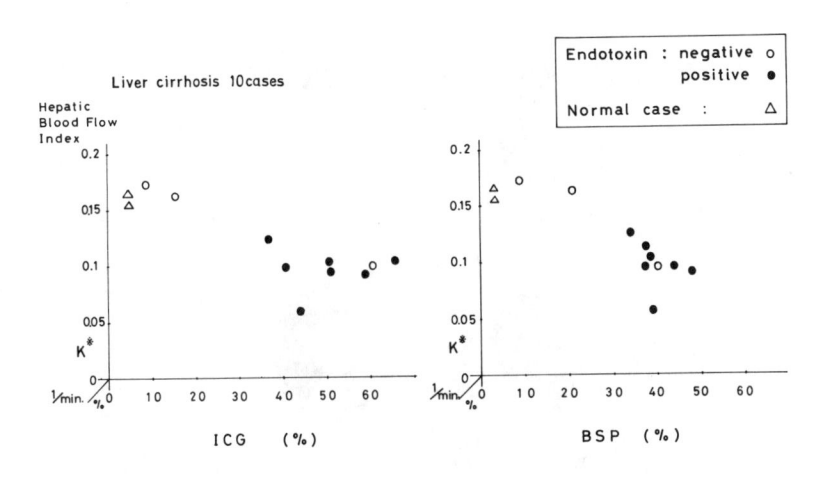

Fig. 3. Relationship between endotoxemia and hepatic blood flow index(K) by ^{99m}Tc-phytate or percent ICG(15 min.) and BSP (45 min.) retentions.

* Constant $K = \dfrac{0.693}{T\frac{1}{2}}$

	Liver size			Liver uptake			Spleen uptake		
endotoxin	normal	atrophy		normal	low	very low	low	high	very high
		slight	marked						
negative 16	3	7	6	6	6	4	2	12	2
positive 19	4	4	11	9	6	4	4	10	4

Table 1. Relationship between endotoxemia and ^{99m}Tc-phytate liver scintigram in 35 cases of liver cirrhosis.

Endotoxemia: Clinical Aspects

	Ascites(+)	Long term endotoxemia	Ascites(−)	Long term endotoxemia
LC	53 / 27 (51%)	14/27 (52%)	54 / 20 (37%)	6/20 (30%)
LC·HCC	58 / 30 (52%)	12/30 (40%)	12 / 5 (42%)	2/5 (40%)
Total	111 / 57 (51%)	26/57 (46%)	66 / 25 (41%)	8/25 (32%)

▧ Endotoxin positive

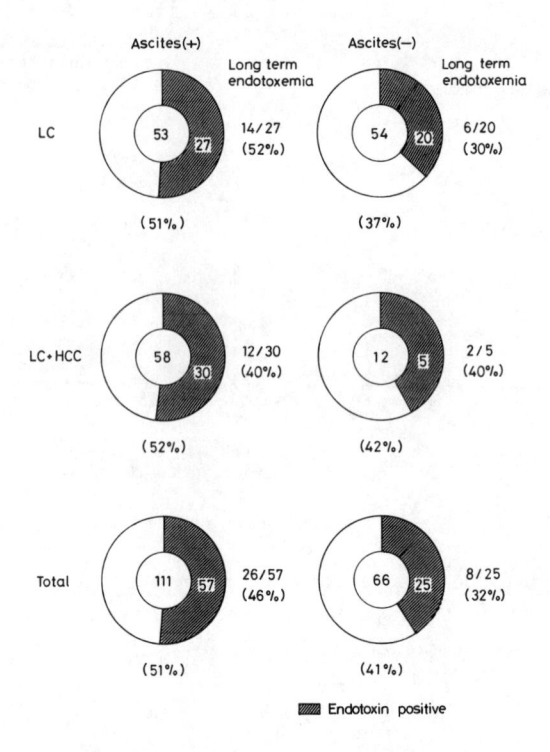

Fig. 4. Frequency of endotoxemia in liver cirrhosis with or without ascites.

Endotoxin is more often detected in the ascitic fluid than in the serum (Fig. 5). The reason for this is probably due to few cells such as Kupffer cells detoxifying endotoxin in the abdominal cavity. This is imortant that endotoxin is frequently detected in the ascitic fluid. In uncontrolable ascites we often use a Le Veen's tube, of which serious complication is disseminated intravascular coagulation. Thus, endotoxin in the ascitic fluid must be checked when use of this tube is indicated. The relationship between endotoxemia and duration of ascites shown in Figure 6. indicates that ascites is hard to be controlled in cases with endotoxemia. Cirrhotic patients with endotoxemia also show to be short-lived. It is well known that most of cirrhotic patients develope esophageal varices and their rupture is one of the main causes of death in liver cirrhosis.

All cases with long term endotoxemia have episodes of upper GI tract hemorrhage , and in Stage 3, which is most advanced stage of esophageal varix by endoscopy, endotoxemia is most often observed with higer frequency of upper GI hemorrhage (Fig. 7). Table. 2 shows a correlation between frequency of peptic ulcer or erosion in the stomach or duodenum and endotoxemia in cirrhotic patients. Gastroduodenal ulcer or erosion is much often found in endotoxin positive patients. Thus, endotoxin could be considered as one of the main causes for rupture of esophageal varix and ulcer or erosion seen in the stomch or duodenum of cirrhotic patients.

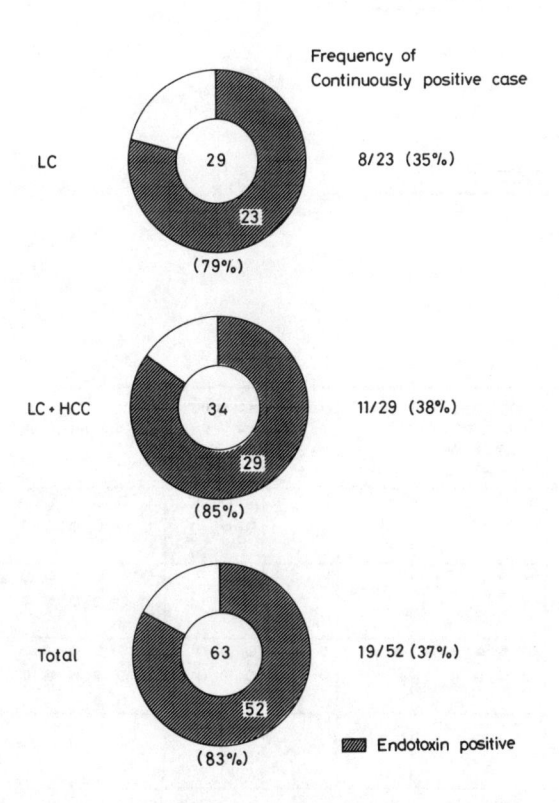

Fig. 5 . Frequency of endotoxin in the ascitic fluid in liver cirrhosis. LC=liver cirrhosis; HCC=hepatocellular carcinoma.

Endotoxemia: Clinical Aspects

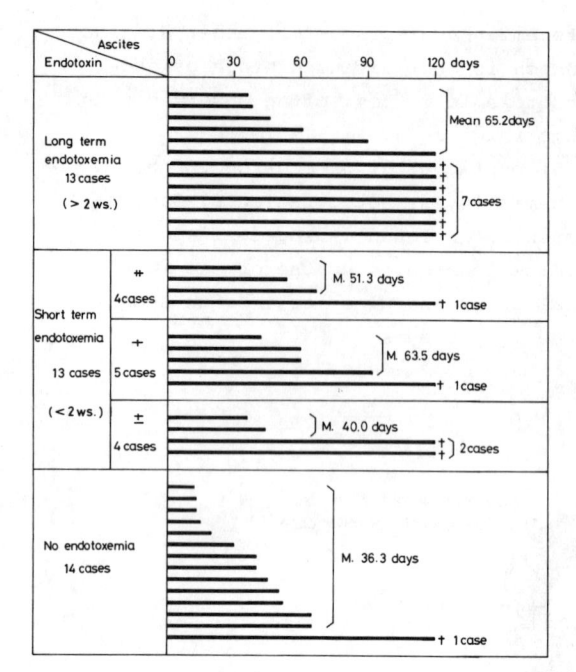

Fig. 6. Relation between endotoxemia and duration of ascites in liver cirrhosis.

Fig. 7. Correlation between stages of esophageal varices and episodes of upper GI hemorrhage or endotoxemia.

●: Hemorrhage(+) in 3 months after or before the endoscopic examination
○: Hemorrhage(−)

Table. 2. Correlation between gastro-duodenal lesions and endotoxemia in 76 cases of liver cirrhosis.

endotoxin in plasma	Long term endotoxemia ╫ - ±	Short term endotoxemia ╫	+	±	No endotoxemia
Gastric ulcer	3	0	1	3	2
Scar of the ulcer	0	0	0	3	2
Duodenal ulcer	1	1	2	1	1
Erosion	4	1	3	4	4
No ulceration	4	1	3	8	25
Total	12		30		34

66.6% / 62.0% / 26.5%

4. Treatment of endotoxemia in liver diseses

Endotoxemia is a serious complication in liver diseases which often needs emergency treatments. Table 3 summarizes clinical managements for endotoxemia in the liver disese.

Table. 3. Clinical managements for endotoxemia in the liver disease

1). Decreasing gut absorption
   Antibiotics (polymyxin B, neomycin, paramomycin)
   Lactulose
   Cholestylamine
   Bile acids
2). Removal from circulation
   Direct hemoperfusion with activated charcol
   Plasma exchange
3). Protect against endotoxin damages
   Steroids etc.

335

Lactulose, nonabsorbable agent which lowers the pH values of the colon and inhibits bacterial growth, is commonly used for treatment or prevention of ammonemia, and this agent is also indicated for endotoxemia. As shown in Table 4, good results were obtained for treatment or prevention of endotoxemia from cases in which lactulose was used for long period of time in comparison with those without use of lactulose. Oral administration of polymyxin B for three days was also effective for endotoxemia, but no effect by kanamycin (Table 5). It is known that polymyxin B acts not only as bactericide, but also inactivate endotoxin itself. Thus, in vitro study was carried out on inactivation of endotoxin by polymyxin B. As shown in Table 6, polymyxin B inactivates endotoxin, but no effects were obtained by kanamycin. We have had no serious side effects in using this agent by oral administration. Table 7 shows endotoxemia in fulminant hepatitis in association with its prognosis. It is important to state that plasma exchange or direct hemoperfusion were performed in all survived cases with endotoxemia.

Those mentioned evidences suggest that endotoxemia is closely associated with various manifestations seen in liver diseases and also with their prognosis. Thus, the treatment of endotoxemia, one of the serious complications in the liver disease, is inevitably important.

Table 4. Effect of lactulose on endotoxemia in liver cirrhosis and hepatoma with liver cirrhosis

Effect	LC (61 cases)		LC + Hepatoma (50 cases)	
	38 cases Lactulose(+)	23 cases Lactulose(−)	29 cases Lactulose(+)	21 cases Lactulose(−)
Improved Endotoxin (#),(+),(±) → (−)	14 (36.8%)	0	11 (37.9%)	0
No change Endotoxin (−) → (−) (#),(+),(±) → (#),(+),(±)	22 (57.9%)	10 (43.5%)	17 (58.6%)	12 (57.1%)
Aggravated Endotoxin (−) → (±),(+),(#)	2 (5.3%)	13 (56.5%)	1 (3.5%)	9 (42.9%)

Table 5. Effects of polymyxin B or kanamycin on endotoxemia in various liver diseases.

Effect of Polymyxin B

Group	Serum Endotoxin	Liver cirrhosis (10 cases)	Hepatoma (10 cases)	Obstructive Jaundice (2 cases)
A	(±)⟶(−) / (+)⟶(−)	4/10 (40%)	4/10 (40%)	0/2 (0%)
B	(−)-(+)-(−)	2/10 (20%)	2/10 (20%)	2/2 (100%)
C	⟶(−)⟶(−)	4/10 (40%)	4/10 (40%)	0/2 (0%)

Effect of Kanamysin

Group	Serum Endotoxin	Liver cirrhosis (6 cases)
A	(±)⟶(−) / (+)⟶(−)	0/6 (0%)
B	(−)-(±)-(−)	3/6 (50%)
C	⟶(±)⟶(±)	3/6 (50%)

Table 6. Inactivation of endotoxin by polymyxin B or kanamycin in vitro.

		Endotoxin (mg/dl)				Control
		1×10^{-5}	10^{-6}	10^{-7}	10^{-8}	
Saline		++	+	±	−	−
Polymyxin B (mg/dl)	0.1	+	±	−	−	−
	0.01	++	+	−	−	−
	0.001	++	+	±	−	−

		Endotoxin (mg/dl)				Control
		1×10^{-5}	10^{-6}	10^{-7}	10^{-8}	
Saline		++	+	±	−	−
Kanamycin (mg/dl)	1.0	++	+	±	−	−
	0.1	++	+	±	−	−
	0.01	++	+	±	−	−
	0.001	++	+	±	−	−

Table 7. Endotoxemia in 30 cases of fulminant hepatitis and their prognosis

	Survived	Dead	Total
Et* (−)	5/30 (16.7%)	7/30 (23.3%)	12/30 (40%)
Et (+)	**4/30 (13.3%)	14/30 (47.7%)	18/30 (60%)
Total	9/30 (30%)	21/30 (70%)	

*Et : Endotoxin
**: Plasma exchange or Direct hemoperfusion were performed in all survived cases with positive Et.

5 References

(1) J. P. Nolan, A. I. Leibowitz, Gastroenterology 75 (1978) 445-449.

(2) J. P. Nolan, M. V. Ali, J. Med, 4 (1973) 28-38.

(3) D. S. Camara, J. A. Carana, M. Montes, J. P. Nolan, Hepatology 1 (1981) 500A.

(4) W. D. Salmon, P. M. Newborne, J. Nutr. 76 (1962) 483-486.

(5) W. Mori, J. Shiga, A. Kato, Virchows Arch. Path. Anat. Hist. 382 (1979) 179-189.

(6) J. P. Nolan, D. S. Camara in H. Popper, F. Schaffner (Ed.), Progress in Liver Diseases Vol. VII, Grune & Stratton, Inc, New York 1982, p. 361.

(7) J.P. Nolan, M. V. Ali, Lancet 1 (1974) 999

(8) K. Tanikawa, H. Setoyama, Saibo (Japanese) 15 (1982) 546-549.

(9) M. S. Cooperstock, R. P. Tucker, J. F. Boublis, Lancet 1 (1975) 1272-1274.

Complement Levels in Endotoxemia

Tetsuro Takaoka, Akishige Nakamura, Nagao Shinagawa, Jiro Yura.

First Department of Surgery, Nagoya City University, Medical School, 1-1 Kawasumi, Mizuho-cho, Mizuho-ku, Nagoya, JAPAN

Abstract

It is important to study immunological behavior during severe infections. Serum complement levels were studied during endo- toxemia which included both experimental and clinical endotoxin shock. In endotoxemia a decrease in total hemolytic complement activity (CH_{50}) was demonstrated very clearly experimentally, but not as clearly clinically. The degree of decrease in CH_{50} and survival time was found to be somewhat endotoxin dose-dependent experimentally. In endotoxin shock a decrease in CH_{50} was demonstrated experimentally and clinically. It was shown both experimentally and clinically that the prognosis was favorable in cases where CH_{50} levels returned to normal and rose after shock. Prognosis was unfavorable in cases where CH_{50} levels fell and decreased. These findings suggest that in cases of endotoxemia and endotoxin shock sequential determination of CH_{50} is considered to be useful in judging the severity of the disease, as well as in predicting the prognosis.

1 Introduction

With the recent increase in infections due to gram negative bacilli, the incidence of endotoxemia and endotoxin shock related to such infections also shows an increasing tendency which poses a great problem in all clinical fields. A great many experimental studies have been performed on the endotoxin responsible for this shock, but many problems have remained unsolved concerning the immunological behavior when endotoxin is in the blood. In view of these facts, we studied experimentally and clinically complement levels during endotoxemia and endotoxin shock.

2 Changes of Serum Complement Levels in Experimental Endotoxemia

1) Materials and Methods

The test animals used were male white rabbits weighing 2.5 to 3.0 kg. The endotoxin used was E.coli, 026:B6 lipopolysaccharide (Difco), which was dissolved in 2 ml of distilled water and given in doses of 10 mg/kg, 5 mg/kg, 3 mg/kg, 0.5 mg/kg and 0.05 mg/kg, five minutes after dissolving. It was injected in one dose into the ear vein of each rabbit. Determination of total hemolytic complement activity (hereafter referred to as CH_{50}) was carried out by Okada's method (1). Due to the recent attention directed at the dissociation of serum CH_{50} from plasma CH_{50}, we measured both values. However, as no significant difference was found between these two titers, only serum complement titers are given in the following description.

2) Results

Because of a wide normal range of CH_{50} in normal rabbits, the CH_{50} level before treatment is taken to 100%. Changes in CH_{50} levels with the passage of time are expressed in percentage.

(a) Group 1 received 10 mg/kg of endotoxin (Fig. 1).

Fifteen minutes after injection of endotoxin, a decrease in CH_{50} was found in four of five rabbits of this group. Although a transient rise was observed in three rabbits at thirty and sixty minutes after injection, CH_{50} levels decreased with time in all rabbits thereafter. Death occurred three to nine hours following injection of endotoxin.

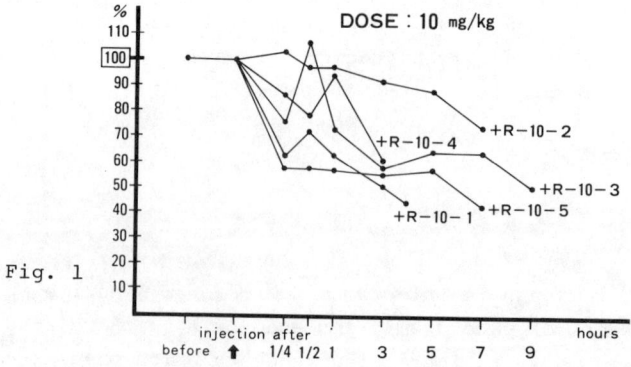

Fig. 1

Changes of Serum Complement Levels after Injection of E. coli Endotoxin

(b) Group 2 received 5 mg/kg of endotoxin (Fig. 2).

The same changes of CH_{50} were observed in these rabbits. Death occurred four to nine hours following the injection of endotoxin. In all but one level (obtained for R-4 thirty minutes later), CH_{50} was lower than initial levels.

Fig. 2

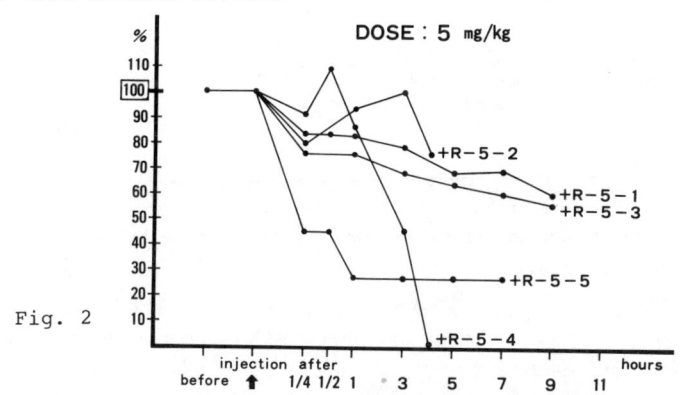

Changes of Serum Complement Levels after Injection of E. coli Endotoxin

(c) Group 3 received 3 mg/kg of endotoxin (Fig. 3).

Fifteen minutes after injection CH_{50} was lowered, with a decrease in four of five rabbits in this group. In the subsequent course, R-1 and R-4 showed a slight rise in CH_{50} at three and five hours respectively, and all rabbits in this group died twelve to fourteen hours after endotoxin injection.

Fig. 3

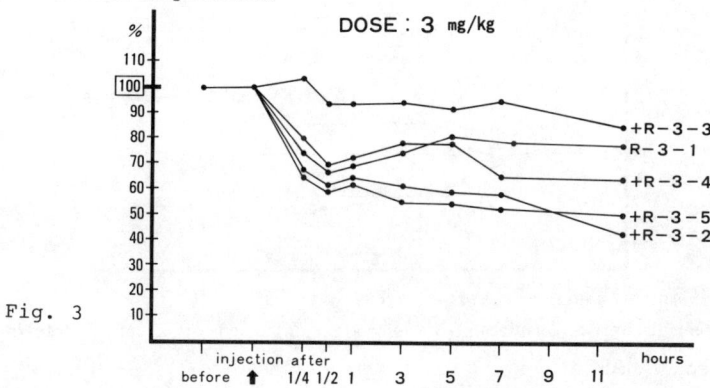

Changes of Serum Complement Levels after Injection of E. coli Endotoxin

(d) Group 4 received 0.5 mg/kg of endotoxin (Fig. 4).

Fifteen minutes after injection, CH_{50} was decreased in three of five rabbits. However, unlike groups given more than 3 mg/kg of endotoxin, two rabbits in this group showed higher CH_{50} than initial levels. Death in this group occurred approximately twenty four to forty hours after injection. Their survival time was remarkably prolonged when compared with the previous three groups.

Fig. 4

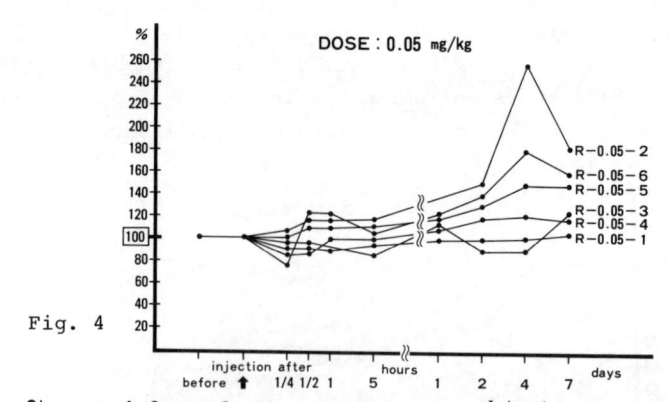

Changes of Serum Complement Levels after Injection of E.coli Endotoxin

(e) Group 5 received 0.05 mg/kg of endotoxin (Fig. 5).

Fifteen minutes after injection the degree of decrease in CH_{50} was smaller in five of six rabbits when compared with the previous four groups. The remaining rabbit (R-6) showed a decrease of 77%. Moreover, within one hour CH_{50} either returned to the initial levels or was elevated. Subsequently, in all cases, a gradual increase in CH_{50} was seen accompanied by a prolonged survival time in comparison with the previous four groups. All rabbits in this group survived thirty days or more.

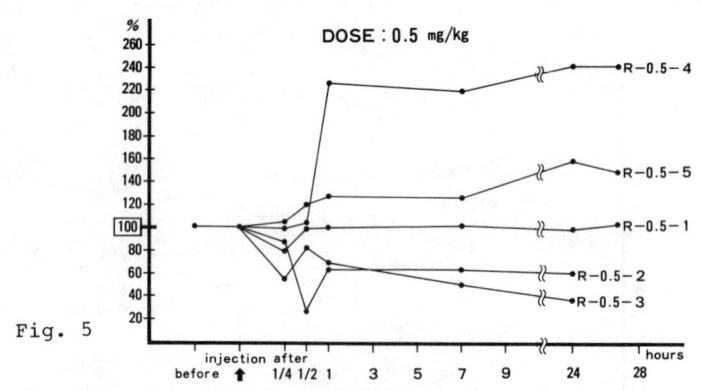

Fig. 5

Changes of Serum Complement Levels after Injection of E.coli Endotoxin

(f) Control Group

Five rabbits serving as controls received 2 ml of distilled water injected into the ear vein. In these animals fluctuations in CH_{50} (due to procedures such as blood sampling) were studied and it was found that CH_{50} was consistently between 91% and 106%. Changes shown were not more than ± 10%.

3) Summary

Before injection of endotoxin, if CH_{50} level was taken to 100%, a remarkable decrease was seen in groups receiving more than 3 mg/kg of endotoxin. In the groups receiving 0.5 mg/kg and 0.05 mg/kg, CH_{50} level showed only a slight decrease and rabbit survival time was longer than in the previous three groups (Fig. 6).

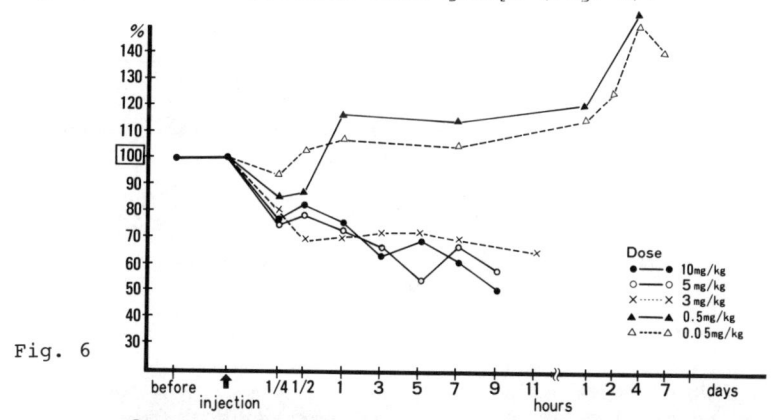

Fig. 6

Changes of Serum Complement Levels in Experimental Endotoxin Shock.

Our findings show that changes in CH_{50} at fifteen minutes after injection of endotoxin and survival time were somewhat dose-dependent. In addition, these results indicated that after endotoxin injection, those cases where CH_{50} levels returned to normal and rose with the passage of time had a favorable prognosis, while in cases where CH_{50} levels decreased or showed a transient rise followed by a fall, had an unfavorable prognosis.

3 Changes of Serum Complement Levels in Clinical Endotoxemia

1) Subjects and Methods

Studies were done on forty-eight patients. Severe surgical infection was determined by the following: 1) fever over 37.5°C, 2) leukocytosis, 3) an infectious focus, and 4) resistance of endotoxemia or bacteremia to antibiotics. Clinical endotoxin shock was determined by a fall of blood pressure to less than 60 mmHg and endotoxemia. Endotoxin was measured by Limulus lysate method. CH_{50} was measured by Mayer's method and C_3, C_4, and C_5 by single radial immunodiffusion method. As in the experimental studies, both serum and plasma complement levels were measured. Serum complement levels only are shown in this paper because in our forty-eight patients, no significant difference was found between these two results.

2) Results

(a) CH_{50} levels in severe surgical infections (Fig. 7, 8).

Blood samples were drawn during the critical period of severe infection. CH_{50} on an average was within normal range at the time of severe infection. There were no differences in CH_{50} between the positive endotoxin group and the negative endotoxin group, and all were within normal range. However, the patients with clinical shock and endotoxemia showed lower levels of CH_{50} than did the patients without clinical shock. CH_{50} in the patients with clinical shock was below the normal range. On the other hand, compared with the prognosis of patients with endotoxemia, CH_{50} levels in the unfavorable prognosis group was lower than CH_{50} in the favorable prognosis group. CH_{50} in the unfavorable prognosis group was below the normal range.

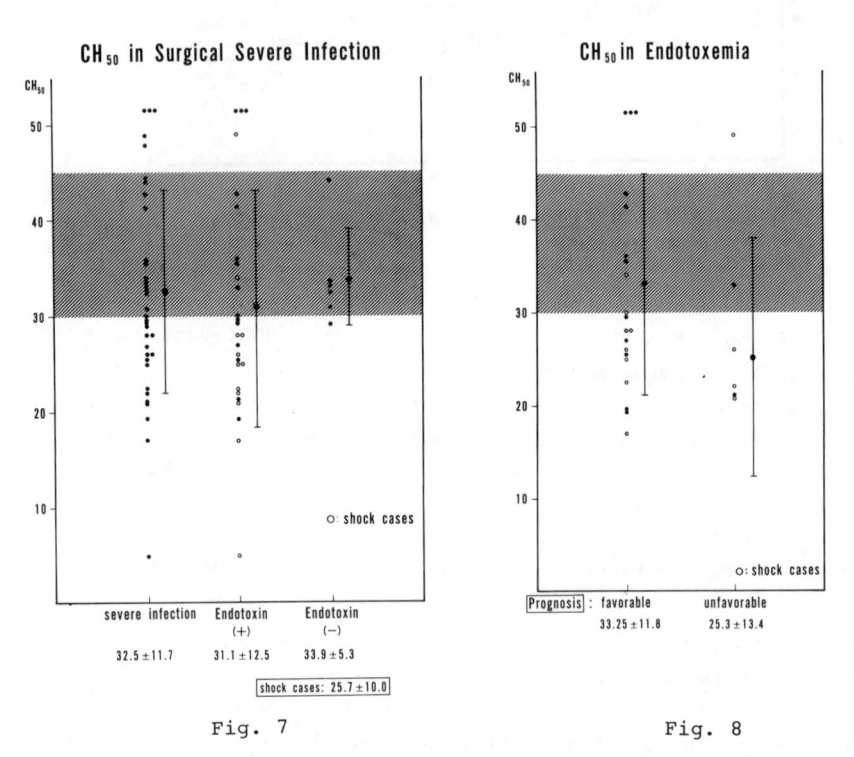

Fig. 7

Fig. 8

(b) C_3, C_4, and C_5 in severe surgical infections (Figs. 9, 10).

In the severe infections, C_3 was below the normal range, but C_4 and C_5 were within the normal range on the average. Compared with the positive endotoxin group and the negative endotoxin group, there was no difference. Compared with the prognosis in the patients with endotoxemia, C_3 and C_4 showed no difference, but C_5 was slightly different. C_5 in the favorable prognosis group was shown higher than in the unfavorable prognosis group, but all were within normal range on the average.

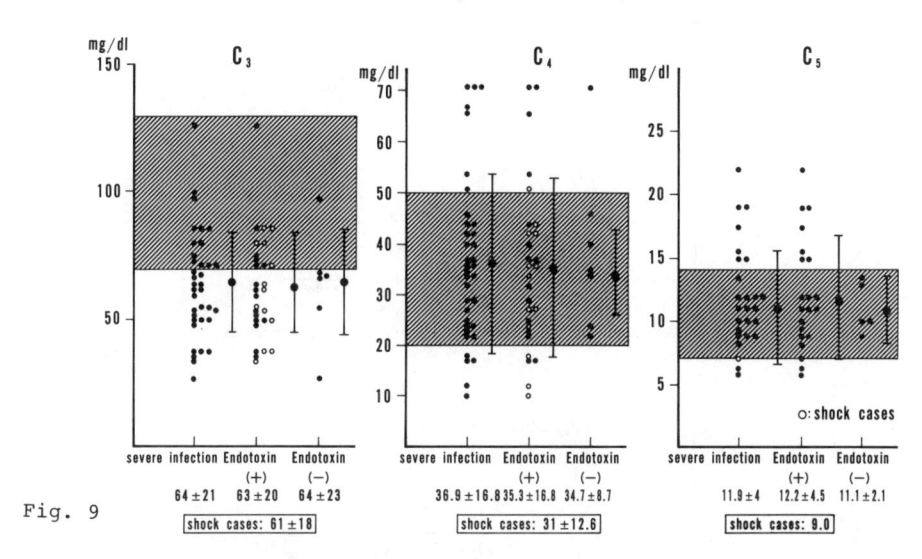

C_3, C_4, C_5 in Surgical Severe Infection

Fig. 9

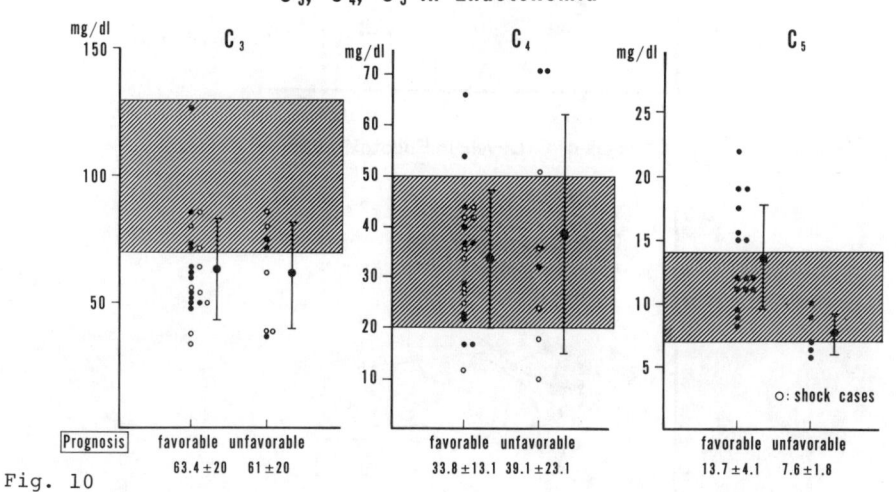

	severe infection	Endotoxin (+)	Endotoxin (−)
C_3	64±21	63±20	64±23
	shock cases: 61±18		
C_4	36.9±16.8	35.3±16.8	34.7±8.7
	shock cases: 31±12.6		
C_5	11.9±4	12.2±4.5	11.1±2.1
	shock cases: 9.0		

o: shock cases

C_3, C_4, C_5 in Endotoxemia

Fig. 10

Prognosis	favorable	unfavorable
C_3	63.4±20	61±20
C_4	33.8±13.1	39.1±23.1
C_5	13.7±4.1	7.6±1.8

o: shock cases

(c) Sequential CH_{50}, C_3, and C_4 in clinical endotoxin shock.

The favorable prognosis group is shown in Fig. 11. In two cases, CH_{50} was within the normal range, but in another four cases, CH_{50} was below the normal range during shock. In all of the cases after the day of shock, CH_{50} levels entered the normal range or showed a tendency to rise. The sequential C_3 and C_4 exhibited the same pattern as that of CH_{50} in all patients (Figs. 12, 13).

Endotoxemia: Clinical Aspects

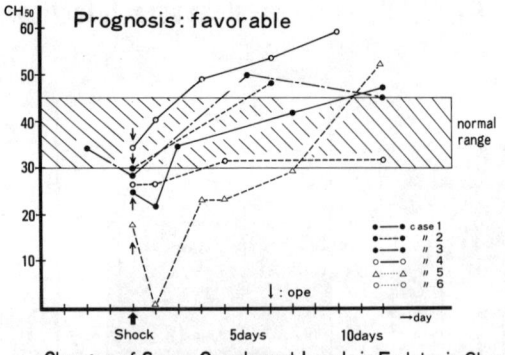

Fig. 11   Changes of Serum Complement Levels in Endotoxin Shock.

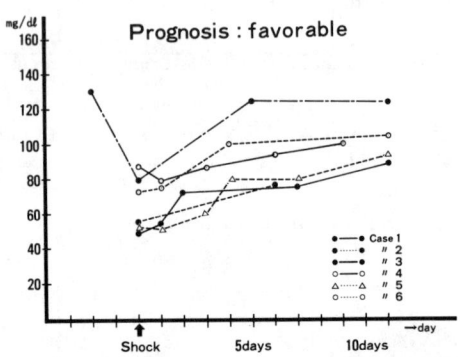

Fig. 12   Changes of C_3 Levels in Endotoxin Shock.

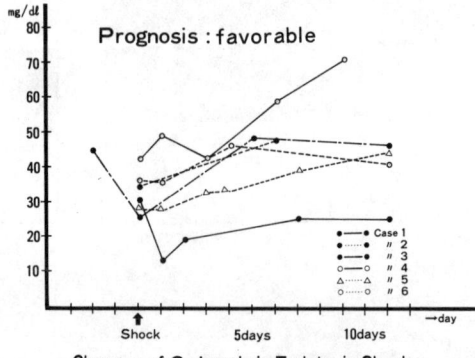

Fig. 13   Changes of C_4 Levels in Endotoxin Shock.

In the group with unfavorable prognosis (Fig. 14), three cases had CH_{50} levels that were below the normal range and showed a tendency to decrease. In another patient, CH_{50} levels were higher than the normal range and a transient rise was followed by a fall, and then showed a marked decrease with recurrence of shock. All the patients died. Sequential C_3 and C_4 exhibited the same pattern as that of CH_{50} in all patients (Figs. 15, 16).

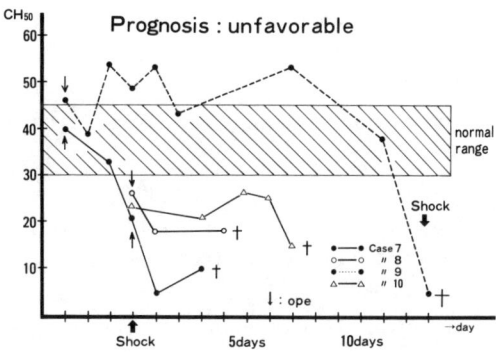

Fig. 14 Changes of Serum Complement Levels in Endotoxin Shock.

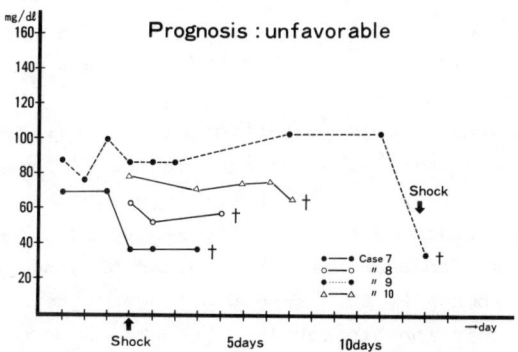

Fig. 15 Changes of C_3 Levels in Endotoxin Shock.

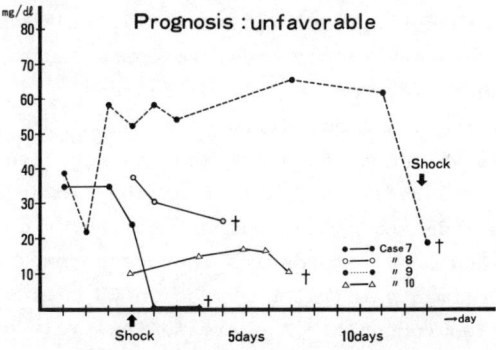

Fig. 16 Changes of C_4 Levels in Endotoxin Shock.

3) Summary

Compared with the normal range of CH_{50}, C_3, C_4, and C_5 in severe infections, there were no differences in CH_{50}, C_4, and C_5. However, C_3 was, on an average, below the normal range. Between the positive endotoxin group and the negative endotoxin group no significant differences of CH_{50}, C_3, C_4, and C_5 were found. Although, when clinical shock occurred, CH_{50} levels were lower than in the no shock group and lower than the normal range. We compared CH_{50}, C_3, C_4 and C_5 in endotoxemia with the prognosis. CH_{50} and C_5 in the favorable prognosis group were higher than levels in the unfavorable prognosis group. We measured sequential CH_{50}, C_3, and C_4 during and after shock. The prognosis was favorable in cases in which levels returned to normal and showed a tendency to rise, while the prognosis was unfavorable in cases in which levels showed a tendency to decrease.

4 Discussion

Various studies have been carried out on the effects of endotoxin on the serum complement system. Gilbert et al (2) showed that serum complement levels decreased when endotoxin was injected in large doses, and also in doses less than LD_{50}.

In our experimental studies, injection of endotoxin obtained from E.coli was also found to cause a decrease in CH_{50}. The degree of decrease in CH_{50} at fifteen minutes after injection was dose-dependent to some extent. In cases where prognosis was unfavorable, CH_{50} showed a tendency to decrease with the passage of time, while in cases with a favorable prognosis, CH_{50} showed a tendency to increase. There was a close relationship between the changes in CH_{50} levels after injection of endotoxin and the period of survival.

In clinical studies, McCabe (3) reported that in cases of bacteremia due to the gram-negative bacillus, but with no complications, C_3 levels did not differ significantly from those of control patients. However, in bacteremia accompanied by shock, C_3 levels were lower than those of control patients.

In our patients, there was no difference in CH_{50}, C_3, C_4, and C_5 between the positive endotoxin group and the negative endotoxin group. When shock occurred, CH_{50} was remarkably lower than in the no shock group. We made sequential determinations of complement levels after shock and obtained results that were comparable to those of animal experiments in which a rise in CH_{50} after shock indicated a favorable prognosis. A decrease in levels of CH_{50} after shock indicated an unfavorable prognosis. Complement levels differ according to underlying disease (benign or malignant), surgical manipulation (4), or treatment (e.g. surgery), and the age of the patient. Therefore, it does not seem proper to compare CH_{50} levels of shock patients with those of control patients. It is most important to measure the complement levels sequentially in order to predict the severity and prognosis.

The reason for the lack of difference of the complement levels between the positive endotoxin group and the negative endotoxin group were as follows: 1) the patients have an infectious focus before endotoxemia, 2) the immunological reaction occurred naturally, and 3) immunological status between product and consumption of CH_{50} has little balance before severe endotoxemia. When the activation and consumption of CH_{50} was stimulated immediately by a large amount of endotoxin and bacteria, the balance was broken and the complement levels decreased markedly.

We conclude from these findings that in cases of endotoxemia, sequential determination of complement levels is considered useful in judging the severity of the disease as well as in predicting the prognosis.

5 References

(1) H. Okada, Clin. Immunol. 5 (1973) 951-963.

(2) V. E. Gilbert, A. I. Braude, J. Exp. Med. 116 (1962) 477-490.

(3) W. R. McCabe, New Eng. J. Med. 288 (1973) 21-23.

(4) T. Takaoka, Med. J. Nagoya City Univ. 28 (1977) 378-414.

Endotoxemia and Blood Coagulation: Procoagulant Activity of Mouse Bone Marrow Cells

Michimasa Hirata, Masao Yoshida, Nobuko Tsunoda, Katsuya Inada

Department of Bacteriology, School of Medicine, Iwate Medical University, Morioka, 020 Iwate, Japan

Abstract

Bone marrow cells from mouse given endotoxin (post-LPS-cells) had significant procoagulant activity (p-activity). The activity was observed as early as 4 hours after endotoxin injection and the time course of the generation of p-activity coincided with those of the cytotoxicity in the cells and the decrease in nucleated cell counts in marrow due to the migration into the circulation. The increase in either the cytotoxicity or p-activity was demonstrated in proportion to the doses of endotoxin injected. Cytotoxicity and p-activity in post-LPS-cells were observed by injection of as little as 5 µg endotoxin. P-activity of post-LPS-cells was considered to be tissue thromboplastin. P-activity was found mainly in both adherent and granulocyte cell fractions. P-activity was destroyed by heating at 56° C for 10 min. P-activity of intact cells was higher than that of sonic lysate of the cells. Most of p-activity was found in the 14,500 x g sediment (membrane fragments) of the sonic lysate.

Abbreviations : P-activity:Procoagulant activity; DIC: Disseminated intravascular coagulation ; PTT: Partial thromboplastin time ; MBT: Mouse brain thromboplastin.

1 Introduction

Endotoxin injection triggers intravascular coagulation in intrinsic and extrinsic pathways. The intrinsic clotting is initiated by the activation of factor XII (Hageman factor) with exposed collagen which results from damage of vascular endothelial cells or with damaged leukocytes. In the extrinsic system, tissue factor (procoagulant) triggers the blood coagulation by activating the factor X in the presence of factor VII. It has been shown that monocytes and granulocytes isolated from peripheral blood are the principal sources of p-activity induced by endotoxin, and the activity plays an important role in the manifestation of disseminated intravascular coagulation (DIC) (1, 2).

The bone marrow reactions, i.e. the cytotoxic damage of granulocytes, the decrease in nucleated cell counts in mouse bone marrow resulted from the migration of the granulocytes into the circulation, have been studied in our laboratory (3 - 5).

In order to evaluate the role of bone marrow cells in pathogenesis of the DIC in endotoxemia, bone marrow cells were investigated as to their ability to generate the p-activity since bone marrow is the source of all kinds of blood cells which are supplied into circulation and to other organs.

2 Materials and Methods

<u>Animals.</u> Male ddY mice, randomly bred, weighing 30 - 40g and 10 to 12 weeks-old, were obtained commercially from Shizuoka Farms (Shizuoka, Japan). <u>Endotoxins.</u> The endotoxin of Salmonella typhimurium LT2 was prepared according to the method of Westphal et al. E.coli O111:B4 (Westphal type, Difco Lab.) was also used. <u>Plasma samples.</u> Citrated mouse plasma was prepared according to a previous paper (6). Normal human plasma (Ortho Diagnostics), and congenital deficiency plasmas (Dade) in factor XII, VIII and VII were obtained commercially. Simplastin (General Diagnostics) and acetone powder of mouse brain thromboplastin were used as standard thromboplastin preparations. <u>Preparations of bone marrow and spleen cells.</u> Bone marrow and spleen cells were obtained according to a previous paper (6). Mononuclear cells were isolated by centrifugation over Sodium-metrizoate-Ficoll (d=1.090) at 22°C for 20 min at the interface. Adherent- and non adherent-cells were isolated by incubation of mononuclear cells in RPMI medium for 1 hour in 16 mm- diameter plastic tissue culture plates at 37°C and 5% CO_2. <u>Coagulation studies.</u> P-activity of bone marrow or spleen cells was expressed as clotting time of mouse plasma (partial thromboplastin time = PTT, sec) and thromboplastin units (the rabbit

brain thromboplastin standard of 1.125 mg was assigned a value of 1,000,000 units). Bone marrow reactions. Bone marrow reactions (cytotoxicity and decrease in nucleated cell counts in bone marrow) were investigated according to Yoshida et al. (3, 7).

3 Results

1) Time course studies.

Time course studies were performed to determine the peak time as to the cytotoxicity, the decrease in nucleated cell counts and p-activity of bone marrow cells after i.p. injection of 50 µg of Salmonella typhimurium endotoxin (Fig. 1). For the coagulation study, sonic lysate of the cells (4×10^6 unfractionated whole nucleated cells) and mouse plasma were used. A significant p-activity in the cells was demonstrated as early as 4 hours and the activity persisted until 24 hours after endotoxin. This pattern coincided with those of the cytotoxicity and the decrease in nucleated cell counts in marrow.

Cytotoxicity — Cell suspension of humerus was stained with trypan blue and the cells stained were counted.

Nucleated cell number — Absolute numbers of nucleated cells per 2 femora were counted.

Procoagulant activity (PTT) — Mouse plasma (0.1ml) was preincubated with sonic lysate (4×10^6 cells/ml) at 37°C for 3 min, and clotted by adding 0.1 ml of 25 mM-$CaCl_2$-phospholipid.

ddY mice in groups of 5 each were given i.p. 50 µg of S.typhimurium endotoxin.

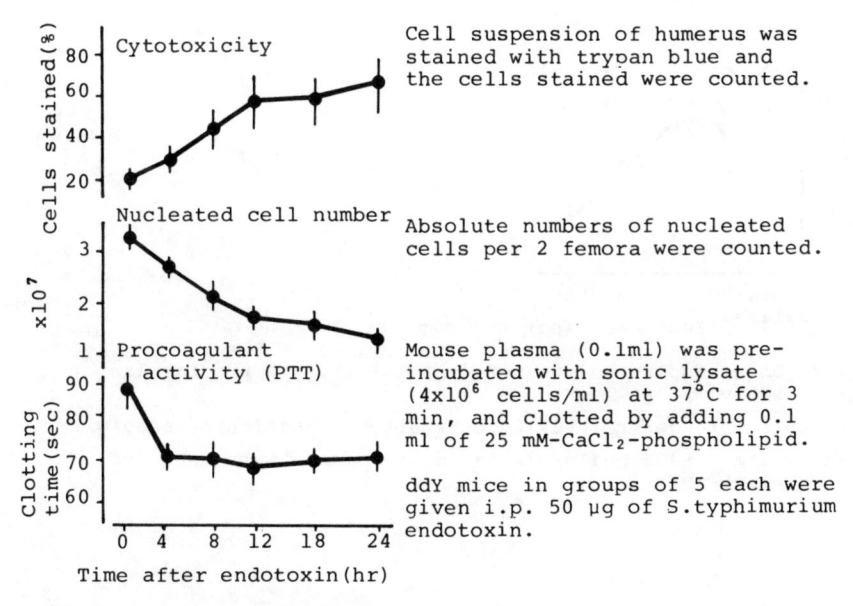

Fig. 1. Bone marrow reactions and procoagulant activity
 -Time course-

2) Dose response.

Mice in groups of 5 to 10 each were given graded doses of endotoxin, ranging from 0.5 to 50 µg per mouse, and the cytotoxicity and p-activity were investigated 18 hours after injection. As shown in Fig. 2, the increase in either the cytotoxicity or p-activity was demonstrated in proportion to the doses of endotoxin injected and the two changes in post-LPS-cells were observed even by injection of 5 µg endotoxin. There existed a definite correlation between cytotoxicity and p-activity (n=20, r=0.8452, p<0.001).

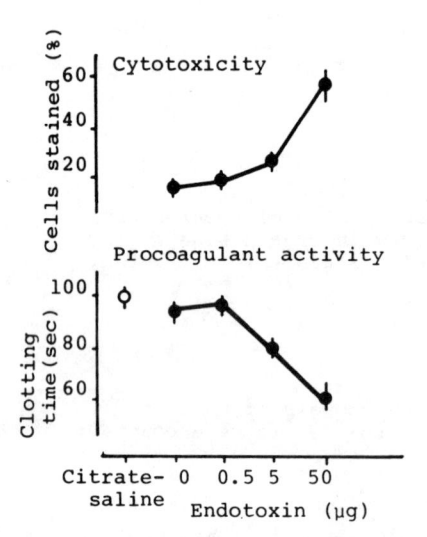

Fig. 2. Dose response curves of cytotoxicity and procoagulant activity.

Both tests were determined 18 hours after injection of endotoxin.
Sonic lysate (2x10^7 cells/ml) and mouse plasma were used for clotting test (PTT).

3) Nature of p-activity.

Two-fold dilutions of post-LPS-cell suspension or mouse brain thromboplastin solution (MBT) were made, and each dilution was added to mouse plasma, then clotting times were estimated. As illustrated in Fig. 3, post-LPS-cells significantly shortened the clotting time compared to normal-cells at each dilution. Parallelism between the two regression lines for post-LPS-cells and MBT was recognized since the mean regression coefficient of the regression line for post-LPS-cells (-0.2462) did not differ from that for MBT (-0.2607). Post-LPS-cells did shorten the clotting time of plasma deficient in factor XII or VIII necessary for the activation of intrinsic coagulation pathway, but did not shorten that of plasma deficient in factor VII necessary for the activation of extrinsic coagulation pathway. (Table 1 and reference 6).

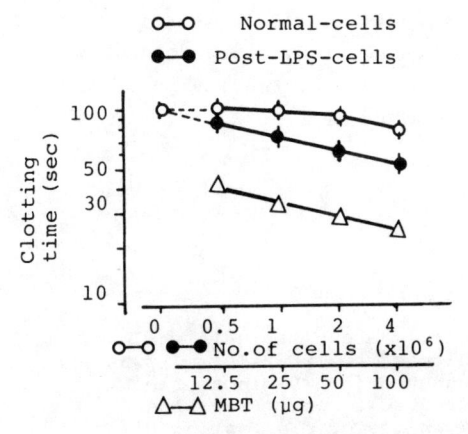

Fig. 3. Parallelism between regression line for post-LPS-cells and mouse brain thromboplastin (MBT).

Post-LPS-cells were harvested 18 hours after i.p. injection of 50 µg of endotoxin.

Sonic lysate and mouse plasma were used for clotting test.

Table 1. Influence of procoagulant activity of post-LPS-cells(a) on clotting time of factor VIII-deficient plasma.

		Clotting time (APTT, sec)(b)	
		Exp. 1	Exp. 2
Saline		96.1 ± 1.0	
Bone marrow cells after endotoxin	0 hr	101.2 ± 0.8	105.0 ± 5.4
	2	96.1 ± 2.5	81.1 ± 5.4(c)
	4	92.4 ± 1.6(c)	57.1 ± 7.7(c)
	18	83.2 ± 4.5(c)	62.6

(a) Post-LPS-cells were harvested 2, 4 or 18 hours after i.v. injection of 100 μg endotoxin.
(b) Sonic lysate of the cells ($2 \times 10^6 - 1 \times 10^7$) was added to factor VIII-deficient plasma and APTT was recorded.
(c) $p < 0.025$ compared with normal-cells (0 hr).

4) Thermostability of p-activity.

Fig. 4 shows the effects of heat treatment of post-LPS-cells(spleen) on p-activity. Sonic lysate of whole nucleated cells (2×10^7/ml) harvested 4 hours after i.v. injection of 100 μg of E.coli endotoxin was incubated for 10 min at the indicated temperature. P-activity of post-LPS-cells was stable at 45°C but destroyed at over 56°C.

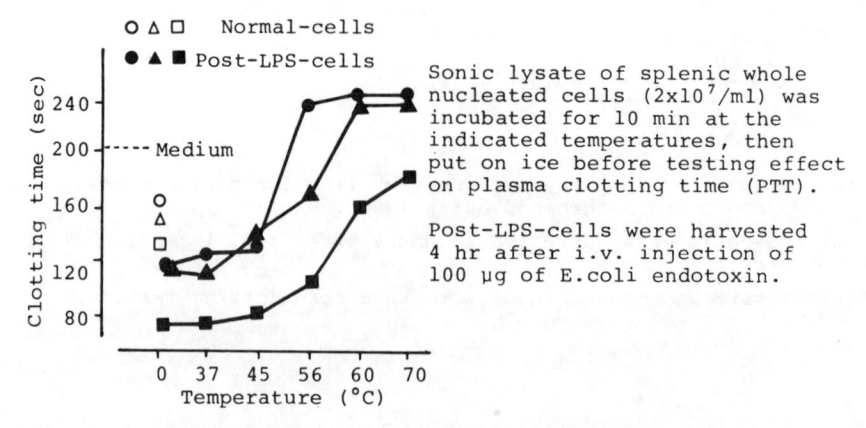

Fig. 4. Thermostability of the procoagulant activity.

5) P-activity of cell populations.

Four hours after i.v. injection of 100 µg of E.coli endotoxin, post-LPS-cells (bone marrow) were collected and p-activity was compared with normal-cells (Table 2). Thromboplastin units of 1×10^6 unfractionated whole nucleated cells increased 3.4-fold from 279 units (normal-cells) to 951 units (post-LPS-cells). Thromboplastin units of mononuclear cells, adherent cells, and granulocytes increased 2.1-, 3.2-, and 1.8-fold, respectively. P-activity was found mainly in both adherent (monocytes) and granulocytes cell fractions.

Table 2. Procoagulant activity of cell populations of mouse bone marrow after endotoxin injection.

Cell population	Thromboplastin units/10^6 intact cells		
	Normal-cells	Post-LPS-cells (a)	Index
Whole nucleated cells	279 ± 25	951 ± 62	3.4
Whole mononuclear cells (b)	340 ± 61	701 ± 131	2.1
Granulocytes	287 ± 75	528 ± 145	1.8
Non-adherent cells (c)	165 ± 20	219 ± 13	1.3
Adherent cells (c)	164 ± 2	520 ± 60	3.2

(a) Post-LPS-cells were harvested 4 hr after i.v. injection of 100 µg of E.coli endotoxin.
(b) Mononuclear cells were isolated by centrifugation over Sodium-metrizoate-Ficoll (d=1.090) at 22°C for 20 min at the interface.
(c) Adherent- and non-adherent cells were isolated by incubation of mononuclear cells in RPMI medium for 1 hr in 16 mm-diameter plastic tissue culture plates at 37°C and 5 % CO_2.

6) P-activity of intact cells and sonic lysate.

Table 3 shows the p-activity of intact cells and sonic lysate of the cells. Sonic disruption tended to prolong the clotting time and to decrease in the thromboplastin units in both the normal-cells and post-LPS-cells. P-activity of intact cells was higher than that of sonic lysate of the cells.

Table 3. Procoagulant activity of intact cells and sonic lysate.

		Procoagulant activity(a)	
		Intact cells	Sonic lysate
Normal-cells	PTT	118.3 ± 3.7	141.8 ± 9.9
	Units	117.0 ± 12.1	66.0 ± 14.0
Post-LPS-cells (b)	PTT	86.9 ± 3.3	110.9 ± 7.1
	Units	334.0 ± 48.7	154.6 ± 37.5
	Index	2.9	2.4

(a) Procoagulant activity was expressed as PTT (sec) and thromboplastin units per 1×10^6 whole nucleated cells. PTT of medium control was 138.5 ± 3.4 sec.

(b) Post-LPS-cells were collected 18 hours after i.p. injection of 100 µg of E.coli endotoxin.

7) Localization of p-activity.

As shown in Table 4, p-activity of several fractions obtained from differential centrifugation of sonic lysate of the spleen cells. Whole nucleated cells were disrupted and sedimented successively at the indicated gravities, each sediment resuspended in predetermined volume was tested for p-activity. Most of the total p-activity was found in the 14,500 x g sediment which contained membrane fragments. The p-activity of this sediment in post-LPS-cells was also higher than that in normal-cells as seen in intact cells.

Table 4. Sedimentation of the procoagulant activity.

	Disrupted(a)	400xg	14,500xg	40,000xg	40,000xg sup	Total
Normal-cells	4228	129	2167	306	228	2830
Post-LPS-cells(b)	9387	390	6331	902	314	7937

(a) Sonic vibrations for 30 - 40 sec.
(b) Collected 4 hr after i.v. injection of 100 µg of E.coli endotoxin.

Splenic whole nucleated cells (5 mice) were disrupted and sedimented successively at the above gravities, each sediment resuspended in predetermined volume was tested for procoagulant activity.

Procoagulant activity is expressed in units of standard thromboplastin per 10^7 cells or cell fractions.

4 Discussion

A variety of in vitro stimuli including endotoxin (1, 2), antigen-antibody complexes (8), lectin (8), complement C5a (9) and plasma lipoprotein (10) have been reported to generate p-activity of thromboplastin character in leukocytes. As to endotoxin-induced p-activity in leukocytes, several authors have demonstrated that monocytes and granulocytes are the cells responsible for the generation of p-activity (1, 2, 11). This activity is thought to play a major role in the manifestation of DIC in endotoxemia. Tissue factor (p-activity) has been found in highest concentration in the plasma membranes of endothelial cells and many other cell types (12).

In the present paper, a single dose of endotoxin is demonstrated to generate p-activity in the bone marrow cells. Post-LPS-cells possessed a greater p-activity than normal-cells and activity was found in both the intact cells and 14,500 x g sediment. These findings suggest that the activity is present in the cell membranes, which is activated as a result of the interaction of the cells with endotoxin in vivo.

The increased p-activity was found in the post-LPS-cells collected as early as 4 hours after injection of endotoxin and the time course of procoagulant generation in the cells was similar to those of cytotoxicity and decrease in nucleated cell numbers in the marrow.

The granulocytes in the marrow migrate into peripheral blood to restore granulocytopenia and then to induce granulocytosis.

It was suggested that cytotoxic damage of bone marrow cells due to endotoxin is responsible for the induction of p-activity of the cells, since (a) bone marrow cells of mouse given endotoxin (post-LPS-cells) had a significant p-activity and the time course of procoagulant induction was similar to that of cytotoxicity, and (b) there existed a definite correlation between cytotoxicity and p-activity.

P-activity of post-LPS-cells was considered to be tissue thromboplastin, since (a) parallelism between the two regression lines for post-LPS-cells and MBT was recognized, and (b) post-LPS-cells did not shorten the clotting time of plasma deficient in factor VII required for extrinsic clotting but did shorten that of plasma deficient in factor XII or VIII required for the intrinsic clotting system.

Adherent cells and granulocytes had an increased p-activity in post-LPS-cells compared with normal-cells. From these results, it seems that monocytes and granulocytes in the marrow are the cells responsible for the generation of p-activity in endotoxemia.

It should be stressed that bone marrow cells (monocytes and granulocytes) are one of the sources of p-activity (tissue thromboplastin) which activates the extrinsic coagulation pathway, and the cells with

```
                    ┌──────────┐
                    │Bacterial │
                    │endotoxin │
                    └──────────┘
          ┌──────────┐        ┌──────────┐
          │Intrinsic │        │Extrinsic │
          │ system   │        │ system   │
          └──────────┘        └──────────┘
   ┌────────────┐  ┌──────────┐           ┌──────────────┐
   │Endothelial │  │Leukocyte │ Migration │Cytotoxic damage│
   │ damage     │  │ damage   │ ◀──────── │of bone marrow │
   └────────────┘  └──────────┘           │ cells         │
         │              │                 └──────────────┘
      XII ──▶ XIIa      ▼
                 ┌──────────────────┐
       XI ──▶ XIa│Tissue thromboplastin│
          Ca++   └──────────────────┘
       IX ──▶ IXa                    Cationic (acid
                                     soluble) protein
      Ca++, VIII              VII
      platelet-3
       X ──▶ Xa
                    Ca++, V
                    phospholipid

      Prothrombin ──▶ Thrombin
             Fibrinogen ──▶ Fibrin
```

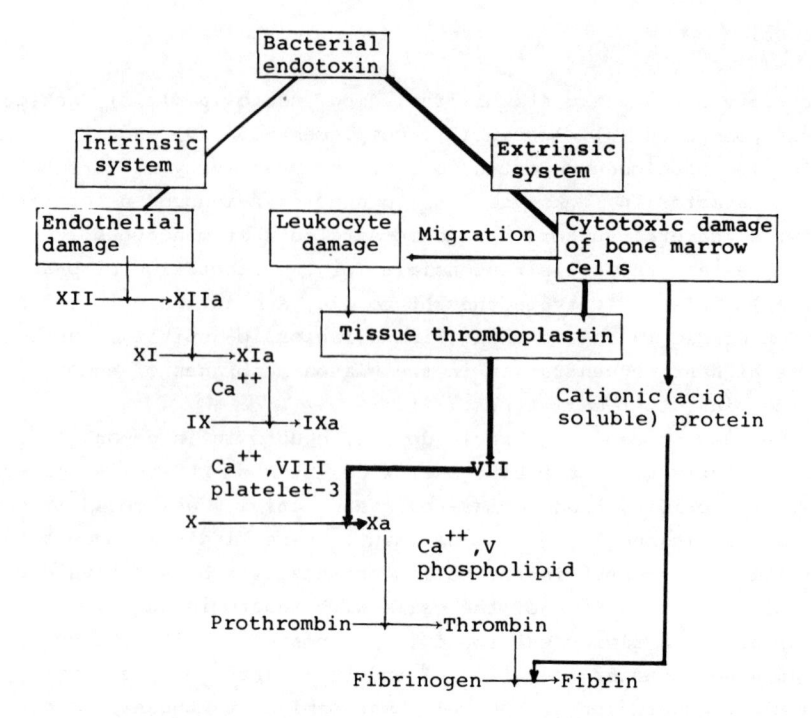

Fig. 5. Blood coagulation in endotoxemia.

an enhanced p-activity either in marrow or in circulation would therefore cause DIC in endotoxemia (Fig. 5).

As to the role of granulocytes in endotoxemia, they have been shown to contain or to produce many active materials in addition to the procoagulant. For example, protease from granulocytes activates clotting factor VII (13), granulocytes generate thromboxane A_2 which causes vascular injury and DIC (14), and cationic protein (acid soluble protein) which precipitates soluble-fibrin has been isolated from granulocytes (15). We could also obtain cationic protein from bone marrow or spleen cells (16).

It has been demonstrated that the bone marrow is most sensitive to endotoxin, and granulocytes are the predominant cells in the bone marrow and these cells are most sensitive to endotoxin-induced cytotoxicity (17). This is the reason why we stress the role of bone marrow cells in DIC in endotoxemia.

This work was supported by a grant from Iwate Medical University-Keiryokai Research Foundation (1981-1982) and a grant-in-aid for Fundamental Scientific Research from the Ministry of Education of Japan (1982).

5 References

(1) R. P. A. Rivers, W. E. Hathaway, W. L. Weston,
     Br. J. Haematol. 30 (1975) 311-316.

(2) R. G. Lerner, R. Goldstein, G. Cummings,
     Thromb. Res. 11 (1977) 253-261.

(3) M. Yoshida, F. Parant, M. Hirata, L. Chedid,
     Japan. J. Med. Sci. Biol. 25 (1972) 243-247.

(4) M. Yoshida, M. Hirata, M. K. Agarwal, Animal, Plant, and Microbial Toxins, Plenum, New York 1 (1976) pp. 509-520.

(5) M. Hirata, J. Infect. Dis. 132 (1975) 611-616.

(6) M. Hirata, K. Inada, N. Tsunoda, M. Yoshida in M. K. Agarwal (Ed.), Bacterial Endotoxins and Host Response, Elsevier/North-Holland Biomedical Press 1980, pp. 255-272.

(7) M. Hirata, M. Yoshida, K. Inada, S. Yaegashi,
     Japan. J. Med. Sci. Biol. 31 (1978) 184-188.

(8) H. Prydz, T. Lyberg, P. Deteix, A. C. Allison,
     Thromb. Res. 15 (1979) 465-474.

(9) T. W. Muhlfelder, J. Niemetz, D. Kreutzer, D. Beebe, P. Ward,
     S. I. Rosenfeld, J. Clin. Invest. 63 (1979) 147-150.

(10) B. S. Schwartz, G. A. Levy, L. K. Curtiss, D. S. Fair,
     T. S. Edginton, J. Clin. Invest. 67 (1981) 1650-1658.

(11) G. A. Levy, T. S. Edginton, J. Exp. Med. 151 (1980) 1232-1244.

(12) J. Niemetz, J. Clin. Invest. 51 (1972) 307-313.

(13) J. Kaparati, K. Varadi, S. Elodi, At the 19th Congress of the
International Society of Haematology (Budapest, August 1-7, 1982).
pp. 362 (abstract).

(14) P. J. Spagnuolo, J. J. Ellner, A. Hassid, M. J. Dunn,
J. Clin. Invest. 66 (1980) 406-414.

(15) J. Hawiger, R. D. Collins, R. G. Horn,
Proc. Soc. Exp. Biol. Med. 131 (1969) 349-353.

(16) M. Hirata, N. Tsunoda, K. Inada, M. Yoshida, At the 5th Congress
of the Japanese Society on Thrombosis and Hemostasis (Tokyo, Japan,
1982), pp. 82 (abstract).

(17) M. Yoshida, M. Hirata, R. Satodate, J. Chiba,
Japan. J. Med. Sci. Biol. 27 (1974) 114-117.

Limulus Test
Endotoxin Standard Preparation

The Limulus Coagulation System Sensitive to Bacterial Endotoxins

Sadaaki Iwanaga, Takashi Morita, Toshiyuki Miyata, Takanori Nakamura, Masuyo Hiranaga and Sadami Ohtsubo

Department of Biology, Faculty of Science, Kyushu University 33, Fukuoka 812, Japan

Abstract

The amebocytes circulating in horseshoe crab (Limulus) hemolymph contain a coagulation system, which participates both in hemostasis and in defence against invading microorganisms. This coagulation system consists of several protein components and is highly sensitive to Gram-negative bacterial endotoxins. The so-called "Limulus test" for detection of endotoxins is based on the endotoxin-induced coagulation reaction, using the amebocyte lysate. Although the overall molecular events of this system have not yet been elucidated, a few coagulation factors, such as proclotting enzyme and coagulogen, from the amebocyte lysate have been purified and characterized biochemically. The previous works from our laboratories have also established the gelation mechanism of coagulogen to coagulin, the whole amino acid sequences of Tachypleus and Limulus coagulogens and a new method for assaying endotoxins, using chromogenic (Boc-Leu-Gly-Arg-p-nitroanilide) or fluorogenic (Boc-Leu-Gly-Arg-4-methylcoumarin-7-amide) substrate of the clotting enzyme. Moreover, our latest studies have shown that the amebocytes contain at least two independent coagulation pathways, the endotoxin-mediated and $(1\rightarrow 3)$-ß-D-glucan-mediated pathways, both of which result in the transformation of coagulogen to coagulin. In the course of these studies, we also found a potent anticoagulant, named anti-LPS factor, which inhibits specifically the endotoxin-mediated activation of factor C. Based on these results, we propose here a tentative mechanism for the Limulus hemolymph coagulation cascade, as follows:

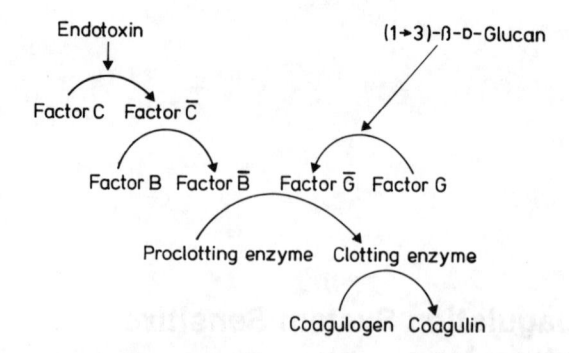

```
           Endotoxin                    (1→3)-β-D-Glucan
               ↓                               ↓
        Factor C   Factor C̄              ↓      ↓
                 ⌒   ⌒                   ↓      ↓
             Factor B  Factor B̄    Factor Ḡ   Factor G
                        ⌒    ⌒    ⌒
                    Proclotting enzyme  Clotting enzyme
                                       ⌒   ⌒
                                  Coagulogen  Coagulin
```

Tentative mechanism of the Limulus coagulation cascade.

1 Introduction

Limulus hemolymph contains only one type of cell, called amebocytes or granular hemocytes (1-3). The amebocytes contain a number of dense granules and exposure of the cells to bacterial endotoxin results in aggregation associated with striking degranulation and clot formation. This endotoxin-induced clotting phenomenon has been known as a possible defence mechanism, serving to immobilize invading gram-negative bacteria (4, 5). Because of this unique property and its extreme sensitivity to endotoxins, it is of great interest to investigate the overall molecular events of the clotting system in Limulus amebocytes, in comparison with those of mammalia.

This paper concerns mainly the biochemical studies on intracellular clotting system closely related to the function of this hemocytes. First of all, we would like to mention about the primary structure of a clottable protein, named coagulogen, and its structural change during gelation (6-9). Secondly, we will describe the purification and properties of proclotting enzyme (10-12). Thirdly, we shall show a new glucan-mediated pathway associated with Limulus clotting system (13). Finally, we shall deal with an anticoagulant, which inhibits specifically the endotoxin-mediated activation of the clotting system (14).

2 Collection of hemolymph from horseshoe crabs

The hemolymph is collected by inserting a needle into a joint between the cephalothorax and the abdominal region. Fifty to 150 ml of the hemolymph per individual is drawn and amebocytes are obtained by centrifugation. Limulus hemolymph thus obtained is shown in Fig. 1.

We will see the amebocytes at the bottom in the first tube. A light blue plasma fraction contains mainly hemocyanin and lectins. The second tube shows a clot formed in the presence of endotoxins, so call= ed lipopolysaccharides, and this clot seems to be stabilized on pro= longed incubation, as shown in the third tube. We have recently deter= mined the amino acid sequence of this clottable protein, named coagulo= gen.

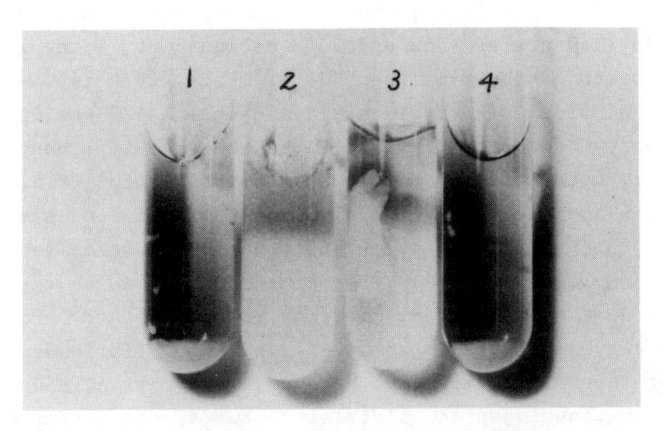

Fig. 1. Hemolymph collected from Japanese horseshoe crab, Tachypleus tridentatus. 1. Ten min after collection; 2. Clot formed by addition of endotoxin (E. coli 0111-B4) into the hemolymph; 3. A stabilized clot after allowed to stand for 1 hr; 4. The same as tube 1.

3. Complete amino acid sequences of Tachypleus and Limulus coagulogens

Figure 2 shows the complete amino acid sequence of two coagulogens, isolated, respectively, from the amebocyte lysates of Japanese and American horseshoe crabs, Tachypleus tridentatus and Limulus polyphemus (8).

```
L.p.  G-D-P-N-V-P-T-C-L-C-E-E-P-T-L-L-G-R-K-V-I-V-S-Q-E-T-K-D-K-I-E-E-A-V-Q-A-
T.t.  A-D-T-N-A-P-I-C-L-C-D-E-P-G-V-L-G-R-T-Q-I-V-T-T-E-I-K-D-K-I-E-K-A-V-E-A-

L.p.  I-T-D-K-D-E-I-S-G-R-G-F-S-I-F-G-G-H-P-A-F-K-E-C-G-K-Y-E-C-R-T-V-T-S-E-D-
T.t.  V-A-Q-E-S-G-V-S-G-R-G-F-S-I-F-S-H-H-P-V-F-R-E-C-G-K-Y-E-C-R-T-V-R-P-E-H-

L.p.  S-R-C-Y-N-F-F-P-F-S-H-F-H-P-E-C-P-V-S-V-S-A-C-E-P-T-F-G-Y-T-T-S-N-E-L-R-
T.t.  S-R-C-Y-N-F-P-P-F-T-H-F-K-L-E-C-P-V-S-T-R-D-C-E-P-V-F-G-Y-T-V-A-G-E-F-R-

L.p.  I-I-V-Q-A-P-K-A-G-F-R-Q-C-V-W-Q-H-K-C-R-A-Y-G-S-N-F-C-Q-R-T-G-R-C-T-Q-Q-
T.t.  V-I-V-Q-A-P-R-A-G-F-R-Q-C-V-W-Q-H-K-C-R-    F-G-S-N-S-C-G-Y-N-G-R-C-T-Q-Q-

L.p.  R-S-V-V-R-L-V-T-Y-D-L-E-K-G-V-F-F-C-E-N-V-R-T-C-C-G-C-P-C-R-S-
T.t.  R-S-V-V-R-L-V-T-Y-N-L-E-K-D-G-F-L-C-E-S-F-R-T-C-C-G-C-P-C-R-S-F
```

Fig. 2. Amino acid sequence homologies between Limulus (L. p.) and

Tachypleus (T. t.) coagulogens. The peptide bonds cleaved by
Limulus clotting enzyme are shown by arrows and identical
residues in the sequences are framed (9).

Both proteins consist of 175 residues and contain a total of 16 half-cystines, which are linked with disulfide bridges. Five of them are clustered in the COOH-terminal region. The complete arrangement of these disulfide linkages are now under investigation. Tachypleus coagulogen consists of three polypeptide segments, that is, A-chain (18 residues), peptide C (28 residues) and B-chain (129 residues). The clotting enzyme described later cleaves the Arg-Gly linkage and Arg-Thr linkage, both located at the NH_2-terminal region (Fig. 2). This Arg-Gly linkage cleaved by the clotting enzyme is the same type as that cleaved by α-thrombin in the transformation of mammalian fibrinogen to fibrin. Moreover, the COOH-terminal octapeptide sequences of A-chain and peptide C exhibit great homology for each other, and their sequences are very similar to that of Rhesus monkey fibrinopeptide B (7). The most interesting finding, however, is a partial sequence homology between coagulogen and platelet factor 4, the latter of which is known as antiheparin (8). When the sequences are aligned for maximum homology, coagulogen is found to contain a number of similar partial sequences with platelet factor 4 (8, 15). Especially, the first eleven NH_2-terminal residues are very similar to those of platelet factor 4. When the sequence of Tachypleus coagulogen is compared with that of Limulus polyphemus coagulogen, the overall sequence of the latter is very close to that of the former, indicating 70 percent sequence homology. It should be noted that in coagulogen molecule the α-helix region, which is predicted by the method of Chou and Fasman, is found mainly in the peptide C segment and that the ß-sheet and reverse turn regions are distributed in the B-chain segment. The 16 half-cystines of these coagulogens are in the same linear position, suggesting a very similar conformation for each other. Moreover, the COOH-terminal tripeptide regions of A-chain and peptide C, Leu-Gly-Arg and Ser-Gly-Arg, are completely conserved in both coagulogens. Therefore, these oligopeptide sequences immediately preceding the bond to be cleaved may be required for the clotting enzyme to split the Arg-Gly and Arg-Thr linkages so as to initiate clot formation. To realize the hypothesis, we synthesized a number of peptidyl-p-nitroanilides, which contain the tripeptide sequence of Leu-Gly-Arg and Ser-Gly-Arg found in coagulogen, and tested their susceptibilities to Limulus clotting enzyme.

Table 1 shows kinetic parameters of Limulus clotting enzyme towards various synthetic peptide substrates. The apparent Michaelis constants for synthetic substrates appear to be relatively good, since their Km values fall in the μmolar range. Especially, Boc-Leu-Gly-Arg-pNA and Bz-Val-Gly-Arg-pNA seem to have greatest affinity to the clotting enzyme. Using this chromogenic substrate, we next studied on proclotting enzyme, a precursor of the clotting enzyme, contained in the amebocytes.

Table 1. The apparent Km and Vmax values of Tachypleus clotting enzyme for peptidyl-pNAs (16).

Substrate	Km (M)	Vmax μmoles/min/A_{280}=1.0	Vmax/Km
Tos-Ile-Glu-Gly-Arg-pNA	2.1×10^{-4}	2.7	13
Boc-Val-Leu-Gly-Arg-pNA	5.4×10^{-5}	3.2	59
Boc-Leu-Gly-Arg-pNA	6.5×10^{-5}	4.2	65
Boc-Val-Ser-Gly-Arg-pNA	1.4×10^{-4}	3.3	24
Boc-Ser(OBzl)-Gly-Arg-pNA	4.9×10^{-5}	3.3	67
Bz-Val-Gly-Arg-pNA	4.6×10^{-5}	3.3	72
H·D-Val-Leu-Gly-Arg-pNA	1.5×10^{-4}	0.9	6

Boc, tert-butoxycarbonyl; Bz, benzoyl; Tos, p-toluenesulfonyl; H·D, D-amino acid residue.

4. Purification and properties of Tachypleus proclotting enzyme

Figure 3 summarizes the purification procedures of Tachypleus proclotting enzyme from the lysate. The proclotting enzyme was purified by a heparin-Sepharose CL-6B column followed by chromatographies on DEAE-Sepharose CL-6B, Sepharose CL-6B, DE-52 and Sephacryl S-300.

```
Amebocyte lysate of Tachypleus tridentatus
                ↓
        Heparin-Sepharose CL-6B
                            ──── under sterilized condition ────
                ↓
        DEAE-Sepharose CL-6B
                ↓
           Sepharose CL-6B
                ↓
                DE-52
                ↓
           Sephacryl S-300
                ↓
       Purified Proclotting Enzyme
```

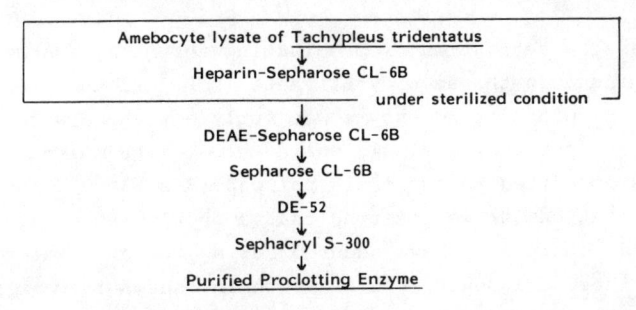

Fig. 3. Procedures for purification of the proclotting enzyme from the Tachypleus amebocyte lysate.

Through these procedures, about 2 mg of the purified material was obtained from one liter of hemolymph, and about 100-fold purification was achieved (15). The purified proclotting enzyme gave a single band on disc-gel electrophoresis at pH 8.3. Furthermore, a single peak with the amidase activity due to clotting enzyme was found at the same position as the protein band stained with Coomassie brilliant blue (Fig. 4B). On SDS-gel electrophoresis with and without reducing agent, the preparation gave a single molecular species, suggesting that the protein consists of a single polypeptide chain (Fig. 4A).

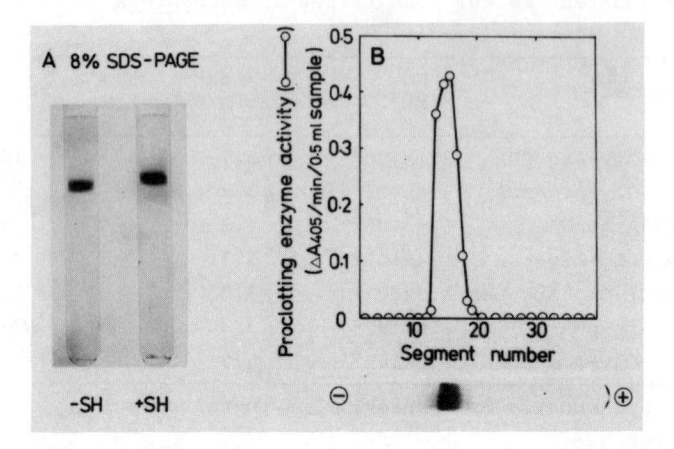

Fig. 4. Polyacrylamide gel electrophoresis of the purified Tachypleus proclotting enzyme. A: SDS-gel (8 %) electrophoresis (5 μg sample) was performed in the presence (right) or absence (left) of 2-mercaptoethanol. B: Disc-gel (8 %) electrophoresis was made at pH 8.3. One of two gels was sliced into 1.5 mm segments and each segment was soaked in 0.5 ml of 0.05 M Tris-HCl buffer, pH 8.0, containing 0.2 M NaCl and 0.4 % bovine serum albumin, and kept overnight at 4°C. The extracts (75 μl) were assayed for proclotting enzyme (—o—). The other gel was stained with Coomassie brilliant blue R-250.

Figure 5 shows the activation of proclotting enzyme by factor B, which was recently found in the amebocyte lysate (17). In the presence of active factor $\overline{B}$, proclotting enzyme was fully activated within 15 min, and during these periods, a single chain enzyme with molecular weight of 54,000 was converted to two chain polypeptides, 31,000 (heavy chain) and 27,000 (light chain), suggesting that a peptide bond cleavage along the polypeptide chain must have occured during the activation. Moreover, one of these fragments, the heavy chain, seems to corresponding to the fragment with a molecular weight of 30,000, which is derived from the purified clotting enzyme (10). These results indicate that the heavy chain portion in proclotting enzyme must contain a "catalytic site" for the enzyme activity. Tachypleus proclotting enzyme has a minimum molecular weight of 54,000, and contains 6 percent of hexo= samines. This protein appears not to be vitamin K-dependent protein like vertebrate prothrombin family, since its γ-carboxyglutamic acid content is very low. Moreover, no direct activation of the purified proclotting enzyme by bacterial endotoxins was observed. These proper= ties quite differ from those reported on Limulus polyphemus proclotting enzyme (17). Based on these results, we concluded that the endotoxin-mediated coagulation system must consist of a multi-enzyme system.

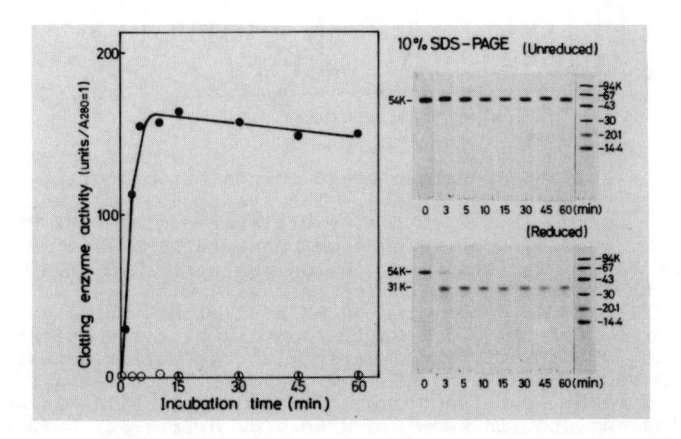

Fig. 5. Time course for activation of the proclotting enzyme by the purified active factor $\bar{B}$ (15). Proclotting enzyme (99 µg/ml) was incubated with (●) and without (○) purified active factor $\bar{B}$ (21.8 µg/ml), in a total volume of 1.0 ml. At the indicated time, 10 µl aliquots were withdrawn and diluted with 160 µl of 0.1 M Tris-HCl, pH 8.0, containing 0.053 % bovine serum albumin. The diluted mixture was assayed for clotting enzyme activity, using Boc-Leu-Gly-Arg-pNA as substrate (16). The other aliquots were subjected to SDS-gel electrophoresis in the absence or presence of 2-mercaptoethanol.

5. Purification and properties of coagulation factors from Tachypleus amebocyte lysate.

  To isolate coagulation factors associated with the Limulus clotting system sensitive to endotoxins, the Tachypleus amebocyte lysate was fractionated on a pyrogen-free dextran sulfate-Sepharose CL-6B column. This column was recently found to give a better separation for the proclotting enzyme, factor G (described later), factor B and anti-LPS factor (described later) rather than that of a heparin-Sepharose CL-6B column previously used (12, 15). The elution profile is shown in Fig. 6.

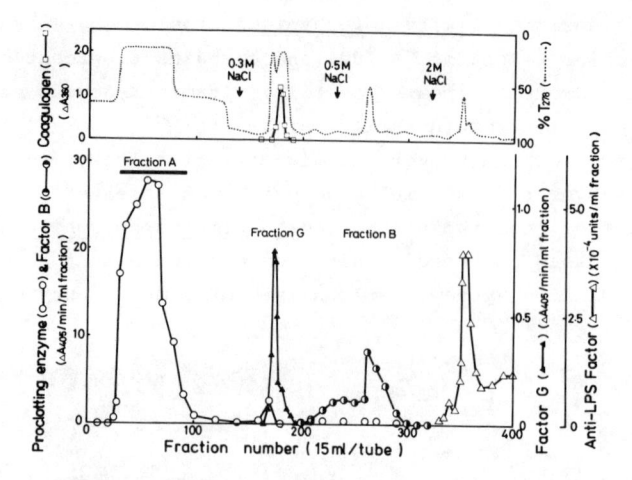

Fig. 6. Dextran sulfate-Sepharose CL-6B column chromatography of Tachypleus amebocyte lysate. The lysate (630 ml) was applied to the column (5 x 23.5 cm), equilibrated with 0.02 M Tris-HCl buffer, pH 8.0, containing 1 mM benzamidine-HCl, 1 mM EDTA and 50 mM NaCl. The stepwise elution was performed at 4°C firstly with the equilibration buffer, secondly with 0.02 M Tris-HCl buffer, pH 8.0, containing 50 mM NaCl alone, thirdly with the buffer containing 0.3 M NaCl, fourthly with the buffer containing 0.5 M NaCl, and finally with the buffer containing 2.0 M NaCl. Fractions of 15 ml were collected at a flow rate of 105 ml/h. The fractions indicated by a solid bar were pooled, diluted in two-fold with 0.02 M Tris-HCl buffer, pH 8.0, containing 1 mM benzamidine-HCl and 1 mM EDTA. The activities of proclotting enzyme (○), factor G (▲), factor B (●), coagulogen (□), and anti-LPS factor (△) were assayed and the detailed procedures were described in the previous paper (18).

The proclotting enzyme described in the previous section was found in fraction A, as revealed by the amidase activity towards Boc-Leu-Gly-Arg-pNA after treatment with endotoxin in the presence of fraction B. The fraction A did not show any amidase activity even in the presence of endotoxin. On the other hand, fraction B, which contained factor B (15), was fractionated further, using a pyrogen-free Sepharose CL-6B column. The elution profile is shown in Fig. 7. On this column, the factor B activity, which was measured in the presence of proclotting enzyme and endotoxin, was separated into two peaks. Moreover, one of the peaks eluted in fraction Nos. 70-85 (fraction I) developed the amidase activity towards Boc-Val-Pro-Arg-pNA, after treatment with only endotoxins, as shown in Fig. 7. The other peak corresponding to fraction Nos. 86-90 (fraction II), however, did not show such amidase activity and the component in these fractions required both the pro=

clotting enzyme and endotoxin to express the amidase activity towards Boc-Leu-Gly-Arg-pNA. These results were in good agreement with that reported previously (19), and suggested that a factor sensitive to endotoxin, tentatively named factor C in the previous paper (15, 19), had been eluted in fraction I (Fig. 7).

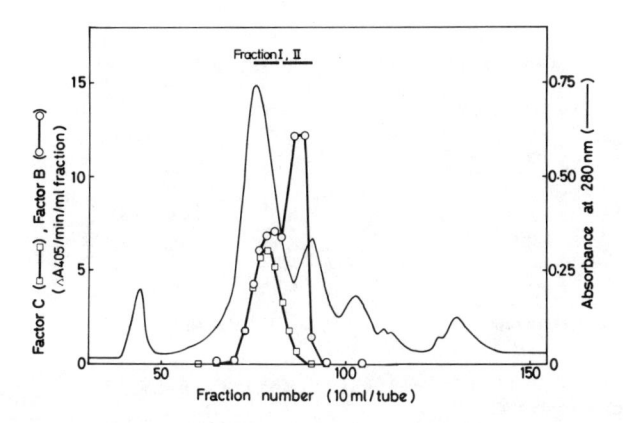

Fig. 7. Further fractionation of fraction B (Fig. 6) on a pyrogen-free Sepharose CL-6B column. A pooled fraction B was applied to the column (3.5 x 120 cm), equilibrated with 0.05 M Tris-HCl, pH 8.0, containing 0.2 M NaCl. The elution was made with the equilibration buffer at 4°C.

The purification of this factor C from fraction I was achieved by additional three steps with chromatography on dextran sulfate-Sepharose CL-6B, Sephacryl S-300 and CM-Sepharose CL-6B columns (18). The final preparation gave a single protein band on SDS-gel electrophoresis. On the other hand, the further purification of factor B was performed with column chromatographies of benzamidine-CH-Sepharose followed by dextran sulfate-Sepharose CL-6B and CM-Sepharose CL-6B. However, as judged with a SDS-gel electrophoretic pattern, the final preparation still contained a few of minor protein bands, in addition to a single major band of factor B.

Figure 8 shows SDS-gel electrophoresis of the activated factor $\bar{B}$ and factor $\bar{C}$. These active forms were prepared, respectively, from fractions I and II by using an affinity column of benzamidine-CH-Sepharose, after activations of factors C and B with endotoxin alone in the former and with factor $\bar{C}$ in the latter. As shown in Fig. 8, factor $\bar{B}$ gave a single major and minor protein bands on SDS-gel in the absence of 2-mercaptoethanol, and the major band had a molecular weight of 57,000. On reduction, its molecular weight decreased to about 31,000, suggesting that factor $\bar{B}$ consists of two polypeptide chains bridged by disul=

fide(s). On the other hand, factor $\overline{C}$ gave a single band on unreduced SDS-gel and had a molecular weight of about 130,000. On reduced SDS-gel, it gave three protein bands with the molecular weights of about 80,000, 40,000 and 10,000. This result suggests that factor $\overline{C}$ consists of at least three chains bridged by disulfide(s).

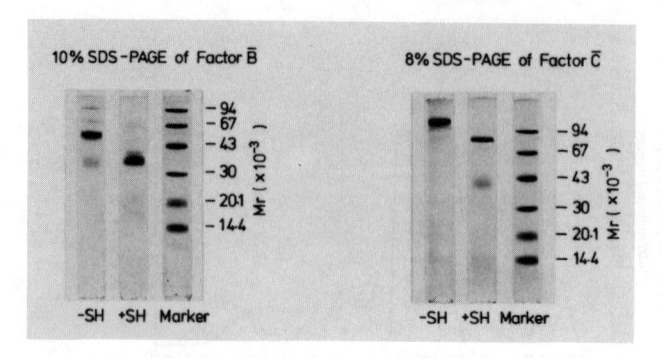

Fig. 8. SDS-polyacrylamide gel electrophoresis of the purified factor $\overline{B}$ and factor $\overline{C}$.

The substrate specificities of factor $\overline{C}$ towards peptidyl-MCA substrates are shown in Table 2. Of the 6 peptide substrates examined, factor $\overline{C}$ hydrolyzed Boc-Val-Pro-Arg-MCA, Bz-Ser-Ser-Arg-MCA and Bz-Thr-Ser-Arg-MCA more efficiently, and its specificity differed from Limulus clotting enzyme which prefers the peptide substrate of Boc-Leu-Gly-Arg-MCA.

Table 2. Amidase activity of factor $\overline{C}$ toward peptidyl-MCA substrates

	Substrate*	µmoles AMC released/min/A_{280}=1
1	Boc-Val-Pro-Arg-MCA	4.78
2	Boc-Leu-Gly-Arg-MCA	0.82
3	Bz-Ser-Ser-Arg-MCA	5.48
4	Bz-Thr-Ser-Arg-MCA	5.93
5	Bz-Thr-Thr-Arg-MCA	2.03
6	Boc-Met-Thr-Arg-MCA	0.40

* The concentration of substrate was 0.1 mM. MCA, 4-methylcoumaryl-7-amide; AMC, 7-amino-4-methylcoumarin; Boc, N-tert=butoxycarbonyl; Bz, benzoyl.

To elucidate the relationship among proclotting enzyme, factor B and factor C so far purified, a recombination experiment was performed, using endotoxin as a mediator for the Limulus coagulation system. The results are shown in Table 3. A strong amidase activity towards Boc-Leu-Gly-Arg-pNA was developed only in the reaction mixture containing above three components (exp. 1), and there was no amidase activity in the combination of proclotting enzyme and factor B (exp. 2) or factor C (exp. 3). These results suggest that factor C must be an endotoxin sensitive component and that the activated factor $\overline{C}$ induces the activation of factor B. Thus, bacterial endotoxin mediates the activation of factor C and its active form converts factor B to factor $\overline{B}$, which then activates the proclotting enzyme to clotting enzyme (18). Therefore, this cascade system existed in Limulus amebocytes may provide an extremely high sensitivity of the lysate to endotoxins.

Table 3. Recombination experiments of proclotting enzyme, factor B and factor C

Experiments	Amidase activity (nmoles pNA released · min^{-1})	
	Without LPS	With LPS
1 ProCE+B+C	0.1	14.9
2 ProCE+B	0	0
3 ProCE+C	0	0.5
4 B+C	0	0
5 ProCE	0	0
6 B	0	0
7 C	0	0.1

ProCE, Proclotting enzyme; B, Factor B; C, Factor C. The incubation mixture containing 30 μl of proclotting enzyme (0.96 mg/ml), 20 μl of factor B (A$_{280}$=0.2) and 5 μl of factor C (A$_{280}$=0.016), in a total volumn of 200 μl, was preincubated at 37°C. Then, E. coli 0111-B4 endotoxin (40 ng) was added and incubated for 15 min. After incubation, the amidase activity developed was measured using Boc-Leu-Gly-Arg-pNA as substrate.

6. A new (1→3)-ß-D-glucan-mediated pathway found in Limulus amebocytes

In the course of these studies, Dr. A. Kakinuma, the Central research Division, Takeda Chemical Industries, informed us that, beside endotoxin, a water-soluble antitumor carboxymethylated ß-D-glucan activates the Limulus coagulation system and forms a clot. This result prompted us to examine which components associated with the coagulation system is activated by ß-glucan. In this experiment, we used fractions A, G and B, which were separated by a heparin-Sepharose column (14). The fractions A, G and B were separately collected and their amidase activities and clot-forming abilities were measured after treatments of the pooled fraction with ß-glucan, instead of endotoxin. As shown in Table 4, the amidase activity and clot-forming ability were found in fraction G, indicating that the fraction G contains a glucan-sensitive factor, tentatively named factor G (Exp. 2). Moreover, a mixture of fraction G with fraction A (Exp. 4) or fraction B (Exp. 6) in the presence of ß-glucan showed a stronger amidase activity, as compared with fraction G alone. However, there was neither amidase activity nor clot-forming ability in the combination of fractions A and B (Exp. 5).

Table 4. Recombination experiments of fractions A, G and B in the presence of carboxymethylated (1→3)-ß-D-glucan(CMPS)

Experiment	Fraction	Amidase activity (μmoles pNA released/20min/ml) ($\times 10^{-4}$)	Clot-forming ability
1	Fraction A	0	−
2	Fraction G	135	+
3	Fraction B	0	−
4	Fraction A+G	400	+
5	Fraction A+B	0	−
6	Fraction G+B	278	+
7	Fraction A+G+B	600	+

The reaction mixture containing 50 μl each of fraction, CMPS (final, 24 ng/ml), 0.4 mM Boc-Leu-Gly-Arg-pNA substrate, 80 mM Tris-HCl buffer, pH 8.0, and 5.2 mM MgCl$_2$, in a total volume of 250 μl, was incubated at 37°C for 30 min. In recombination experiments, 50 μl each of fractions A, G or B was mixed for each other and the amidase activity was measur= ed (16). A clot-forming ability of each fraction was tested by using a highly purified coagulogen as follows. The reaction mixture contain= ed 50 μl each of fractions A, G and B or their combined mixture, 30 μl CMPS (400 ng/ml), 100 μl Tachypleus coagulogen (2 mg/ml) and 50 μl 0.4 M Tris-HCl buffer, pH 8.0, containing 26 mM MgCl$_2$, in a total volume of 230 μl. When a clot appeared within 1 hr at 37°C, it was judged that the sample showed a positive reaction.

The maximum amidase activity was obtained in the mixture of fractions A, G and B (Exp. 7). The results suggest that a ß-glucan sensitive factor, factor G, must be contained in fraction G and that the factor induces the activation of the known proclotting enzyme eluted in fraction A. Based on these results, we would suggest a new glucan-mediated coagulation pathway in Limulus amebocytes, as shown in the abstract. Thus, ß-D-glucan activates factor G and the active factor $\overline{G}$ converts the known proclotting enzyme to clotting enzyme, which then catalyzed the transformation of coagulogen to coagulin gel. Therefore, Limulus amebocytes seem to contain two independent coagulation path= ways, endotoxin-mediated and glucan-mediated, both of which result in the clot-formation (12, 13). The latter pathway seems to corresponding to an alternative pathway found in the mammalian complement system, as shown in Fig. 9.

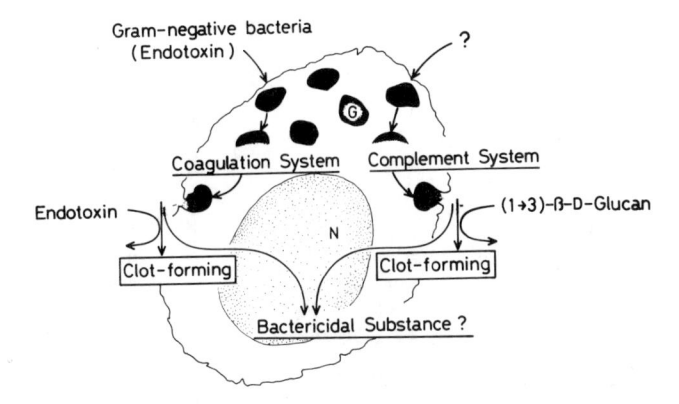

Fig. 9. Schematic visualization of two independent coagulation path=
ways, endotoxin-mediated and ß-D-glucan-mediated pathways,
found in Limulus amebocytes. One of bactericidal substances
appears to be anti-LPS factor described in this paper.

7. Biochemical characterization of anti-LPS factor found in Limulus amebocytes

As reported previously, the amebocyte lysate contained a factor which disturbs the endotoxin-mediated coagulation cascade (14). We named tentatively this principle as anti-LPS factor. This factor was found in the last fraction, which was eluted with 2 M NaCl on a dextran sulfate-Sepharose CL-6B column (Fig. 6). The anti-LPS fraction was purified further by gel-filtration on a pyrogen-free Sephadex G-50 column. Figure 10 shows the elution profile and the SDS-gel electro= phoretic patterns in the anti-LPS fractions. The fractions with anti-LPS factor activity were collected and finally purified by a CM-Sepharose column, as shown in Fig. 11. The anti-LPS factor appeared as a single peak and the inhibitory activity coincided with a peak detect= ed by the absorbance at 230 nm. Moreover, the protein in this fraction showed a single band on SDS-gel electrophoresis in the presence of reducing agent, and also on a disc-gel at pH 4.0. The molecular weight of anti-LPS factor estimated by SDS-gel electrophoresis had about 15,000. The amino acid composition of the purified anti-LPS factor is shown in Table 5. The material contained relatively large amounts of threonine, serine, glycine, leucine and lysine. Hexosamine content was less than 0.1 percent, suggesting that it consists of a simple basic protein.

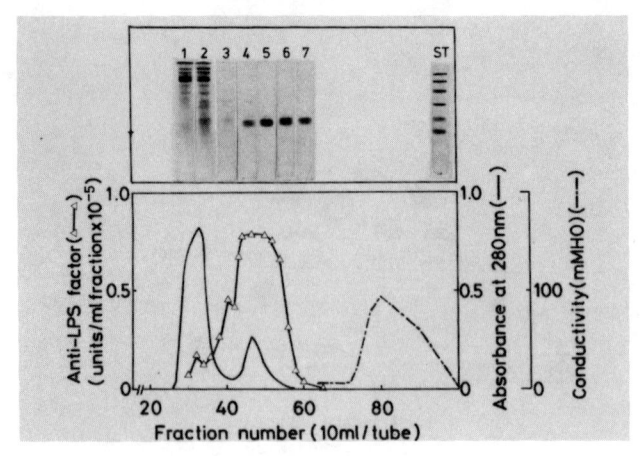

Fig. 10. Gel-filtration of the anti-LPS factor fraction on a pyrogen-free Sephadex G-50 column. The anti-LPS factor fraction obtained from a dextran sulfate-Sepharose CL-6B column (Fig. 6) was applied to the column (3.5 x 76 cm) and eluted by 20 mM HCl at 4°C.

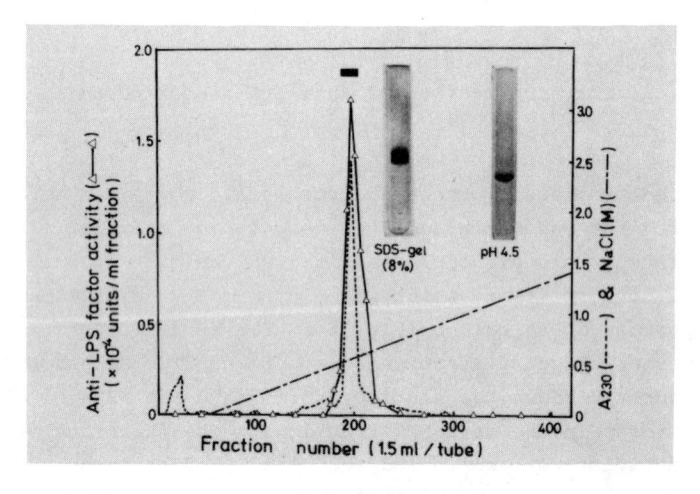

Fig. 11. Chromatography of anti-LPS factor on a CM-Sepharose column. Elution was performed with a linear salt gradient to 1.5 M NaCl at a flow rate of 10 ml per hr. The fractions indicated by a solid bar were collected. The activity of anti-LPS factor was assayed by the previous method (14).

The effect of the anti-LPS factor on endotoxin- and glucan-mediated activation of Limulus coagulation system is shown in Table 6. In these experiments, a recombination system with proclotting enzyme and factor B or a system with proclotting enzyme and factor G was used,

Table 5. Amino acid composition of the isolated anti-LPS factor

Component	Residues per molecule
Asp	5.3 (5)
Thr	9.6 (10)
Ser	13.1 (13)
Glu	14.9 (15)
Pro	3.4 (3)
Gly	9.8 (10)
Ala	8.7 (9)
Cys/2	2.3 (2)
Val	6.7 (7)
Ile	5.5 (6)
Leu	10.1 (10)
Tyr	3.8 (4)
Phe	6.3 (6)
Lys	12.3 (12)
His	3.5 (3)
Trp	6.2 (6)
Arg	6.8 (7)
Total	128
Glucosamine (%)	0.1
Galactosamine (%)	0.1

Amino acid and hexosamine compositions were calculated from extrapolated or average values estimated on samples of 24, 48 and 72 hr hydrolyzates.

Table 6. Effect of anti-LPS factor on LPS-and CMPS-induced activation of Limulus coagulation system

Exp.	Sample	Trigger	Amidase activity Without anti-LPS	With anti-LPS	Inhibition (%)
			μmoles pNA released/30 min/ml (x 10^{-4})		
1	Amebocyte lysate	LPS	682	115	83.1
2	Fr A+B	LPS	335	30	91.0
			μmoles pNA released/17 min/ml (x 10^{-4})		
3	Amebocyte lysate	CMPS	288	315	0
4	Fr A+G	CMPS	295	316	0
			μmoles pNA released/10 min/ml (x 10^{-4})		
5	Active clotting enzyme		165	170	0
6	Activated factor B + Proclotting enzyme (Fraction A)		210	220	0

The reaction mixture containing 20 μl <u>Tachypleus</u> amebocyte lysate or the mixtures containing 50 μl each of fractions obtained from a heparin-Sepharose column (14) and LPS (E. coli 0111-B4, final 24 ng/ml) were incubated at 37°C for 30 min in the presence of the anti-LPS factor fraction (30 μl) in Fig. 10. After incubation, the LPS-induced amidase activity was measured (16). The same experiments as above were made using CMPS (final, 24 ng/ml)

in addition to Limulus amebocyte lysate. In experiment 1, the appear=
ance of LPS-induced amidase activity due to the clotting enzyme in the
lysate was strongly inhibited in the presence of anti-LPS factor. The
same inhibitory effect of anti-LPS factor on the recombination system
with fractions A and B was observed (Exp. 2). However, there was no
inhibitory effect of anti-LPS factor of the glucan-induced activation
(Exp. 3). Moreover, anti-LPS factor did not any effect on the activi=
ties of clotting enzyme and factor B in their activated forms (Exp. 5).
These results indicate that anti-LPS factor inhibits only the activa=
tion of Limulus coagulation system mediated with LPS but not with ß-
glucan. The inhibitory site of anti-LPS factor in the coagulation
cascade seems to be on the activation of factor C mediated with endo=
toxin. Although we need further studies to know a possible mode of
action of this factor, some interesting result as to the biological
activity has been obtained in collaboration with Drs M. Niwa and K.
Ohashi. That is a bactericidal action of the isolated anti-LPS factor
on the growth of Salmonella minnesota R595. The growth of this bacte=
ria in the presence of anti-LPS factor is strongly inhibited. This
result may suggest that anti-LPS factor plays an important role in
biological defence of Limulus against microorganisms. To establish
this hypothesis, further investigation will be required.

Acknowledgments: We wish to express our thanks to Dr. M. Niwa, Osaka
City University Medical School, for helpful discussion, and to Dr. S.
Sakakibara and his group for the kind supply of some of the chromogenic
and fluorogenic peptide substrates used here. We also thank Miss
Kazuko Usui for her assistance of amino acid analysis and Miss Mizumo
Akiyoshi for her expert secretarial assistance. This work was support=
ed in part by Grants-in-Aid for Scientific Research from the Ministry
of Education, Science and Culture of Japan, and by Toray Science foun=
dation for 1981.

8. References

(1) Dumount, J. H., Anderson, E. and Winner, G., J. Morph., 119 (1966)
    181-208.

(2) Ornberg, R. L. and Reese, T. S., Prog. Clin. Biol. Res., 29 (1970)
    125-130.

(3) Niwa, M. and Waguri, O., Seikagaku (in Japanese) 47 (1975) 1-13.

(4) Levin, J. and Bang, F. B., Bull Johns Hopkins Hosp., 115 (1975) 265-274.

(5) Mürer, E. H., Levin, J. and Holme, R., J. Cell Physiol., 86 (1975) 533-542.

(6) Nakamura, S., Iwanaga, S., Harada, T. and Niwa, M., J. Biochem., 80 (1976) 1011-1021.

(7) Nakamura, S., Takagi, T., Iwanaga, S., Niwa, M. and Takahashi, K., Biochem. Biophys. Res. Commun. 72 (1976) 902-908.

(8) Takagi, T., Hokama, Y., Morita, T., Iwanaga, S., Nakamura, S. and Niwa, M., Prog. Clin. Biol. Res., 29 (1979) 169-184.

(9) Miyata, T., Hiranaga, M., Umetsu, M. and Iwanaga, S., Ann. N. Y. Acad. Sci., in press.

(10) Nakamura, S., Morita, T., Harada-Suzuki, T., Iwanaga, S., Takahashi, K. and Niwa M., J. Biochem., 92 (1982) 781-792.

(11) Nakamura, T., Ohtsubo, S., Tanaka, S., Morita, T. and Iwanaga, S., Seikagaku, 54 (1982) 822.

(12) Ohki, M., Nakamura, T., Morita, T. and Iwanaga, S., FEBS Letters, 120 (1980) 217-220.

(13) Morita, T., Tanaka, S., Nakamura, T. and Iwanaga, S., FEBS Letters, 129 (1981) 318-321.

(14) Tanaka, S., Nakamura, T., Morita, T. and Iwanaga, S., Biochem. Biophys. Res. Commun., 105 (1982) 717-723.

(15) Morita, T., Nakamura, T., Ohtsubo, S., Tanaka, S., Miyata, T., Hiranaga, M. and Iwanaga, S., Proteinase Inhibitors: Medical and Biological Aspects (eds. N. Katsunuma et al.) Japan Sci. Soc. Press, Tokyo/Springer-Verlag, Berlin, in press.

(16) Harada-Suzuki, T., Morita, T., Iwanaga, S., Nakamura, S. and Niwa, M., J. Biochem., 92 (1982) 793-800.

(17) Tai, J. Y., Seid, R. C jr., Huhn, R. D. and Liu, T. Y., J. Biol. Chem., 252 (1977) 4773-4776.

(18) Nakamura, T., Hiranaga, M., Ohtsubo, S., Miyata, T., Morita, T. and Iwanaga, S., Proceedings of the 30th Symposium on Toxins, pp 16-20, Atami city, 1983.

(19) Ohki, M., Nakamura, T., Morita, T. and Iwanaga, S., Proceedings of the 27th Symposium on Toxins, pp 49-54, Tokyo, 1980.

Basic and Applied Studies of the Limulus Test

Makoto Niwa*, Masaru Umeda** and Kunihiro Ohashi*

* Department of Bacteriology and ** Department of Urology, Osaka City University Medical School, Asahi-machi, Abeno-ku, Osaka 545, Japan.

Abstract

   Basic studies on the biochemical mechanism of the gelation reaction induced by endotoxin in amebocyte lysates from the horseshoe crab led us to introduce quantitative methods for microassays of endotoxin, such as a clot protein method and a chromogenic substrate method. With the chromogenic substrate method it was possible to determine endotoxin concentrations as low as 5-10 pg/ml with satisfactory accuracy and reproducibility. The mechanism and specificity of the Limulus amebocyte lysate (LAL) test were investigated by the quantitative methods. The LAL gelation activities of various endotoxins and related substances were approximately parallel to their pyrogenic activities. The lipid moiety of lipopolysaccharide was found to be responsible for the LAL activity. In an attempt to prepare an endotoxin adsorbing material, polymyxin B was covalently attached to Sepharose beads (Polymyxin-Sepharose). It was demonstrated that Polymyxin-Sepharose effectively adsorbed endotoxin.

1. Introduction

   Standardization of endotoxin requires the most sensitive and accurate assay method. Bang and Levin (1) discovered that Limulus amebocyte
_____
Abbreviations: LAL: Limulus amebocyte lysate; LPS: Lipopolysaccharide; Tris: Tris-hydroxymethylaminomethane; Abs: Absorbance; pNA: p-nitroaniline; Px: Polymyxin B; Px-Seph: Polymyxin-Sepharose.

lysate (LAL) form a firm gel in the presence of a minute amount of endotoxin. The gelation reaction, so-called the Limulus test, is now extensively employed as a convenient microassay for endotoxin in medical and pharmaceutical fields. Although the Limulus test is very sensitive and simple, it is rather qualitative than quantitative, and a lack of reproducibility has been experienced. Since 1972 we have investigated, collaborating with Drs. Iwanaga, Nakamura and Morita, the biochemical mechanism and the specificity of the gelation reaction. We have already established the biochemical mechanism of the gelation process and the amino acid sequence of a clottable protein, designated coagulogen, from the Japanese horseshoe crab (Tachypleus tridentatus) (2,3,4). Coagulogens from other three species of the horseshoe crab (Limulus polyphemus, Tachypleus gigas and Carcinoscorpius rotundicauda) were found to have homology in their amino acid sequences with that of Tachypleus tridentatus (4). In all of the four species, clotting enzyme in LAL sequentially activated through Factor C and Factor B by endotoxin splits coagulogen, releasing a peptide C, and proteolysed coagulogen forms a firm gel. The amount of gel can be determined as insoluble clot protein (5). Based on the amino acid sequence of the split sites of coagulogen, Iwanaga et al. (2,6,7) introduced new chromogenic substrates for the clotting enzyme and elaborated a sensitive and quantitative method for endotoxin assays. The new chromogenic method and the clot protein method have enabled us to perform the most accurate quantitation of endotoxins. We report here the feasibility of these quantitative methods and potential applications in endotoxin studies. This paper also includes adsorption characteristics of a novel endotoxin binding substance, Polymyxin-Sepharose.

2. Materials and Methods

LAL: The amebocyte lysate from Tachypleus tridentatus was prepared by the method previously described (8). Commercial LAL preparations, Pregel^R, Pyrodick^R (a kit for pNA method) and Toxicolor^R (a kit for diazo-dye method) were the products of Seikagaku Kogyo Co. Ltd., Tokyo. Lyophilized LAL from Limulus polyphemus (Wako Pure Chemicals Co. Ltd., Osaka) was also employed. Chemicals: A synthetic chromogenic substrate, Boc-Leu-Gly-Arg-pNA, was the product of the Protein Research Foundation, Minoh, Osaka. Polymyxin B sulfate (Taito-Pfeizer Co. Ltd., Tokyo) and CNBr-activated Sepharose 4B (Pharmacia, Sweden) were used for the preparation of Polymyxin-Sepharose. Endotoxins: Endotoxin preparations used were as follows; E. coli O113 LPS (phenol extract), E. coli UKTB LPS (Japanese Reference LPS, reported by Akama et al. in this

Symposium), E. coli O111: B4 LPS (Difco Laboratories, Detroit, U.S.A.), Salmonella enteritidis endotoxin (aqueous ether extract provided by Dr. E. Ribi, NIH, U.S.A.), Salmonella minnesota R595 glycolipid (phenol-chloroform-petroleum ether method (9)). The native polysaccharide from E. coli O113 was given by E. Ribi (10). All the endotoxin preparations were reconstituted and diluted with sterile, pyrogen-free water for injection (Ohtsuka Pharmaceutical Co. Ltd., Tokyo). Assay methods: All glassware was rendered pyrogen-free by heating at 250° for 1 hr in an oven. The clot protein method was performed as follows; The reaction mixture containing 0.1 ml each of LAL, Tris-HCl buffer (pH 7.2, 0.12 M) and 0.2 ml sample was incubated at 37° for 60 min. The gel formed was washed with 3 ml of an ice-cold buffer (Tris-HCl, pH 7.2, 5 mM-$CaCl_2$, 1 mM-NaCl, 0.15 M) three times by centrifugation. The amount of insoluble clot was determined by Lowry's method and expressed as bovine serum albumin equivalent (5). The technique of the chromogenic substrate method has already been described (6,7). Briefly, reaction mixtures consisted of 100 μl each of LAL, 2 mM Boc-Leu-Gly-Arg-pNA, Tris-HCl (pH 8.0, 0.2 M + $MgCl_2$, 0.04 M) and samples were incubated at 37° for 20-60 min. After terminating the reaction by adding 0.6 N acetic acid, the liberated pNA was measured at 405 nm in a photometer (Hitachi, Model 624). In some cases, pNA was converted to a diazo-dye by coupling with $NaNO_2$, ammonium sulfamate and n-1-naphthyl-ethylenediamine according to the method of Takagi et al. (11). The absorbance of the red-colored dye was determined photometrically at 545 nm. Preparation of Polymyxin-Sepharose: Polymyxin B was covalently immobilized to Sepharose 4B beads by the procedure shown in Fig. 1. Usually about 40 mg polymyxin/g of Sepharose could be coupled. As a control CNBr-activated Sepharose treated by the same procedure without polymyxin was used.

```
CNBr activated Sepharose 4B
   │  washed with 1 mM HCl
   │     "      "   Coupling Buffer(pH 8.3,NaHCO₃ 0.1 M,NaCl 0.5 M)
   ├─ + Polymyxin B sulfate    100 mg/g Sepharose
   │     Shake at room temperature for 2 hr
   ├─ + Bloking Agent( Glycine,pH 8.0,0.2 M or Tris-HCl,pH 8,o.1 M)
   │     stand at 4° for 16 hr
   │     washed with Coupling Buffer
   │        "      "   Acetate Buffer( pH 4.05 + NaCl 0.5 M )
   │        "      "   Coupling Buffer
POLYMYXIN-SEPHAROSE
```

Fig. 1. Preparation of Polymyxin-Sepharose

3. Results

Fig. 2 illustrates the biochemical basis of the LAL gelation. The initial stage of activation of Factor C by endotoxin is followed by sequential activation of Factor B and a proclotting enzyme. The activated clotting enzyme splits coagulogen at the site Arg^{18}-Thr^{19} and Arg^{46}-Gly^{47} liberating a peptide C. The remaining portion of coagulogen polymelizes and thus form a gel. The conventional LAL test, in which the firmness of gel is scored, routinely detect 0.1 ng/ml of endotoxin. The amount of gel can be determined as insoluble clot protein. The activated clotting enzyme, acting as an amidase, liberates pNA from the chromogenic substrate Boc-Leu-Gly-Arg-pNA. Depending on the potency of LAL and the reaction time, the amount of liberated pNA is a function of endotoxin concentrations.

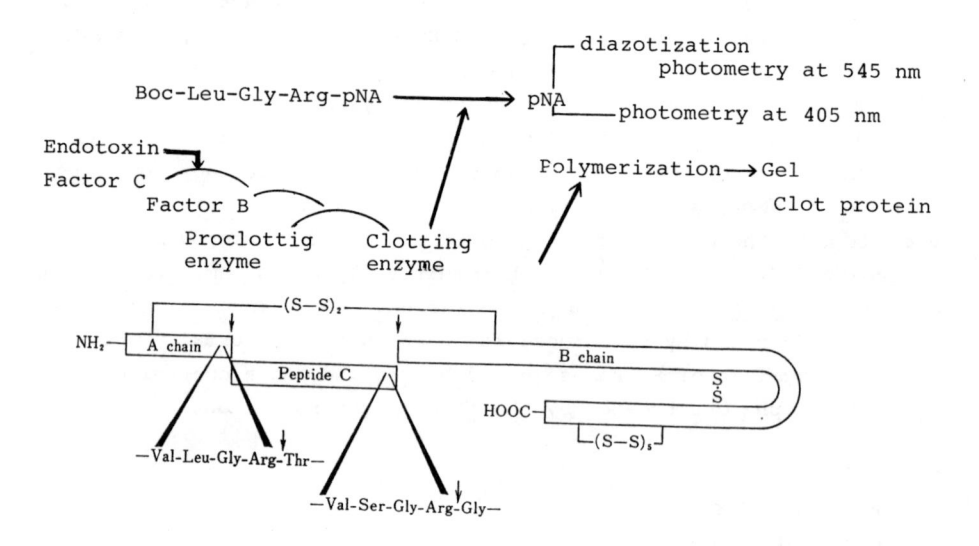

Fig. 2. Biochemical basis of Limulus test

A representative standard curve of the pNA method for E. coli UKTB LPS are given in Fig. 3. In the optimal condition, even 10 pg/ml of the endotoxin could be clearly distinguished from the blank (usually Abs at 405 nm of the blank was about 0.03). The intra-assay variability with an endotoxin sample varied about 5% and the inter-assay reproducibility, depending on the potency of LAL used, was fairly good. The measurement of the absorbance at 405 nm is interfered by yellow colored samples such as plasma and serum. To avoid such interference the released pNA can be converted to a red-colored diazo dye by the

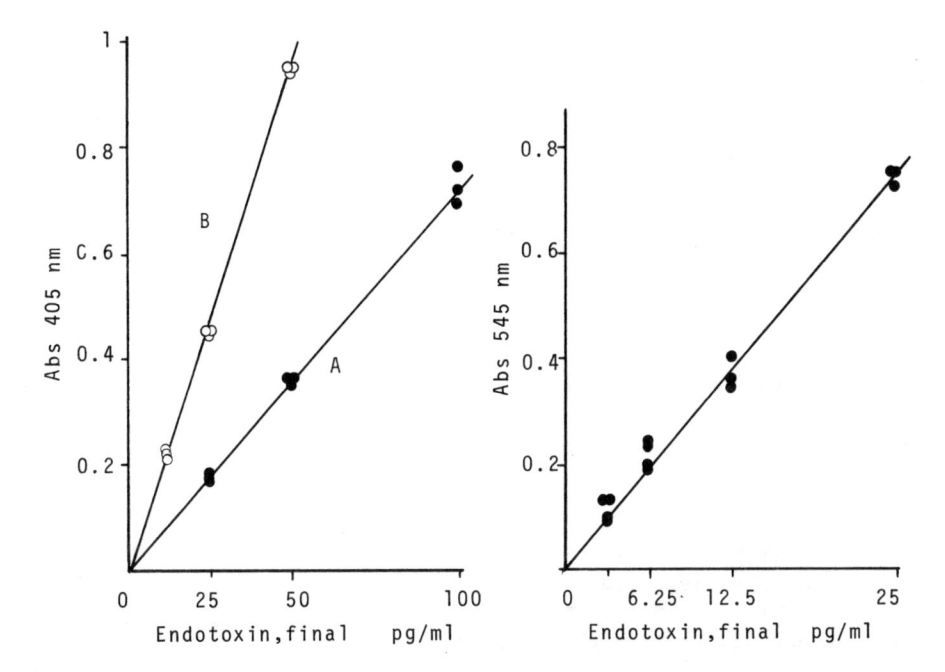

Fig. 3. Standard curve of pNA method.
  A: E. coli UKTB LPS (100 µl) was added to Pyrodick (100 µl). 37°, 30 min.
  B: E. coli UKTB LPS and Wako LAL. 37°, 40 min.

Fig. 4. Standard curve of diazo-dye method. E. coli O111: B4 endotoxin (100 µl) was added to Toxicolor (100 µl). 37°, 30 min.

diazotization method. The sensitivity of the diazo-dye method was higher than that of the pNA method. Endotoxin as low as 5 pg/ml or less could be determined (Fig. 4). The amount of liberated pNA revealed a good correlationship with the amount of the clot protein when the same endotoxin was assayed by these methods.

A variety of endotoxins and related substances were compared in their gelation inducing activities using the clot protein method. For a quantitative comparison we have calculated a G_{50} value which was defined as: G_{50} = the concentration (µg/ml) of sample which gives 50% of the maximum clot with the same batch of LAL. As summarized in the Table, the gelation inducing potencies of S. minnesota R595 glycolipid and E. coli UKTB LPS were very high, whereas preparations such as E. coli O113 native polysaccharide and Poly I: Poly C which elicit only feeble endotoxic activities were also very weak in their gelation inducing activities. A significant correlation between the gelation and the pyrogenic activities were found. Among the endotoxin preparations tested Salmonella minnesota R595 glycolipid and E. coli UKTB LPS were the most potent ones. Similar results were obtained using the chromogenic substrate method (7).

Table. Comparisons of gelation potencies and pyrogenic activities of various preparations.

Preparation	G_{50} µg/ml	MPD* µg/kg
S. minnesota R595 glycolipid	1.3×10^{-5}	4×10^{-3}
E. coli UKTB LPS	6.3×10^{-5}	1×10^{-3}
Shigella K3 endotoxin	4.0×10^{-3}	3×10^{-3}
E. coli O111: B4 LPS	2.4×10^{-2}	1×10^{-2}
Ps. aeruginosa N10 OEP	1.7×10^{-1}	3×10^{-1}
Poly I: Poly C	2.8	(40 +)**
E. coli O113 native polysaccharide	27	(1,000 +)**

G_{50} = the concentration which gives 50% of the maximum clot.
*: the minimum pyrogenic dose which gives 0.6° temperature rise in rabbits.
**: pyrogenic with the dose indicated.

The pNA releasing activity was induced by all the active endotoxins in the range of 1 ng/ml to few pg/ml, whereas more than 1 µg/ml was required with native polysaccharides from E. coli O113 and E. coli O111: B4 (7). As reported by Morita et al. (12), (1-3)-β-D-glucan activates the clotting enzyme in LAL through an alternative pathway, but the optimal concentration was 10-100 ng/ml.

Several substances other than endotoxin have been reported to be Limulus positive. Because of the proteolytic nature of the LAL reaction, proteases such as trypsin cleave the coagulogen molecule. Peptidoglycans and muramyl peptides from various strains of gram-negative bacteria were claimed to cause gelation of LAL. However, these substances were Limulus positive at much higher concentrations (1 µg/ml or more). Thus the LAL test may be specific for assaying endotoxin.

Nowadays it has been established that almost all the biological activities of endotoxin are attributed to the lipid moiety. This is the greatest contribution to biology and medicine made by Westphal's "Lipid A theory". The lipid moiety has been shown to play a very important role in inducing the LAL activation. In 1969, Niwa et al. (13) demonstrated that the fatty acyl ester in the endotoxin molecule which is labile to mild alkaline treatments was essential for eliciting

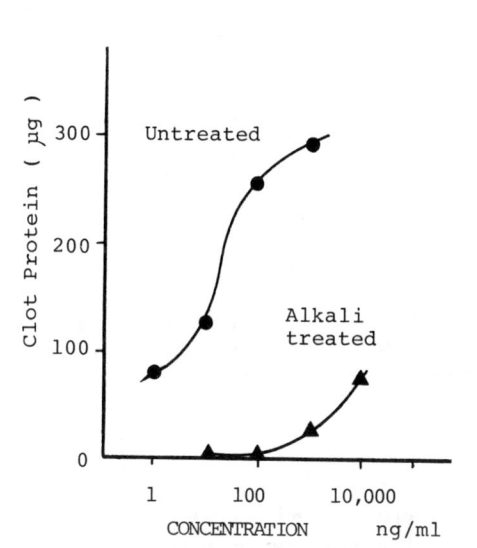

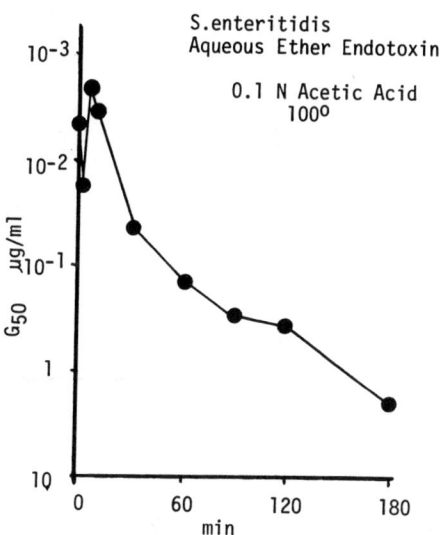

Fig. 5. Effect of alkaline treatment of E. coli O113 LPS on the gelation activity. E. coli O113 LPS was treated with 0.03 N NaOH in 95% ethanol at 30° for 60 min.

Fig. 6. Effect of acid hydrolysis of S. enteritidis endotoxin on the gelation activity. It was hydrolysed with 0.1 N acetic acid at 100°.

pyrogenicity and chick embryo lethality of endotoxins. The mild alkaline treatment also abolished the gelation inducing activity of E. coli O113 LPS as shown in Fig. 5. Salmonella enteritidis endotoxin (aqueous ether extract) was hydrolysed with 0.1 N acetic acid at 100°. After neutralization and dilution the gelation inducing activities were measured. Fig. 6 illustrates the time course of the hydrolysis. A gradual and consistent loss of the gelation inducing activity was observed. These results suggest that the fatty acyl ester moiety and the acid labile components in the endotoxin molecule are essential for the activation of the LAL coagulation system.

The chromogenic substrate method allows precise quantitations of endotoxin in basic as well as applied studies which were not achieved by the conventional Limulus test. For example, we have investigated effects of polymyxin B on the manifestation of endotoxic activities. The polymyxins are a family of antibiotics which contain a cationic cyclopeptide with a fatty acid chain. Polymyxin B is known to bind endotoxin and thus neutralize the biological activities of endotoxin (14). We have confirmed that polymyxin B prevented the toxicity of endotoxin in mice and that the Re glycolipid from S. minnesota R595 inhibited the antimicrobial activity of polymyxin B. Polymyxin B also

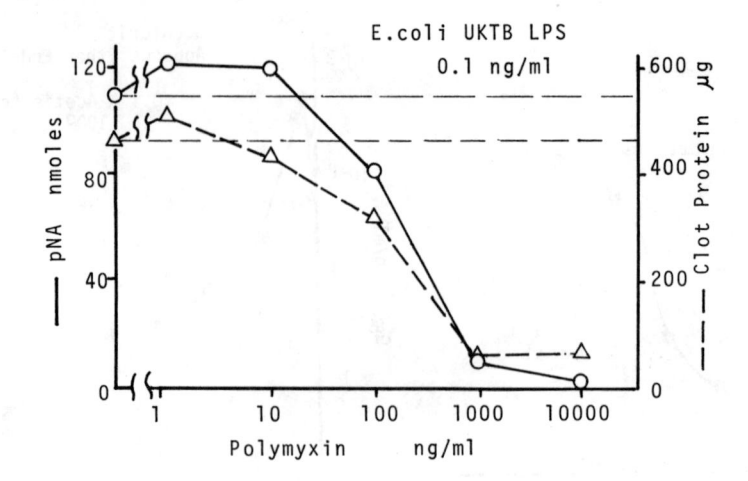

Fig. 7. Inhibition of the gel formation and the pNA releasing activity of LAL by polymyxin B. Varying concentrations of polymyxin B were mixed with the endotoxin and added to the reaction mixtures. Clot protein and pNA were determined after 60 min. and 30 min. respectively.

inhibited the activation process of the LAL system. As can be seen in Fig. 7, polymyxin B inhibited the clot formation and the pNA releasing activity of LAL when it was added with endotoxin simultaneously. Nakamura et al. (15) have demonstrated that the clotting enzyme which had been activated by endotoxin was not inhibited by polymyxin even at 50 µg/ml. These data indicate that polymyxin B prevent the initial activation stage by complexing with endotoxin and/or by interfering Factor C.

Endotoxin contamination provides numerous troublesome problems in manufacturing parenteral drugs and vaccines, and any slight contamination of endotoxin in hemodialysis systems should be eliminated. In order to remove endotoxin from contaminated dialysis fluids or blood, columns filled with activated charcoal are tried to employ in hemodialysis systems with little success. Nolan et al. (16) reported that a cholestyramine resin, Dowex 1-X2, adsorbed ^{51}Cr-labelled endotoxin. However the adsorption characteristics of such anion-exchange resin as well as of activated charcoal are non-specific. It may reasonably be expected that the polymyxin immobilized on an adequate carrier serves as an effective and more specific adsorbent for endotoxin.

In an attempt to remove endotoxin from contaminated fluids, we prepared a novel endotoxin adsorbing material, Polymyxin-Sepharose.

Polymyxin B was covalently immobilized on Sepharose 4B beads by the
CNBr coupling method. It was found that the immobilized polymyxin
was able to adsorb endotoxin. Polymyxin-Sepharose (about 10 mg containing 364 µg of immobilized polymyxin) was mixed with S.
minnesota R595 glycolipid (80 µg) dissolved in a buffer (Tris-HCl, pH 7.2,
0.01 M) and centrifuged. Almost all the endotoxin was adsorbed on
the adsorbent, and no endotoxin was found in the supernatant (17).
Because a known amount of the endotoxin added to the supernatant was
recovered quantitatively, polymyxin which inhibit the endotoxin assay
did not liberated from Polymyxin-Sepharose. The apparent affinity
constant between S. minnesota R595 glycolipid and Polymyxin-Sepharose
estimated by the Scatchard plot was found to be approximately 4×10^8.
No adsorption of the endotoxin by control Sepharose was observed.
Human plasma containing E. coli O113 LPS was passed through a small
column filled with Polymyxin-Sepharose or control Sepharose and eluted
with a pyrogen-free phosphate buffer (pH 7.0, 0.1 M). Eluates were
assayed for endotoxin by the chromogenic substrate method. Endotoxin
was found in the eluate from the control column, but no endotoxin was
found in the eluates from the Polymyxin-Sepharose column (Fig. 8).

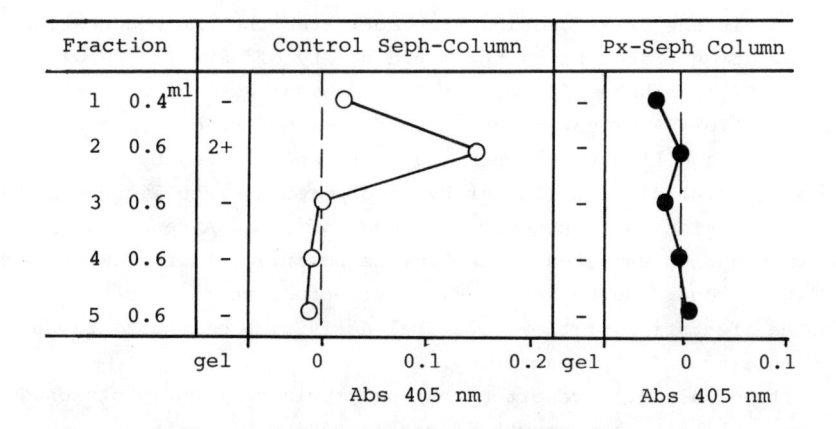

Fig. 8. Adsorption of endotoxin by Polymyxin-Sepharose. Column size = 0.6 × 4 cm. Human plasma containing E. coli O113 LPS (0.2 ml of 10 ng/ml endotoxin plasma) was charged. Eluent: 0.1 M phosphate buffer, pH 7.0. The eluates were diluted and assayed for endotoxin by the chromogenic substrate method.

Polymyxin B is known to combine not only to endotoxin but to some phospholipids. We observed that Polymyxin-Sepharose beads adsorbed human erythrocytes irrespective of their A, B, O blood type, HeLa cells and Vero cells, whereas the control Sepharose did not. Polymyxin-Sepharose inhibited the growth of E. coli O113 in Penassay media when the inoculum was less than 2×10^7 cells, while a heavier inoculum (2×10^8 cells) allowed the microbial growth in the presence of Polymyxin-Sepharose.

4. Discussion

To date several modifications of the conventional Limulus test, such as turbidimetric methods (18), have been reported to improve the sensitivity and accuracy of the test. However, the gelation reaction involves a complicated process of polymerization of coagulin, and it may be interferred by many substances. The introduction of the chromogenic substrate method by Iwanaga et al. provided the most sensitive and precise assay for the quantitation of endotoxin. By the use of this method as low as few pg/ml of endotoxin can be determined with satisfactory accuracy and reproducibility. A distinct advantage of this method is that the biochemical basis of the reaction has already been elucidated (4,7,15). Further improvements in the sensitivity of the chromogenic substrate method are now attempted.

The constituent of endotoxin which activates the LAL coagulation system was demonstrated to be the lipid moiety of LPS. The Re glycolipid from S. minnesota R595 was the most potent activator of the LAL system. Lipid A preparations derived from gram-negative bacteria also activate the LAL system when they were solubilized by complexing with bovine serum albumin (19) or by incorporating into liposomes (20). In this connection the synthetic analogues of lipid A reported in this Symposium might be expected to be Limulus positive. There remains much to be learnt regarding to the chemical structure which is essential for the manifestation of the biological activities of endotoxin such as pyrogenicity and LAL reactivity.

La Porte et al. (21) reported that polymyxin covalently attached to Agarose inhibited the growth of gram-negative bacteria, but no information about adsorption of endotoxin was presented. Our data indicate that Polymyxin-Sepharose effectively adsorbs endotoxin. The immobilized polymyxin seems to be useful in numerous basic and applied investigations of endotoxin due to its promising potential for adsorption of endotoxin. Surveys for more suitable carriers other than Sepharose are now in progress. The quantitative approach as described in this paper is essential for endotoxin studies and may provide more detailed understanding of the nature of endotoxin.

Acknowledgement

The authors are indebted to Drs. S. Iwanaga, T. Morita (Faculty of Sciences, Kyushu University) and Dr. S. Nakamura (Institute for Primate Research, Kyoto University) for their constant collaboration. This work was supported in part by grants from the Scientific Research Foundation of the Ministry of Education, Japan.

5. References

(1) J. Levin, F. Bang, Bull. Johns Hopkins Hosp. 115 (1964) 265-274.

(2) S. Nakamura, T. Takagi, S. Iwanaga, M. Niwa, K. Takahashi, J. Biochem. 80 (1976) 649-652.

(3) T. Takagi, Y. Hokama, T. Morita, S. Iwanaga, S. Nakamura, M. Niwa, Prog. Clin. Biol. Res. 29 (1979) 169-184.

(4) S. Nakamura, F. Shishikura, S. Iwanaga, T. Takagi, K. Takahashi, M. Niwa, K. Sekiguchi, in T. Y. Liu et al. (Eds.). Frontiers in Protein Chemistry, Elsevier North Holand, Inc. New York, 1980, pp. 495-514.

(5) M. Niwa, T. Hiramatsu, O. Waguri, Jap. J. Med. Sci. Biol. 28 (1975) 98-100.

(6) S. Iwanaga, T. Morita, T. Harada, S. Nakamura, M. Niwa, K. Takada, T. Kimura, S. Sakakibara, Haemostasis, 7 (1978) 183-188.

(7) T. Harada-Suzuki, T. Morita, S. Iwanaga, S. Nakamura, M. Niwa, J. Biochem. 92 (1982) 793-800.

(8) M. Niwa, O. Waguri, Seikagaku, 47 (1975) 1-12.

(9) C. Galanos, O. Lüderitz, O. Westphal, Eur. J. Biochem. 9 (1969) 245-249.

(10) R. L. Anacker, R. A. Finkelstein, W. T. Haskins, M. Landy, K. C. Milner, E. Ribi, P. W. Stashak, J. Bacteriol. 88 (1964) 1705-1720.

(11) K. Takagi, A. Moriya, H. Tamura, C. Nakahara, S. Tanaka, Y. Fujita, T. Kawai, Thrombosis Res. 23 (1981) 51-57.

(12) T. Morita, S. Tanaka, T. Nakamura, S. Iwanaga, FEBS Lett. 129 (1981) 318-321.

(13) M. Niwa, K. C. Milner, E. Ribi, J. A. Rudbach, J. Bacteriol. 97 (1969) 1069-1075.

(14) J. Bader, M. Teuber, Z. Naturforsch. 28 (1973) 422-430.

(15) S. Nakamura, T. Morita, T. Harada-Suzuki, S. Iwanaga, K. Takahashi, M. Niwa, J. Biochem. 92 (1982) 781-792.

(16) J. P. Nolan, J. J. McDevitt, G. S. Goldman, Proc. Soc. Exp. Biol. Med. 149 (1975) 766-770.

(17) M. Niwa, M. Umeda, K. Ohashi, Jap. J. Exp. Med. Biol. 35 (1982) 114-115.

(18) J. H. Jorgensen, G. A. Alexander, Appl. Environ. Microbiol. 41 (1981) 1316-1320.

(19) E. T. Yin, C. Galanos, S. Kinsky, R. A. Bradshaw, S. Wessler, O. Lüderitz, M. E. Salmiento, Biochim. Biophys. Acta, 261 (1972) 284-289.

(20) E. C. Richardson, B. Banerji, R. Seid, J. Levin, C. R. Alving, Infect. Immun. 39 (1983) 1385-1391.

(21) D. C. LaPorte, K. S. Rosenthal, D. R. Storm, Biochemistry, 16 (1977) 1642-1648.

A Candidate for Japanese Reference Lipopolysaccharide in Control of Limulus Amoebocyte Lysate Testing

Kiyoto Akama, Kazuo Kuratsuka, Reiko Homma
Department of General Biologics Control, National Institute of Health, 10-35, Kamiosaki 2-chome, Shinagawa-ku, Tokyo, Japan

Seizaburo Kanoh
Department of Pharmacology, National Institute of Hygienic Science, 1-43, Hoenzaka 1-chome, Higashi-ku, Osaka, Japan

Makoto Niwa
Department of Bacteriology, Osaka City University Medical School, Asahimachi 1-chome, Abeno-ku, Osaka, Japan

Sadaaki Iwanaga
Department of Biology, Faculty of Science, Kyushu University, Hakozaki 6-chome, Higashi-ku, Fukuoka, Japan

Chizuko Nakahara
Seikagaku Kogyo Co. Ltd., 1253, Tateno 3-chome, Higashiyamato, Tokyo, Japan

Abstract

It was attempted to establish a standard for control of the limulus amoebocyte lysate (LAL) testing. The purified LPS preparation was prepared from culture of Escherichia coli UKT-B strain by phenol water extraction followed by ethanol precipitation. For in vitro measurement of endotoxin activity, Synthetic Chromogenic Substrate (SCS) technique was mainly employed, in which linearity of the dose-response lines was observed within a range of the LPS doses against various endotoxin preparations. The chemical analysis data, pyrogenicity and LAL activating activity of the preparation were satisfactory as the material for dried "Reference LPS". Its inter-vial uniformity was confirmed. It was satisfactorily stable unless being dissolved.

1 Introduction

It is a well known fact that main pyrogenic factor in biologics including blood products is endotoxin (lipopolysaccharide, LPS) produced by Gram-negative bacteria. Rabbit pyrogen test has been used for the detection of endotoxin in these products. Recently, limulus amoebocyte lysate (LAL) test has debuted as an effective and attractive method to detect bacterial endotoxin which could be contained in various drugs to be injected or transfused and also in clinical blood specimens.

Some of the LAL reagents currently available in Japan are controlled by the manufacturers to their own satisfaction, though some of those marketed in the USA are fairly well regulated (1). Samples of endotoxin preparations included in the reagent kits are in similar situation in respect to their control. The testing procedures are multifarious due to each of reagent kits. Therefore, it may be troublesome to compare the endotoxin values to each other which were obtained by using different reagent kits.

To estimate the contents of endotoxin in terms of "weight unit" or "weight equivalent", an LPS preparation as pure as possible should be used as a standard material. Therefore a project has been carried out to establish a standard LPS preparation which would be necessary in the control of the quantitative LAL reagents to be used in Japan. Similar attempts seem to have been made in other countries. For instance, Food and Drug Administration (FDA) licenced several marketable reagents which fulfilled US requirements for LAL reagent kit. The USP XX specified in detail the procedure of determining endotoxin by LAL test using US Reference endotoxin established by FDA. Recently, WHO is planning to establish international reference endotoxin for LAL testing, but the attempt does not seem to be rewarded with prompt success.

Although it was anticipated that the establishment of a reference endotoxin was not easy, a great demand for a reference endotoxin in control of the injectables including biologicals as well as in clinical diagnosis encouraged us. The establishment is also desired eagerly for control of various medical drugs which will be produced by DNA recombination technique in near future. In principle, a reference material should be specified in some chemical and physical properties. When the endotoxin contents are to be estimated in weight equivalent to reference LPS, the higher the purity of the reference, the more ideal it would be. Moreover, it is desirable that a reference material can be easily handled in practical use.

The research was carried out by the "Study Group for the Application of LAL Test to the Control of Biological Products", which was supported by a research grant from Japanese Ministry of Health and Welfare.

2 Extraction of LPS from E. coli UKT-B strain

Broth culture of E. coli UKT-B strain was chosen as the starting material as it had been used for preparing endotoxin by Kanoh who is experienced in purification of LPS. LPS was extracted according to the partially modified phenol/water method (2) and purified with 50% alcohol and ultracentrifugation. Outline of the preparation procedures is shown in Table 1. Electronmicroscopic observation of the purified LPS (ultracentrifuged LPS) revealed some aggregates of rod-shaped microparticles. Such aggregated particles were evenly dispersed by ultrasonic treatment, the LAL activating activity being considerably elevated by the ultrasonication. An example is shown in Table 2. The ultrasonic-treated purified LPS preparation was satisfactory in its pyrogenicity in rabbits and LAL activating activity as shown later, it might be suitable for preparing the "Reference LPS".

3 Chemical composition of the purified LPS preparation

The purified LPS was analyzed chemically at the Showa University (Prof. Kasai, N.), Josai University (Prof. Hisatsune, K.) and our laboratories. The results of chemical analyses obtained are given in Table 3. 2-Keto-3-deoxyoctonate (KDO) was quantified by a modified thiobarbituric acid reaction (3). Heptose was estimated by the method of Osborn (4). Neutral sugars and amino sugars were determined by the method of Laine et al. using xylose and mannose as internal standards, respectively (5). Fatty acids were determined by standard methods. Samples were methanolyzed with 5% methanolic HCl under nitrogen for 16hr at 100°C and released methylesters of fatty acids were analyzed on a glass column (2m×4mm) containing 10% Silicon GE (SE-30) on Gas Chrom Q (80-100 mesh). Phosphorus was determined by the modified Fiske-Subbarow method (6).

In addition to the compositions presented in Table 3, the purified LPS contained approximately 9% of water and small amount of metals and unknowns, but negligible amounts of ribonucleic acid.

4 Preparing the candidate Reference LPS

A part of the purified LPS preparation was sonicated, and dispensed into vials so to contain 1μg of LPS and 15mg mannitol per vial. They were freeze-dried and filled with N_2 gas. One thousand vials of the candidate Reference LPS, Lot 8205 was thus obtained.

Reference LPS dissolved easily in water or in buffered saline.

Table 1. Extraction procedure of LPS from E. coli UKT-B (11)

Bacterial suspension ----- 5g of dried bacteria was suspended in 175ml of water and 175ml of 90% phenol was added
    |
    Extraction --------- stirred vigorously for 20min at 65-68°C
    Cooling ----------- with ice-water to less than 10°C
    Centrifugation ----- at 7000rpm for 15min at 4°C
  S   R
        Re-extraction
        Centrifugation
      S   R
        Dialysis ----------- against water for 4 days at 4°C
        Condensation ------- to about 20ml under reduced pressure at 30°C
        Centrifugation ----- at 3000rpm for 10min at room temperature
    S   R
    Extraction ------------- added 10-fold volume of ethanol containing about 100mg of sodium acetate and left overnight at 4°C
    Centrifugation --------- at 7000rpm for 15min at 4°C
  S   R
        Lyophilization = crude LPS

        LPS solution ------- 1g of crude LPS was dissolved in 50ml of water
        Extraction --------- added 50ml of ethanol and left overnight at 4°C
        Centrifugation ----- at 5000rpm for 15min at 4°C
      S   R
        Condensation ------- to about 20ml under reduced pressure at 30°C
        Re-extraction ------ added equal volume of ethanol and left overnight at 4°C
        Centrifugation ----- at 5000rpm for 15min at 4°C
      S   R
        Precipitation ------ added 6-fold volume of ethanol with 200mg of sodium acetate and left overnight at 4°C
        Centrifugation ----- at 7000rpm for 15min at 4°C
  S   R
        Washing ------------ 2 times with ethanol
        Lyophilization = 50% EtOH-purified LPS

        LPS solution ------- 1% solution of 50% EtOH-purified LPS
        Ultracentrifugation- at 105000×g for 4hr at 4°C, 2 times
      S   R
            Lyophilization = Ultracentrifuged LPS (Purified LPS)

Table 2. Enhancing effect of ultrasonication on LAL activating activity of purified LPS

Ultrasonication*	Dose (ng/ml)			
	5×10^{-2}	10^{-2}	5×10^{-3}	10^{-3}
Before	++	+	−	−
After	++	++	+	−

* At 150w, 19.5kHz for 3 min.

Table 3. Chemical composition of LPS from E. coli UKT-B

Saccharides	%	Fatty acids	%	Other constituents	%
Rhamnose	9.8	Lauric (C 12)	2.6	Phosphorus	3.35
Galactose	5.5	Myristic (C 14)	1.9	Phosphoryl-ethanolamine	1.16
Glucose	9.6	β-Hydroxymyristic (3-OH-C 14)	15.9		
Heptose	7.0			4-Amino-4-deoxy-L-arabinose	0.08
Glucosamine	9.6	Palmitic (C 16)	trace		
KDO	5.3	β-Hydroxydodecanoic (3-OH-C 12)	0.4		
Total	46.8		20.8		4.59

5 Rabbit pyrogen test

Pyrogen test in rabbit was carried out according to the "Minimum Requirements for Biological Products" (Ministry of Health and Welfare of Japan, 1982) (7). Three rabbits were used for one dose of the LPS preparation, and the rectal temperatures were measured at every 30 min following the iv injection of LPS solution.

6 Determination of endotoxin activity

For measurement of endotoxin activity in vitro, the Synthetic Chromogenic Substrate method (SCS method) developed by one of the present authors (Iwanaga) was mainly employed, because of its conspicuous performance in quantitative analysis (8, 9). The SCS reagent kit "PYRODICK" was supplied by Seikagaku Kogyo Co. Ltd., Tokyo (10), which is composed of Tachypleus tridentatus (Japanese horseshoe crab) amoebocyte lysate, a chromogenic substrate (Boc-Leu-Gly-Arg-pNA) and a buffer solution. The manufacturer's instructions for use were strictly followed in tests using the kit. The results in the SCS method are expressed by the OD value at 405 nm after incubation at 37°C for 30 min, unless otherwise specified.

The location and the linear range of the dose-response line obtained with the SCS method depend on the incubation time as well as on the dose of LPS. An example is presented in Fig. 1 where a positive endotoxin control (E. coli 0111:B4) for "PYRODICK" reagent was used. The linearity of the dose-response line by the SCS method was found within a certain range of endotoxin doses as shown in Fig. 2.

Dose-response curve and incubation time

Time course and LPS dose

Endotoxin preparation: Positive Control (E. coli 0111: B4 LPS) for "PYRODICK" reagent
PYRODICK reagent: Lot N020

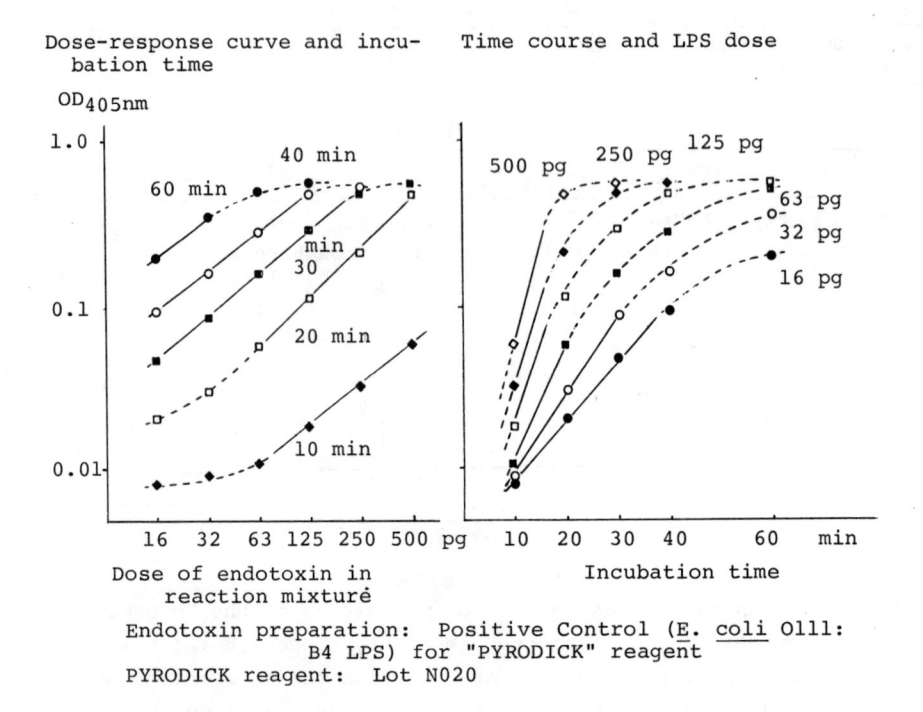

Fig. 1. Effects of endotoxin dose and incubation time on response in the SCS method

A conventional LAL gelation technique was used only for checking the effect of sonication and the activity of LPS in each steps of its extraction.

7 Intra-vial uniformity within a lot of the Reference LPS

The uniformity of the LPS activities within a lot was examined by both in vivo and in vitro tests. Four vials were examined to compare the pyrogenicity. From each vial, three different doses, 1 ng, 10 ng or 100 ng per kg of animal body weight, were injected intravenously into each 3 rabbits, and the rise in temperature in each of the animals was measured. The uniformity in the pyrogenicity within a lot was found as shown in Fig. 3.

The in vitro examination in the SCS method was attempted to confirm the uniformity by using two different LPS concentrations. The examinations 1 and 2 were carried out in triplicate, and the results are summarized in Table 4. No significant inter-vial difference was found in the LAL activating activity.

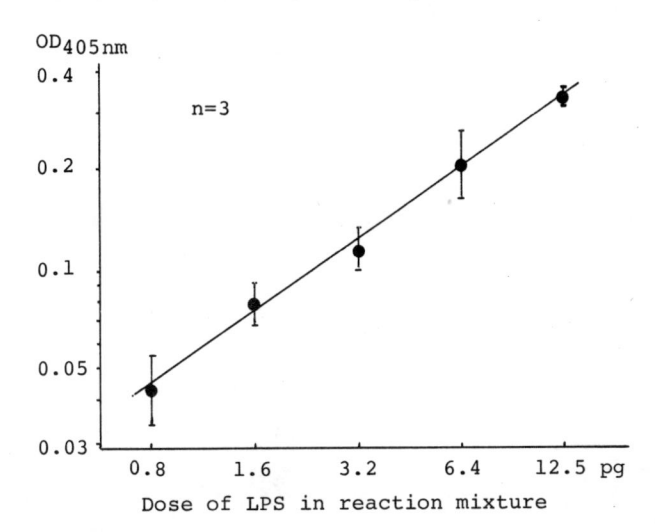

OD_{405nm}

Dose of LPS in reaction mixture

Regression analysis

Factor	SS	DF	MS	F
Regression	1.58483	1	1.53484	1279
Linearity	0.01048	3	0.00349	2.89
Error	0.01208	10	0.00120	
Total	1.55742			

Linearity is not denied (p=0.05)

Fig. 2. Dose-response line with the Reference LPS

8 The minimum pyrogenic dose of the Reference LPS

The data including the one given in Fig. 3 were plotted to draw the dose-response curve. As an example, the one from the results in Fig. 3 is shown in Fig. 4. The linearity of the dose-response relationship was not denied. The mean dose producing 0.56°C rise in temperature, minimum pyrogenic dose, might be estimated by extrapolating its linear part. The minimum pyrogenic dose calculated from the line was found to be 0.25 to 0.35 ng/kg body weight. The pyrogenicity of the Reference LPS was found to be satisfactorily high.

9 Stability of the Reference LPS

Stability of the Reference LPS was assessed in accelarated degrada-

Limulus Test for Endotoxin

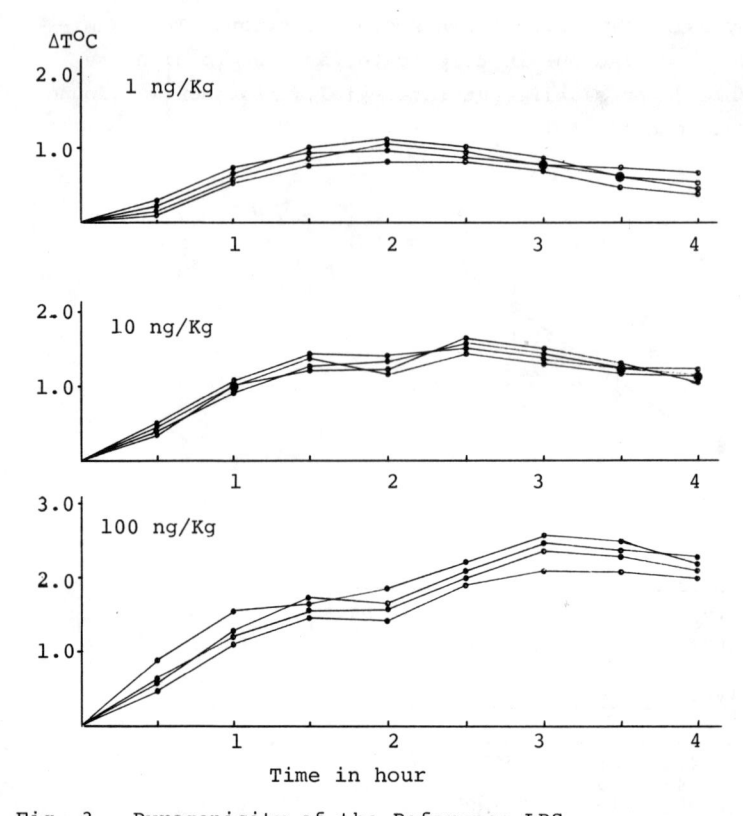

Fig. 3. Pyrogenicity of the Reference LPS

Table 4. Inter-vial uniformity (by SCS method)

Exam. 1 (n=3)

Vial No.	OD$_{405nm}$ (25 pg/ml)	S^2	Confidence interval (p=0.05)
1	0.197	0.000059	0.178 - 0.216
2	0.199	0.000028	0.186 - 0.212
3	0.189	0.000196	0.154 - 0.223
4	0.201	0.000006	0.195 - 0.207
5	0.190	0.000029	0.176 - 0.203

Mean=0.196   Common variance=0.000064

Exam. 2 (n=3)

Vial No.	OD$_{405nm}$ (100 pg/ml)	S^2	Confidence interval (p=0.05)
6	0.409	0.000169	0.377 - 0.441
7	0.401	0.000225	0.364 - 0.438
8	0.401	0.000441	0.349 - 0.453
9	0.420	0.000961	0.343 - 0.497
10	0.393	0.001521	0.296 - 0.490

Mean=0.405   Common variance=0.000663

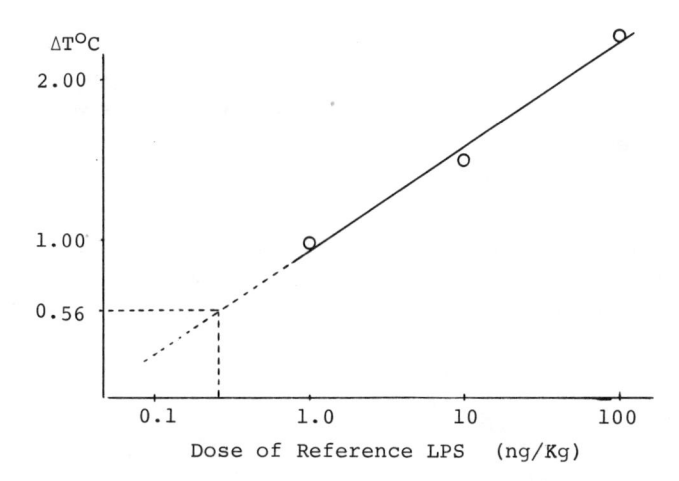

Fig. 4. Linearity of the dose-response line and minimum response dose of the Reference LPS in rabbit pyrogen test

Three rabbits were used for one dose

tion tests.

(1) The dried Reference LPS sealed in vials were heated either in boiling water for 30 min or 120 min, or in water bath at 70°C for 24 or 72 hr. After the heat-treatment, the content of every vial appeared quite well, and was easily dissolved in water (Table 5). The LAL activating activity of the Reference LPS in dried state was unaffected by heating at 100°C for 30 min, but significantly lowered by the heating at the same temperature for 120 min ($\alpha=0.05$).

Table 5. Results of accelarated degradation tests

Treatment	OD_{405nm} Y (n=3)	s^2	Significance in difference ($\alpha=0.05$)
Untreated control	0.294	0.000496	
100°C 30 min	0.280	0.000472	no
120 min	0.275	0.000033	yes
70°C 1 day	0.256	0.000532	no
3 days	0.259	0.000243	no

The activity of LPS was determined at 20 pg in the reaction mixture.

(2) On the other hand, the Reference LPS was not stable once it was dissolved in water. As an example, the results of the stability at concentration of 100 ng per ml are shown in Fig. 5. The relative activity after the treatment at 37°C for 60 min was little lower (0.802)

than unheated control, but after the treatments at 60°C for 30 min and 60 min there was observed marked falls in the relative activities (0.081 and 0.0007) respectively. The activity of further diluted Reference LPS was more affected by the heat-treatment.

The instability of the dissolved Reference LPS was again noticed during the storage at 5°C or at room temperature (25°C) as shown in Fig. 6. The results suggest that the LAL activating activity of dissolved Reference LPS at concentration of 100 ng/ml will be halved by storage in 12.5 days at 5°C or in 1.5 days at room temperature.

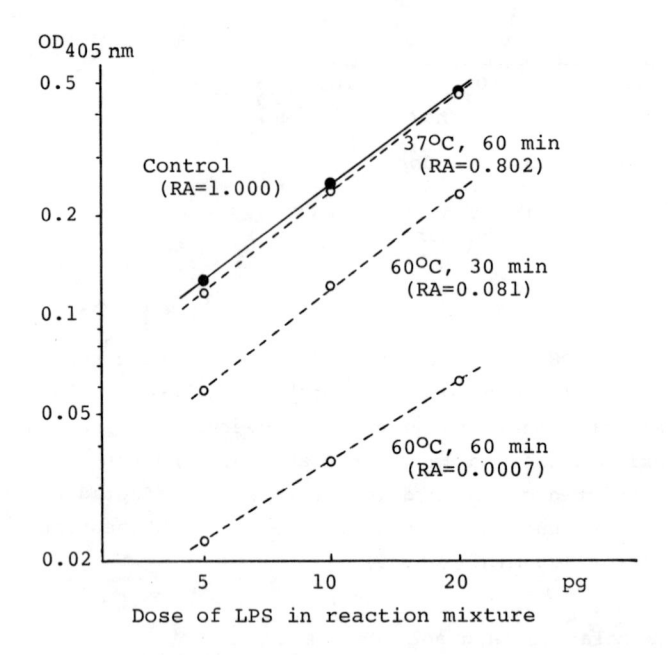

Fig. 5. Stability of the Reference LPS solution (100 ng/ml)

10 Dose-response lines with some LPS preparations

By using the SCS method, the dose-response lines were drawn with each of the LPS preparations available, namely Novo Pyrexal, US Standard Endotoxin EC-5, and various contol endotoxin preparations for LAL testing. Linearity of the lines was not denied for any of the preparations within a certain ranges. The representative lines with the first two preparations and our candidate Reference LPS are shown together in Fig. 7. Further experiences have to be accumulated in testing various crude endotoxin of miscellaneous origins. Parallelism among the dose-response lines will be analysed on the data.

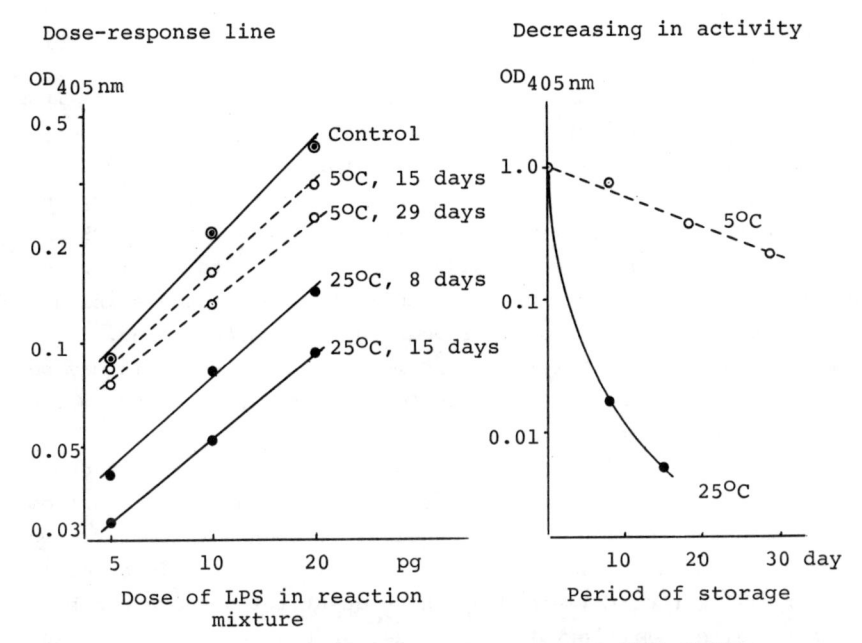

Fig. 6. Stability of dissolved Reference LPS (Lot 8205)
(Stored at concentration of 100 ng/ml)

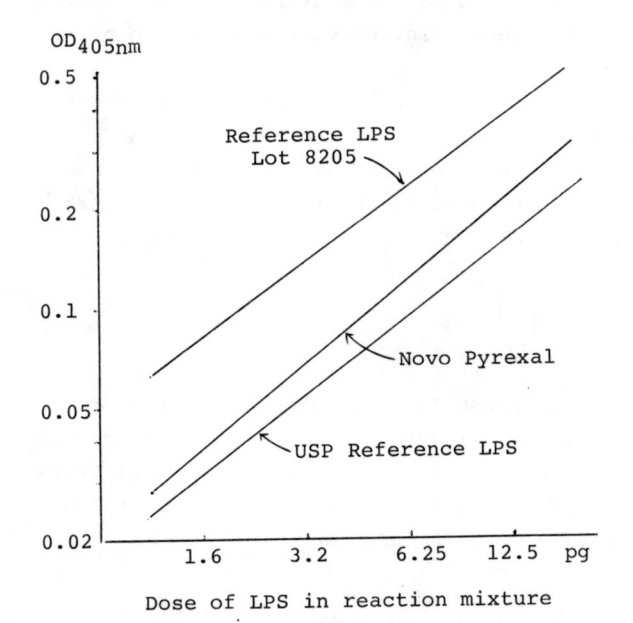

Fig. 7. Dose-response lines with three reference LPS preparations

11 Discussion

The candidate Reference LPS (Lot 8025) are apparently higher in its purity, LAL activating activity and pyrogenicity than USA Reference Endotoxin and the commercially available control endotoxins. The results presented here show that the Reference LPS is satisfactory in respect to inter-vial uniformity. Instability of the Reference LPS in dissolved state would be a great disadvantage for its storage in routine laboratories, in contrast to its stability in dried state which could have been proven by the accelerated degradation technique. Instability of the diluted Reference LPS is out of question. It may be due to inactivation, or possibly be due to polymerization of LPS molecules or to their adsorption to vials.

Linearity of the dose-response line was observed with various endotoxin preparations. However, there might occur somewhat deviation from parallelism with some of the regression lines, though not demonstrated in this study. Then, it will be possible to analyze the differences in regression by using the SCS method by which quantitative data can be obtained with quite a little variance. Thus, should be done in respect to the parallelism among the regression lines with further samples of endotoxin preparations, since it would be the most substantial problem in the determination of relative endotoxin activity that can be present in test samples.

At the beginning of this work, the authors attemptes to prepare two kinds of reference preparations, namely a purified LPS as a standard on the one hand, and a working reference endotoxin for practical use on the other. The product presented here is the former, and the work to establish the latter in suitable lot size is under progress.

12 Acknowledgement

The authors wish to thank Dr. Kazuhito Hisatsune, Department of Microbiology, School of Pharmaceutical Sciences, Josai University, and Dr. Nobuhiko Kasai, Department of Microbial Chemistry, School of Pharmaceutical Sciences, Showa University, for their co-operation in the chemical analyses of the purified LPS preparation.

13 References

(1) National Archives of US, Code of federal regulations, 1982, Part 660.

(2) O. Westphal, O. Luderitz, F. Bister, Z. Naturforschg. 7 (1952) 148-155.

(3) A. Weissbach, J. Hurwitz, J. Biol. Chem. 234 (1959) 705-709.

(4) M. J. Osborn, Proc. Natl. Acad. Sci. USA 50 (1963) 499-506.

(5) R. A. Laine, K. Stelner, S. Hakomori, Methods in membrane biology, Plenum Press, New York 1974, pp. 228-244.

(6) C. H. Fiske, Y. Subbarow, J. Biol. Chem. 66 (1925) 375-400.

(7) Ministry of Health and Welfare, Japanese Government, Minimum Requirement of Biological Products, 1982, 357-358.

(8) S. Iwanaga, T. Morita, T. Harada, S. Nakamura, M. Niwa, K. Takada, T. Kimura, S. Sakakibara, Haemostasis 7 (1978) 183-188.

(9) T. Harada, T. Morita, S. Iwanaga, S. Nakamura, M. Niwa, prog. Clin. Biol. Res. 29 (1979) 209-220.

(10) Y. Fujita, C. Nakahara, Endotoxins and their detection with the limulus amebocyte lysate test, Alan R. Liss, Inc., New York 1982, 173-182.

(11) T. Komuro, Y. Ogawa, H. Niji, H. Kawasaki, S. Kanoh, Jap. J. Pharmacol. 31 (1981) 206P.

Isolation and Purification of a Standardized Lipopolysaccharide from Salmonella abortus equi

Chris Galanos, O. Lüderitz, O. Westphal

Max-Planck-Institut für Immunbiologie, Stübeweg, D-7800 Freiburg, F.R.G.

Abstract

A standardized lipopolysaccharide designated as Novo-pyrexal has been prepared from Salmonella abortus equi. The lipopolysaccharide is free of protein, nucleic acid, and other bacterial contaminants, and it is present in the uniform sodium salt form. The physical, chemical, and biological properties of this preparation are reported here.

Introduction

It is more than 100 years ago since it was recognized that the pathogenicitiy of gram-negative bacteria is closely associated with surface components firmly bound in cell-wall, being released after death and lysis of the cells. For this reason these biologically active substances were named endotoxins, although their precise identitiy remained unknown for many years that followed. The development of procedures for the isolation and subsequent chemical analysis of the endotoxically active component in gram-negative bacteria enabled their chemical identification. Today we know that, chemically, endotoxins are lipopolysaccharides made up according to a common structural principle. They consist of the O polysaccharide, the core oligosaccharide, and the lipid A. The O polysaccharide is highly variable in its structure and composition among gram-negative bacteria and determines the serospecificity of the molecule and of the parent bacterial strain. The structure and composition of the core are less variable, being common to larger groups of bacteria.

Lipid A is a common constituent of all lipopolysaccharides exhibiting large similarities in structure and composition among gram-negative bacteria. (For reviews see 1,2,3). Lipid A is the part of the molecule responsible for toxicity and the other known biological activities of lipopolysaccharides (2).

Of the many lipopolysaccharides investigated so far those of Salmonella have been best characterized, and methods of isolation and chemical analysis modelled for their study have been applied successfully for the investigation of lipopolysaccharides from other genera.

Today, a high standard of lipopolysaccharide isolation and purification methodoloy has been achieved which allows the isolation of pure lipopolysaccharides in physicochemically defined form.

Here we report on the preparation of a standardized lipopolysaccharide derived from S.abortus equi. The preparation is free of proteins and other bacterial cell contaminants and is present in physicochemically defined form as the uniform sodium salt of the lipopolysaccharide. The preparation is also available in the form of sterile ampulled solutions that are suitable for clinical purposes and known as Novo-pyrexal*.

Results

Isolation and purification

A complete account of the methods employed has been given earlier (4). Briefly, S. abortus equi bacteria were cultivated to the late logarithmic phase as described by Schlecht (5), harvested and dried. Isolation of the lipopolysaccharide was carried out by the phenol-water procedure whereby the lipopolysaccharide was obtained in the aqueous phase (together with proteins and other bacterial contaminants). The aqueous phase was then dialyzed against distilled water, freeze-dried, and subjected to the phenol-chloroform-petroleum ether (PCP) extraction procedure whereby the lipopolysaccharide was obtained free of protein and nucleic acid. Subsequent steps in the further purification of the preparation involved ultracentrifugation, electrodialysis and re-extraction by the PCP method, as earlier described, which yielded finally the sodium salt of the S. abortus equi lipopolysaccharide.

Fig. 1 Sedimentation behaviour of the standardized S.abortus equi LPS in the triethylamine form.

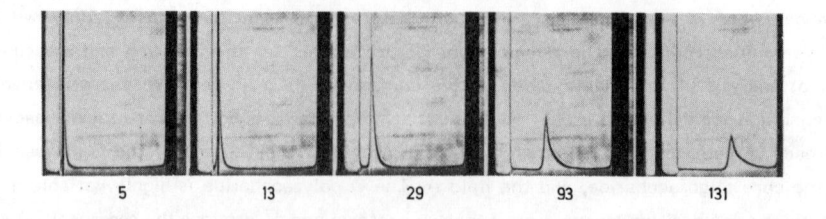

Purity and physicochemical properties

The lipopolysaccharide obtained as described above was found to be free of the usual contaminants that are present in lipopolyaccharides. Thus, nucleic acids and glycans were
*Supplied by Hermal Chemie, Kurt Hermann, Reinbek b. Hamburg.

absent and the protein content measured by amino acid anaylsis was below 0.08%. Further, due to the extensive electrodialysis to which the lipopolysaccharide had been subjected, polyamines (putrescin, spermine, spermidine cadaverine etc.) which are usually present in lipopolysaccharides, ionically bound to negatively charged groups, were not detectable in the present preparation. In the analytical ultracentrifuge the lipopolysaccharide sedimented as a single sharp peak (Fig. 1), which is an additional criterium for the high purity of this material.

Table 1. Sedimentation coefficients of the Salmonella abortus equi lipopolysaccharide in different salt forms

Salt form	S value
Triethylamine	9
Ethanolamine	64
Pyridine	72
Sodium	105
Potassium	135
Putrescine	230
Calcium	partly insoluble

The S. abortus equi sodium salt lipopolysaccharide exhibits high solubility in distilled water, whereby clear solution containing up to 30 mg/ml may be obtained. When converted to other salt-forms (triethylamine, calcium etc.), the lipopolysaccharide exhibits different solubility which is chracteristic for the given salt-form, being highest in the triethylamine form and lowest in the calcium form.

The above differences in solubility are paralleled by corresponding differences in the sedimentation properties of the different salt forms in the analytical ultracentrifuge (Tab. 1). In this respect, the lipopolysaccharide of S. abortus equi behaves like other lipopolysaccharides investigated so far (6).

Chemical structure

The structure of the S. abortus equi lipopolysaccharide as elucidated during the past years is shown in Fig. 2 (7). It consists of an O polysaccharide, the core and the lipid A, a structure present in most lipopolysaccharides that have been analyzed so far. The repeating units of the O-specific chain are represented by pentasaccharides containing D-mannose, L-rhamnose and D-galactose as a linear trisaccharide, the mannose and galactose carrying branched abequose and D-glucose, respectively. Abequose and glucose constitute the sero-specific sugars in the molecule, representing the O factors 4 and 12, respectively.

The strucure and composition of the core polysaccharide are as found for the genus Salmonella (8), containing 2-keto-3-deoxyoctonate (KDO), heptose, glucose, galactose and N-acetylglucosamine.

Fig. 2: Chemical structure of the S. abortus equi lipopolysaccharide

$$\left[\begin{array}{c}\text{Abep}\\\downarrow\alpha 1,3\\\text{D-Manp}\end{array}\xrightarrow{\alpha 1,4}\text{L-Rhap}\xrightarrow{\alpha 1,3}\begin{array}{c}\text{D-Glcp}\\\downarrow\alpha 1,4\\\text{D-Galp}\end{array}\xrightarrow{\alpha 1,2}\right]_{\approx 7}\begin{array}{c}\text{Abep}\\\downarrow\alpha 1,3\\\text{D-Manp}\end{array}\xrightarrow{\alpha 1,4}\text{L-Rhap}\xrightarrow{\alpha 1,3}\text{D-Galp}\xrightarrow{\beta 1,4}$$

O-Specific Chains (Region I)

Core Polysaccharide (Region II)

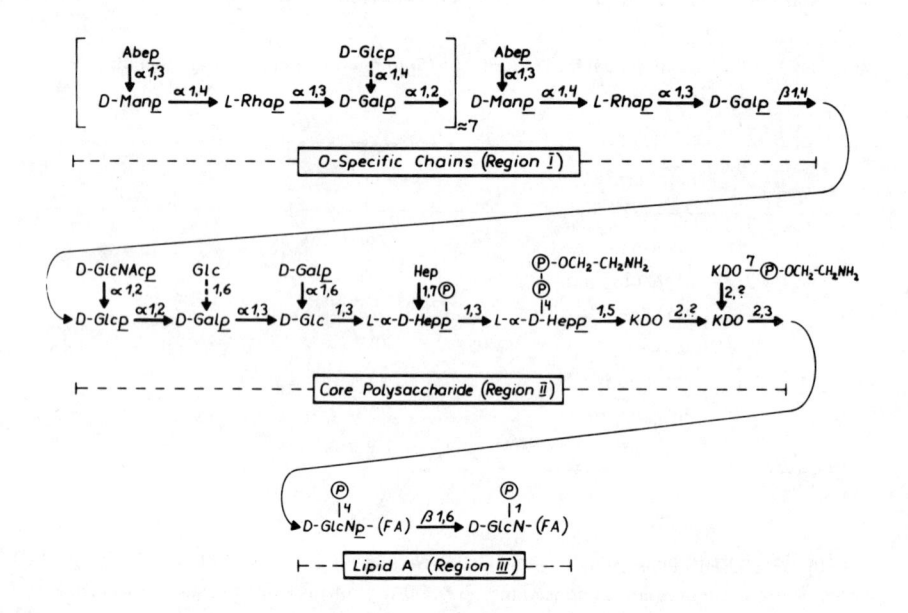

Lipid A (Region III)

Table 2. Chemical composition of the S. abortus equi lipopolysaccharide Na-form

Sugars (unhydro)		Fatty acids (unhydro)	%	Other constituents	%
D-Abequose	13.5	Lauric (C 12)	2.2	Phosphorylresidue	5.0
L-Rhamnose	15.2	Myristic	1.8	Sodium	1.8
D-Mannose	15.7	D-3-hydroxymyristic		4-amino-L-arabinose	0.3
D-Heptose	5.0	(3-OH-C 14)	5.6	Ethanolamine	0.2
D-Galactose	17.8	Palmitic	1.1		
D-Glucose	4.0				
2-Keto-3-deoxy-D-Manno-octonic acid (KDO)	6.8				
D-Glucosamine	4.2				
Total	82.2		10.7		7.3

Total: approx. 100 %

The structure of the lipid A of S. abortus equi is similar to that found for most lipopolysaccharides (9). It consists of a phosphorylated D-glucosamine disaccharide in a β-1-6 linkage carrying amide linked 3-hydroxymyristic acid and ester linked lauric, myristic, palmitic and 3-hydroxymyristic acid.

Other constitutents present in the lipopolysaccharide in low concentrations are 4-amino-L-arabinose, P-ethanolamine and sodium cations.

A complete analysis of the standardized lipopolysaccharide is shown in table 2.

Figure 3: Properties of the lipopolysaccharide of S. abortus equi in different salt forms.

S abortus equi LPS	Sedimentation Coefficient	Solubility (Water)	Lethal Toxicity MICE RATS	Pyrogenicity (Rabbits)	Rate of clearance from the blood	Interaction with C' in vivo and in vitro	Affinity for cells	Mitogenic activity
Triethylamine (TEN)	9.3	↑	↑	↑		(a)	(b)	↑
Pyridine								
Ethanolamine								
Na								
K								
Putrescine	230							
Ca	partly insol.	↓	↓	↓	↓	↓	↓	↓

Arrows show direction of increasing activity.
(a) In the TEN form S form LPS is completely non-anticomplementary.
(b) No difference within the above range of (S) values.

Biological properties of lipopolysaccharides in uniform salt forms

The physical state of a lipopolysaccharide exerts a profound effect on its biological properties (10,11). Investigation of lipopolysaccharides in uniform salt forms which exhibit characteristic

Figure 4: Anticomplementary activity of S. abortus equi in different salt forms.

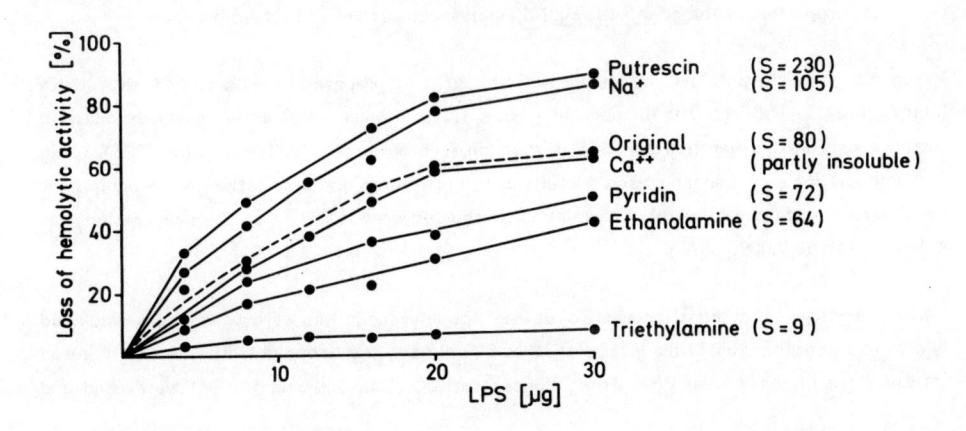

Putrescin (S = 230)
Na+ (S = 105)
Original (S = 80)
Ca++ (partly insoluble)
Pyridin (S = 72)
Ethanolamine (S = 64)
Triethylamine (S = 9)

physicochemical properties revealed that for a given preparation, individual biological activities of different salt forms may vary significantly. Activities such as pyrogenicity and lethal toxicity for mice are higher when the degree of aggregation of the lipopolysaccharide is decreased and are thus expressed optimally in the highly soluble triethylamine form. On the other hand, properties such as anti-complementary activity, interaction with cell membranes or the rapid clearance of the lipopolysaccharide from the blood, require a high degree of aggregation for their optimum expression. Fig. 3 summarizes in a schematic representation the influence of salt form on a number of individual biological activities of lipopolysaccharides.

The above salt form-dependent differences in individual biological activities are also true for the lipopolysaccharide of S. abortus equi. The different salts were obtained by neutralizing the free acid form of the standardized lipopolysaccharide, obtained by electrodialysis with different bases. In accordance with Fig. 3 pyrogenicity and lethal toxicity in mice were highest in the triethylamine salt form while anticomplementary activity, interaction with cell membranes, the rate of clearence from the blood and lethal toxicity in rats were highest in the sodium or calcium form of the lipopolysaccharide. The activities of other salt forms were intermediary to the above, in correspondingly increasing or decreasing order as shown by the arrows in Fig. 3. For mitogenic activity for mouse B cells and for lymulus lyzate gelating activity no significant differences were found among the different salt forms.

As a typical example of aggregation-dependent differences in biological activity, Fig. 4 shows the in vitro anti-complementary activity of the standardized S. abortus equi lipopolysaccharide in different salt forms (10). Here it may be seen that the anti-complementary activity of the different salt forms increases in the order of increasing aggregation (compare sedimentation values in Tab. 1). It may also be seen that in the highly disaggregated triethylamine form anti-complementary activity is completely abolished. Thus, it was possible in this case to abolish selectively one boiological property from the lipopolysaccharide by simple physical means without altering the chemical integrity of the molecule. Differences in the anti-complementary activity of triethylamine and sodium forms were also demonstrated in vivo (12).

Biological properties of the S. abortus equi lipopolysaccharid-Na salt (Novo-pyrexal)

Pyrogenic properties: The pyrogenic activity of the preparation was tested in rabbits (chinchilla, 1.5 - 1.8 kg). The lipopolysaccharide, (0.1, 0.01 and 0.001 µg/kg) was administered intravenously in the marginal ear vein in pyrogen-free phosphate buffered saline (PBS), using five animals for each concentration. Details of the procedure and description of the equipment used were described earlier (4). The minimal pyrogenic dose (MDP-3) of the preparation was calculated to be 0.0008 µg/kg.

Lethal toxicity: The lethal properties of the lipopolysaccharide were tested in 6-week old NMR mice of both sexes. Using groups of 10 or 12 animals, the lipopolysaccharide was injected intraperitoneally in 0.5 ml PBS. Three concentrations (100, 200 and 300 µg) were used and lethality was recorded up to 72 h after injection. The results in Tab. 3 show that 300 µg of the preparation induce about 75% mortality.

Table 3. Lethal toxicity of the S. abortus equi lipopolysaccharide Na salt in mice

Amount of Preparation µg	NMRI (male) dead/total	% Lethality	NMRI (female) dead total	% Lethality
100	0/12	0	2/10	20
200	6/12	50	3/10	30
300	7/12	70	8/10	80

Local Shwartzman reation: Each of 6 rabbits (chinchilla, 1.5 kg) received intradermally 5, 10, 20 and 40 µg of the lypopolysaccharide in 0.1 ml PBS and one intradermal injection of PBS (0.1 ml) as contol. After 24 h all animals received 100 µg S. abortus equi lipopolysaccharide intravenously in 0.5 ml PBS. The results in Tab. 4 show that with 20 and 40 µg strong skin reactions developed in all animals. A moderate reaction was obtained with 10 µg. Even with 5 µg a detectable skin reaction was present in most of the animals. No reaction was seen at the site of injection with PBS (4).

Table 4. Induction of the local Shwartzman reaction by the Salmonella abortus equi lipopoly saccharide Na salt

Amount of Preparation µg	1	2	3	4	5	6
40	+++	+++	+++	+++	+++	+++
20	+++	++	++	+++	+++	+++
10	++	++	++	++	++	++
5	±	+	-	±	±	+
PBS	-	-	-	-	-	-

Extend of Reactions in Rabbit No.

Mitogenic avtivity: Lipopolysaccharides are known to be mitogenic for mouse B lymphocytes. The S. abortus equi preparation described here was found to be mitogenic for mouse B lymphocytes of different mouse strains in concentrations between 5 and 10 µg/10^6 cells (13).

No mitogenic responses were seen in endotoxin non-responder C3H/HeJ mice, indicating the purity of the preparation with respect to protein contamination that is known to exhibit mitogenic activity in the above mice.

Anti-tumor activity: The tumoricidal acivity of the S. abortus equi lipopolysaccharide was investigated on a methylcholantren induced fibrosarcoma in syngeneic CBF_1 mice. Groups of 10 animals (6-8 weeks) were inoculated intradermally with 5×10^5 tumour cells. On day 8, by which time the tumcurs had reached an average size of 7 mm in diameter, the animals received 10 µg of lipopolysaccharide by the intraperitoneal route. Two to 3 days later a strong hemorrhage in the tumour mass developed. The tumours became degenerate and in 70% of the animals disappeared completely 8 to 12 days later (Tab. 5. In the control animals without lipopolysaccharide, tumour growth continued and the animals died about 4 weeks after inoculation (14).

Table 5. Effect of S. abortus equi lipopolysaccharide on the Meth A Fibrosarcoma in CBF_1 mice

Number of mice	Day			Cured animals
	1	8	8	25
10	5×10^5 Meth A Cells i.d.	Tumor Size: 0.6-0.9 cm ⌀	10 µg LPS i.p.	7/10
10			(Control	0/10

Activity of Novo-pyrexal in humans: The sensitivity of humans to endotoxin is very high and for certain parameters (pyrogenicity, induction of leucocytosis) where a limited amount of information has accumulated the response of humans would seem to be comparable to that of rabbits. The effect of Novo-pyrexal on blood leucocytes was tested in 60 healthy volunteers of both sexes ranging between 23 and 55 years of age. All individuals received 0.03 µg (total) of S.abortus equi lipopolysaccharide intravenously in 0.3 ml buffer.

Samples of blood were removed before and at 2, 4 and 6 hours after injection and differential and total blood leucocyte counting carried out according to standard methods. The results are shown in Fig. 5. All individuals responded to the dose employed with a strong leucocytosis reaching maximum values 4 hours after injection and decreasing again two hours later. On the average the increase amounted to 250% of the values before injection. The above increase in leucocytes concerned mainly granulocytes (Fig. 5).

Repeated injections of Novo-pyrexal in the same individuals during a period of 14 days did not lead to tolerance to the granulocytic properties of the preparation (Fig. 6). This is in contrast to earlier observations made with a similar preparation (15) and is explained by the very low content of protein in the Novo-pyrexal preparation. It should be mentioned here that according to personal communications from several clinical sources that have used Novo-pyrexal in humans, in contrast to leucocytosis, tolerance to the pyrogenic properties of the preparation did establish itself following repeated injections in the same individuals.

Figure 5: Induction of granulocytosis by Novo-pyrexal in humans.

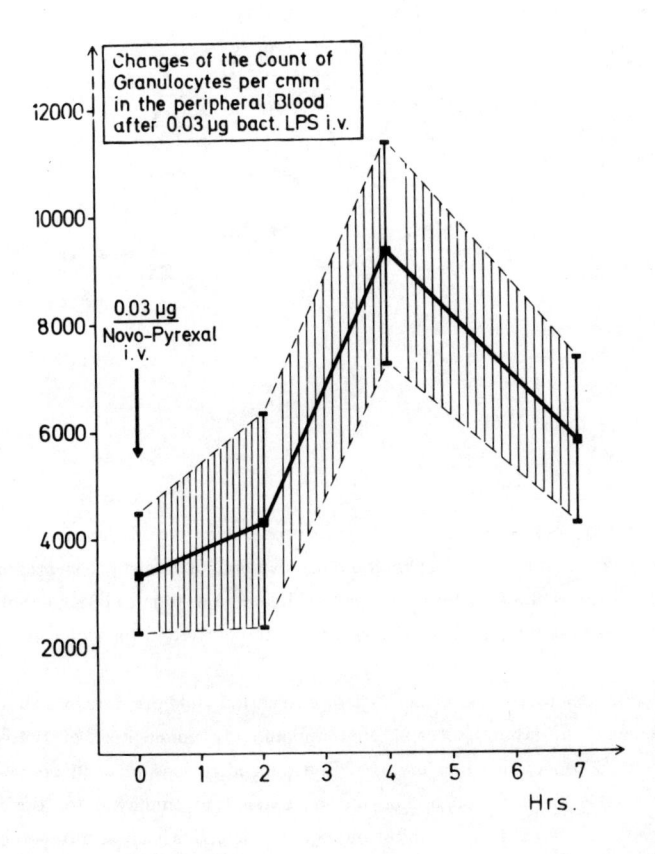

Figure 6: Effect of repeated injections of Novo-pyrexal on leucocytosis in humas.

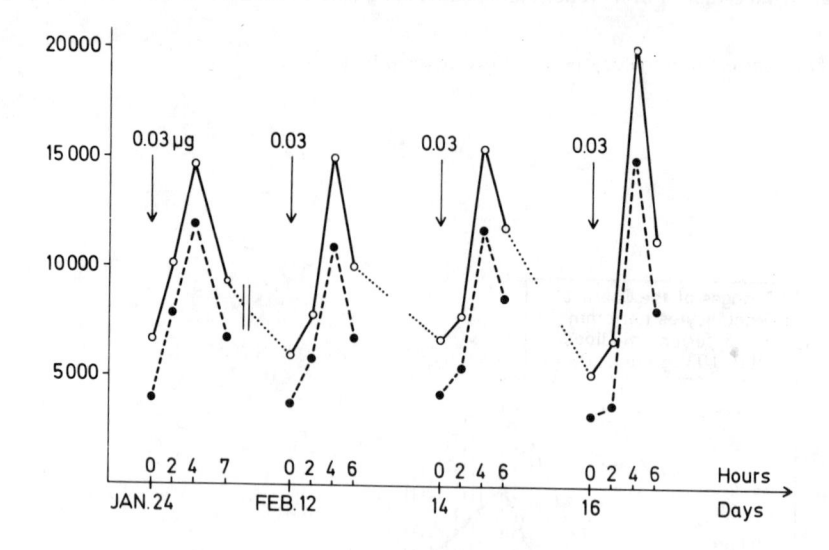

Discussion

In recent years the necessity for a standard endotoxin preparation that is suitable for chemical, biological and also clinical purposes has become more evident. Standard preparations should fulfill certain criteria such as stability and their purity should be as high as possible but in any case they should be free of contaminants that may modify the activities for which they are intended as standards.

Isolation of lipopolysaccharides by the usual methods of extraction yield preparations that always contain variable amounts of other bacterial contaminants the commonest of these being proteins, nucleic acids, phospholipids and glycans. The concentrations of such contaminants may vary from 1 to 3% or more, depending on the extraction employed for their isolation. Bacterial proteins have been shown to exhibit biological properties such as mitogenic activity, however, also in animals that usually do not respond to the mitogenic properties of lipopolysaccharides (16). Bacterial proteins were also shown to sensitize to the lethal effects of lipopolysaccharide in animals that are immunologically primed to the proteins (17). It becomes therefore apparent that while a low degree of contamination of lipopolysaccharides with proteins may have no significant influence on properties such as lethal toxicity or pyrogenicity in normal animals, or on in vitro activities like the limulus lyzate gelation, their effect may be significant on other activities such as mitogenicity, adjuvanticity or on various parameters tested in experimental animals that may be naturally primed to bacterial proteins.

The standardized S.abortus equi lipopolysaccharide (Novo-pyrexal) described her is free of proteins and other macromolecular contaminants. It is further free of polyamines and is present in the uniform sodium salt form, other metal cations being absent. The preparation exhibits constant physicochemical properties being readily soluble in distilled water and buffer solutions. Its chemical structure and composition have been well defined and it exhibits all known characteristic endotoxic properties of lipopolysaccharides. Novo-pyrexal is also available in ampoulled solutions that are suitable also for clinical purposes. Solutions of the preparation are highly stable showing no loss of biological activity over many years.

The standardized S.abortus equi lipopolysaccharide thus fulfills the criteria of purity, stability and biological activity and qualifies for being adopted as a standard endotoxin.

References

(1) O. Lüderitz, C. Galanos, V. Lehmann, M. Nurminen, E.Th. Rietschel, G. Rosenfelder, M. Simon, O. Westphal, J. Infect. Dis. 128 (1973) p. 17-29.

(2) C. Galanos, O. Lüderitz, E.T. Rietschel, O. Westphal, Int. Rev. Biochem. 14 (1977) 239-335.

(3) O. Lüderitz, M.A. Freudenberg, C. Galanos, V. Lehmann, E.Th. Rietschel, D.H. Shaw in S. Razin, S. Rottem (Eds.), Current Topics in Membranes and Transport. Academic Press, Inc., New York, 17 (1982) 79.

(4) C. Galanos, O. Lüderitz, O. Westphal, Zbl. Bakt. Hyg., I. Abt. Orig. A (1979) 226-998.

(5) S. Schlecht, Zbl. Bakt. Hyg., I. Abt. Orig. A 232 (1975) 61-72.

(6) C. Galanos, O. Lüderitz, Eur. J. Biochem. 54 (1975) 603-610.

(7) I. Fromme, O. Lüderitz, A. Nowotny, O. Westphal, Acta Helvet. 33 (1958) 391-400.

(8) O. Lüderitz, C. Galanos, H.J. Risse, E. Ruschmann, S. Schlecht, G. Schmidt, H. Schulte-Holthausen, R. Wheat, O. Westphal, J. Schlosshardt, Ann. N.Y. Acad. Sci. 133 (1966) 349-374.

(9) E.Th. Rietschel, U. Zähringer, H.W. Wollenweber, K. Tanamoto, C. Galanos, O. Lüderitz, S. Kusumoto, T. Shiba in Tu, Habig, Hardegree (Eds.), Marcel Dekker, Inc., New York (1983).

(10) C. Galanos, O. Lüderitz, Eur. J. Biochem. 65 (1976) 403-408.

(11) C. Galanos, Z. Immun.-Forsch. Bd. 149 (1975) 214-229.

(12) M.A. Freudenberg, C. Galanos, Infect. Immun. 19 (1978) 875-882.

(13) J. Andersson, F. Melchers, C. Galanos, O. Lüderitz, J. Exp. Med. 137 (1973) 943.

(14) O. Westphal, unpublished data.

(15) E.M. Eichenberger, H. Schmidhauser-Kopp, M. Fircsay, O. Westphal, Schweiz. med. Wochr. 85 (1955) 1190-1226.

(16) F. Melchers, V. Braun, C. Galanos, J. Exp. Med. 142 (1975) 473-482.

(17) C. Galanos, in preparation.